An Introduction to U.S. Telecommunications Law

Second Edition

For a listing of recent titles in the *Artech House Telecommunications Library*, turn to the back of this book.

An Introduction to U.S. Telecommunications Law

Second Edition

Charles H. Kennedy

Artech House
Boston • London
www.artechhouse.com

Library of Congress Cataloging-in-Publication Data
Kennedy, Charles H.
    An introduction to U.S. telecommunications law / Charles H. Kennedy. — 2nd ed.
    p. cm. — (Artech House telecommunications library)
    Includes bibliographical references and index.
    ISBN 0-89006-380-X (alk. paper)
    1. Telecommunication—Law and legislation–United States.
    I. Title: Introduction to US telecommunications law.   II. Title.   III. Series
KF2765 .K46   2001
343.7309'94—dc21   2001022861

British Library Cataloguing in Publication Data
Kennedy, Charles H.
    An introduction to U.S. telecommunications law. —2nd ed.—
    (Artech House telecommunications library)
    1. Telecommunication—Law and legislation—United States.
    I. Title
    343.7'3'0994
    ISBN 0-89006-380-X

Cover design by Paul Kraytman

© 2001 ARTECH HOUSE, INC.
685 Canton Street
Norwood, MA 02062

All rights reserved. Printed and bound in the United States of America. No part of this book may be reproduced or utilized in any form or by any means, electronic or mechanical, including photocopying, recording, or by any information storage and retrieval system, without permission in writing from the publisher.
    All terms mentioned in this book that are known to be trademarks or service marks have been appropriately capitalized. Artech House cannot attest to the accuracy of this information. Use of a term in this book should not be regarded as affecting the validity of any trademark or service mark.

International Standard Book Number: 0-89006-380-X
Library of Congress Catalog Card Number: 2001022861

10 9 8 7 6 5 4 3 2

For my family

Contents

Contents	*vii*
Preface	*xv*
Introduction	*xvii*
How We Got Here: The Evolution of Telecommunications Law	*xvii*
1913–1968: The Age of Monopoly	*xviii*
1968–1996: The Age of Hybrid Regulation	*xxi*
How Much Freedom for the Established Carriers?	*xxiii*
Regulation of New Entrants	*xxv*
Divided Authority and Differential Regulation	*xxv*
1996 to the Present: The Age of Competition	*xxvi*
Conclusion	*xxviii*
Endnotes	*xxviii*
Part I **Incumbent Local Exchange Carriers**	**1**

		Introduction to Part I	1
		Endnotes	3
1		**How Much May ILECs Charge for Local Exchange Services?**	**5**
	I.	Peninsula Telephone: Local Exchange Service Rates Under Rate-of-Return Regulation	6
		A. Allocating Costs to Local Exchange Service	8
		B. Calculating Peninsula's Revenue Requirement	9
		C. A Word About Depreciation	10
		D. Peninsula's Rate Structure	12
		E. Discriminatory Rates: The Regulator's View	14
		F. Discriminatory Rates: The Economist's View	14
	II.	Peninsula Telephone: Local Exchange Service Rates Under Price-Cap Regulation	15
		Endnotes	18
2		**How Much May an ILEC Charge for Nonbasic Services?**	**23**
	I.	Pricing Nonbasic Services Under Rate-of-Return Regulation	24
	II.	Pricing Nonbasic Services Under Price Caps	27
	III.	Pricing Nonbasic Services Under the Antitrust Laws	29
		Endnotes	31
3		**When Must an ILEC Interconnect Its Facilities with Those of Other Service Providers?**	**35**
	I.	ILEC Services Provided to CLECs	36
		A. Negotiating the Interconnection Agreement Between Peninsula Telephone and PenCLEC	37
		B. Arbitration, Mediation, and Judicial Review	44
	II.	ILEC Services Provided to IXCs	45
		A. Equal Access Defined	46

		B. How Equal Access Has Been Implemented	47
		C. How ILECs Charge for Services Provided to IXCs	49
	III.	ILEC Services Provided to CMRS Providers	53
	IV.	ILEC Services Provided to ISPS and Other Providers of Information Services	54
	V.	The Future of Interconnection and Intercarrier Compensation	55
		Endnotes	55
4		**Special Cases: The Bell Operating Companies**	**61**
	I.	The BOC Restrictions in the Telecommunications Act of 1996	62
		A. The InterLATA Restriction	62
		B. Interconnection with Data Networks and Other Enhanced Service Providers: *The Computer III* Rules	71
		Endnotes	78
5		**Competing Local Exchange Carriers**	**85**
	I.	State Certification of CLECs	86
		A. Election of Facilities-Based or Resale Authority	86
		B. Specification of Services to Be Offered	87
		C. Specification of Area to Be Served	87
		D. Financial Ability to Serve	88
		E. Construction Plans and Environmental Impact Statements	89
		F. Tariff Filings	89
	II.	Ongoing CLEC Regulatory Obligations	89
		A. Contributions to Universal Service and Other Funds	89
		B. Tariffs and Informational Filings	91
		C. Interconnection and Related Obligations	92

III.	CLEC Access to ILEC Facilities, ILEC Services, and Rights-of-Way		93
	A. Interconnection and Reciprocal Compensation		94
	B. Access to Unbundled Network Elements		95
	C. Poles, Conduits, and Rights-of-Way		97
	Conclusion		98
	Endnotes		98

Part II
Non-ILECS: The Competitive Telecommunications Industry — 101

Introduction to Part II — 101

6 Interexchange Carriers — 103

I.	Origins and Growth of the IXC Industry	104
	A. Price-Cap Regulation of AT&T's Rates	106
	B. Classification of AT&T as Nondominant	108
II.	Regulation of IXCs Today	109
	Endnotes	110

7 Pay Telephones and Operator Services Providers — 115

I.	The Business or Premises Owner	116
II.	The Payphone Provider	116
III.	Incumbent Local Exchange Carriers	118
IV.	Operator Service Providers	120
	Endnotes	121

8 Mobile Telephone Companies — 123

I.	Mobile Telephone Technology	123
II.	Licensing Mobile Telephone Companies	125

		A. A Brief History of Mobile Telephone Spectrum Assignment	126
III.		Regulation of Mobile Telephone Licensees	129
IV.		Interconnection Between Mobile Telephone Licensees and Wireline Carriers	131
	A.	Physical Terms of Interconnection	131
	B.	Economic Terms of Interconnection	133
V.		Mobile Satellite Service	134
		Endnotes	135

9 Internet Service Providers — 139

I.		Liability for Harmful Content	140
	A.	Liability of ISPs for Defamation on the Internet	141
	B.	ISP Liability for Copyright Infringement on the Internet	145
	C.	ISP Liability for Trademark Infringement	149
	D.	ISP Liability for Obscenity and Indecency on the Internet	152
	E.	Liability of ISPs for Spamming	157
II.		Jurisdiction over ISP Activities	158
III.		ISPs and the FCC: Regulation of the Internet	161
IV.		Privacy and Data Security on the Internet	162
	A.	Interception of Internet Communications	162
	B.	Access to Computers and Stored Information	163
V.		Problems of Electronic Commerce	170
	A.	Making and Enforcing On-Line Contracts	171
	B.	Protecting Customers' Personal Information	174
		Conclusion	175
		Endnotes	175

10 Universal Service — 185

I.		Explicit High-Cost Support Programs Before the 1996 Act	186
II.		Explicit Support for Low-Income Subscribers	187
III.		Universal Service Requirements of the 1996 Act	188
IV.		The Commission's Post-1996 Universal Service Rules	189
	A.	Defining Universal Service	190
	B.	Eligibility to Receive Universal Service Support	190
	C.	Who Contributes to Universal Service Support?	192
	D.	How Support Payments Are Calculated	193
	E.	The Lifeline and Link-Up Programs	194
	F.	Support for Schools and Libraries	194
		Conclusion	195
		Endnotes	195

11 International Services — 199

I.		The Section 214 Process	199
	A.	Section 214 Applications of U.S. Carriers	200
	B.	Section 214 Applications of Foreign Carriers	201
	C.	Dominant Carrier Safeguards	202
II.		Foreign-Ownership Restrictions	203
III.		The International Settlements Policy	204
		Conclusion	205
		Endnotes	206

Appendix A
The Economic Background of Telecommunications Law — 209

I.		Competition and Monopoly	210
		Consumer Welfare: The Economic Case for Competition	210
	A.	What Is a Competitive Market?	212

	Using Marginal Cost to Find the Most Profitable Output	218
	Minimizing Costs in the Short Run	223
	Minimizing Costs in the Long Run	225
	B. What Is a Monopoly Market?	228
II.	Specific Economic Issues in Telecommunications Law	231
	A. The Puzzle of Natural Monopoly	232
	B. Does Regulation of Rates and Earnings Achieve Economic Efficiency?	234
	C. Efficient Pricing by Carriers	237
	D. Predatory Pricing	242
	E. Measuring a Carrier's Market Power	244
	Conclusion	246
	Endnotes	246

Appendix B
Selected Sections of the
Communications Act of 1934 — 251

About the Author — 359

Index — 361

Preface

The first edition of this book was published in 1994. In that year, basic local telephone service was still a legally-protected monopoly. The Internet, as a mass medium of communication, did not exist. And the business decisions of the Bell operating companies were still largely dictated by a single judge presiding in Washington, D.C.

It is no surprise, therefore, that this second edition is almost a complete rewrite of the first. Subjects that were centerpieces of the first edition, such as the requirements of the AT&T antitrust consent decree, now are peripheral. Subjects that did not appear in the first edition at all, such as content regulation of Internet-based services, are discussed at length. And the Telecommunications Act of 1996 has added so much detail to the statutory framework of telecommunications law that a lengthy appendix, setting out key provisions of the Communications Act and the 1996 amendments, now seems essential to the book's usefulness.

With all of these changes, the purpose of the book is the same as it was in 1994: to introduce some of the principal themes of telecommunications law in a way that assumes no background in the law, economics, or technology of telecommunications. As I said in the preface to the first edition, this book "is intended to be brief, but above all to be understood."

In keeping with its introductory character, this book does not necessarily provide all of the detail that a reader may need to solve a specific legal problem in any of the areas it addresses. Also, the pace of change in telecommunications law is so rapid that no discussion of the subject should be relied upon without a review of post-publication developments. Within these limits, I hope a wide range of readers will find this book useful.

Finally, I want to express my appreciation to Linda Peck, for her able help with the manuscript; to Mark Walsh and Barbara Lovenvirth of Artech House, for their patience and professionalism; to my colleagues at Morrison & Foerster, LLP, for their support; and to my family, for the good humor with which they have endured the writing of yet another book.

Introduction

How We Got Here: The Evolution of Telecommunications Law

This book avoids lengthy, generalized discussions of theory and history, and treats those subjects only as they affect particular legal and business problems. Before we begin, however, a *little* context may prove helpful.

First, we should decide what we mean by "telecommunications." While many formal definitions of this term have been attempted,[1] it is fair to say that any electronic transmission of *information* chosen by the sender, between specific *places* chosen by the sender, qualifies as telecommunications. The medium of transmission may be copper wire, radio, or optical fiber; the transmission may be one-way (as in paging) or two-way (as in an ordinary telephone conversation); the information transmitted may be a voice conversation or bits of data passing between computers. The definition includes ordinary telephone service, mobile telephone service, fax machines, paging, and much else besides; it excludes nonelectronic transmissions (mail, pigeons, smoke signals) and transmissions that do not flow between designated points (radio and television broadcasting). While this definition may not be airtight, it covers most of what people mean when they speak of telecommunications, and most of the services that are regulated as telecommunications [2].

We also should decide what we mean by telecommunications *law*. For our purposes, we may define this term to cover all of the legal rules, from whatever source, that apply specifically to telecommunications companies, or that apply with particular force to telecommunications companies. This definition includes the antitrust laws, the federal Communications Act, the regulatory statutes of the states, and the orders and regulations of the Federal Communications Commission, and the state public utilities commissions. It does not include those areas of the law (securities, taxation, labor law, equal employment opportunity, contract law, and the like) that affect all industries more or less equally.

Finally, and before we describe the particular rules under which telecommunications services are provided, it might be instructive to ask what it is about the telecommunications business that invites so much particularized scrutiny. Why does antitrust enforcement fall more heavily on telephone companies than pipe manufacturers or tae kwon do parlors? Why do 52 regulatory bodies worry more or less constantly about what the telephone companies are doing? And most of all, what are these courts and regulators (and the appallingly complex body of rules they administer) trying to do?

The answer to each of these questions is: it depends. It depends on the service; it depends on who is providing the service; it depends on which court or regulatory body has jurisdiction over the service or its provider; and it depends on when you ask. There are no straightforward answers because telecommunications law today is more the product of its own tangled history than of any unifying principle. Telecommunications law combines a longstanding system of common carrier regulation, rapidly declining in relevance as technology advances, with new rules and policies that have only partly displaced the old. And the continued existence of multiple regulators, with overlapping jurisdictions and differing views of the role of regulation, adds drastically to the confusion.

To understand telecommunications law, therefore, it is necessary to know something about its historical development. We offer a brief version of that history here, and we divide the story into three stages. We call these stages the Age of Monopoly, the Age of Hybrid Regulation, and (somewhat optimistically, as we shall see) the Age of Competition.

1913–1968: The Age of Monopoly

We begin our story at a time when there was nothing so grandiose as "telecommunications." There was only telephone service, and it was provided over expensive, fixed networks made up of mechanical switches and copper wire[3]. From around 1894, when Alexander Graham Bell's patents on the telephone expired, through the first decade of the twentieth century, telephone service in many cities was provided by competing companies with separate networks of copper wire connecting their subscribers to separate exchanges. To many observers, this situation seemed wasteful and contrary to the public interest. One objection was that the enormous investment needed to create and run a wire-based telephone company made the service a natural monopoly. Another was that telephone service should not be left to the free market because it was a business "affected with a public interest." Each of these terms requires a word of explanation.

To say that telephone service is a *natural monopoly* is to say that it can be provided most efficiently by a single firm. For an economist, this means that the average cost of providing the service declines indefinitely as output grows, so that the larger firm (i.e., the firm with the greater output) is always more efficient than the smaller one. For a regulator, it may mean simply that competition *appears* to be messy, inimical to service quality, or otherwise contrary to the public interest. Whatever the rationale, once a service is labeled a natural monopoly the urge takes hold to enshrine in law what economics has ordained, and to forbid competition in the natural monopoly market as wasteful and harmful to the public interest. In the telephone business, such protection from competition became the rule, and was referred to as a carrier's *exclusive franchise* to serve the market in which it operates.

The second principle is both broader and older than the idea of natural monopoly. For centuries governments have decided that some products and services are somehow too important to be entrusted to free enterprise. Some of these products and services (such as mail delivery in most countries) have been provided by government agencies; while other enterprises (ranging from ferry boats to power companies to taxicabs) have been left in private hands, but have been classified as common carriers and subjected to obligations not imposed on ordinary businesses.

Where the common carrier approach is chosen, the typical practice is to license the providers of these services and either protect them from competition altogether (the exclusive franchise again), or limit competition by awarding only as many licenses as the regulators believe the market will sustain. Such limitation of competition in common carrier services typically is rationalized as necessary to ensure that the services will continue to be available, or will be made available to everyone (not just those who are cheapest or most profitable to serve), or will be provided at an acceptable level of service quality.

It is not hard to see how these principles, once applied to particular industries, give rise to elaborate systems of regulation. The exclusive franchise, for example, is a legally protected monopoly that invites abuse of the public through excessive rates and shoddy service. Since the discipline of competition is not available to correct these tendencies, governments step in to scrutinize and control the companies' expenditures, prices, service quality, and other business practices in great detail. Similarly, common carrier status confers obligations to provide service to all that request it, on reasonable and nondiscriminatory terms and conditions. Once these obligations are imposed, detailed rules and enforcement mechanisms are bound to follow.

After the brief era of relatively robust competition in the early twentieth century, the common carrier model emerged as the universal method of regulating telecommunications in the United States. The seminal episode in this evolution was the Kingsbury Commitment—an agreement made in 1913 between the Bell system and the United States Department of Justice. Under that agreement, the Bell system gave up its ownership of Western Union and stopped its aggressive program of acquiring other telephone companies. Bell also agreed to interconnect its network with those of the independent telephone companies that provided service in areas not served by Bell. But Bell made no commitment to permit the independent companies, or any other carrier, to offer competitive services within Bell territory. In fact, the Kingsbury Commitment effectively confirmed an unchallenged Bell system monopoly of telephone service in most of the United States. In exchange for that monopoly, the states and the federal government undertook to regulate telephone service to ensure that it was provided in the public interest. Eventually, the states and the federal government were controlling entry, earnings, and the terms and conditions of service for all telephone companies, Bell and independent alike. AT&T and its integrated Bell system companies were offering most telephone service, both local and long distance, throughout the United States, and independent companies had exclusive franchises to serve areas in which Bell companies did not operate. In no part of the United States was there any competition in the manufacture of telephone equipment or the provision of telephone service.

In the telephone industry, as in railroads, trucking, electrical power, natural gas transmission, and other industries that came to be treated as common carriage, regulation mandated service and pricing decisions that a competitive environment would not support. The regulatory environment that controlled the telephone industry during the several decades when competition was forbidden imposed many practices of this kind, typically in the name of *universal service*.

Universal service, as state and federal regulators gave content to that policy, meant that customers would not be charged the actual cost of providing their service. Instead, some services would be priced well above the associated costs, and the revenues from those services would support other services that would be priced well below cost. These subsidies came to flow in familiar directions: from long-distance service to local service, from business services to residential services, and from urban rate payers to rural rate payers. Once in place, the system of subsidies was an additional reason not to permit competition, since low-cost competition could destroy the telephone companies' higher-priced services and the politically popular subsidies they provided to basic services.

The entrenched system of universal service subsidies became a principal obstacle to competition in the U.S. telecommunications industry. Whenever a would-be competitor proposed to offer a product or service in competition with a Bell product or service, the Bell companies argued that they could meet the competitor's price only by reducing a subsidy that the Bell product or service provided to affordable, basic telephone service [4]. While these claims were difficult to prove, state and federal regulators treated the Bells' arguments with considerable deference. Largely for this reason, competition was slow in coming to the telecommunications industry.

1968–1996: The Age of Hybrid Regulation

In 1968, the FCC ruled that a non-Bell manufacturer could sell the Carterfone—a device for establishing temporary connections between mobile radios and conventional telephones[5]. The following year, the Commission granted the application of MCI to provide long-distance, private-line service to business customers in competition with the Bell system[6]. With these and other market-liberalizing decisions, the Commission began a gradual transition from a monopoly telecommunications industry to a hybrid collection of competitive and monopoly markets.

The long-distance story, which involved a business long regarded as a natural monopoly, was especially dramatic. Starting in the 1960s, some investors and entrepreneurs were willing to bet that with the advent of microwave radio as an alternative to copper wire, long-distance service could be provided in competition with AT&T and its Bell system operating companies. The FCC went along with the gamble but proceeded incrementally, permitting the new carriers to offer only specialized, business-oriented long-distance services that would have a limited impact on AT&T's revenues. In 1977, however, a federal court of appeals expanded the scope of long-distance competition to include ordinary, switched services—mooting the FCC's cautious policy and exposing AT&T to competition in one of its core markets [7].

The FCC's experiments in competition had one immediate result: the telephone companies were subjected to intense and unaccustomed scrutiny under the antitrust laws [8]. A series of actions, brought by private plaintiffs and the Department of Justice, accused AT&T of a variety of misconduct, including denial of reasonable interconnections to its competitors in the long-distance and equipment markets, and pricing its own competitive products and services below their costs in order to drive competitors from the marketplace.

The result of this prolonged legal assault was the watershed event of U.S. telecommunications history—the divestiture of AT&T. After divestiture, the integrated Bell system was succeeded by an AT&T that no longer offered local telephone service, and by seven regional holding companies that were permitted to offer little else. While AT&T now was free to enter almost any business it wished,[9] the Bell operating companies were prohibited by the terms of a consent decree from manufacturing equipment, providing interexchange services, providing information services, or entering any non-telecommunications line of business [10].

After divestiture took effect in 1984, both the telecommunications industry and its legal environment became even more complex.

As for the industry, technology and competition created important markets (such as cellular telephony and on-line information services) that barely existed in 1984; while new carriers began giving customers ways to bypass the local telephone companies, just as MCI and its imitators once had opened the long-distance network to competition.

On the legal side, the industry's continued evolution revealed enormous contradictions among the separate demands of common carrier regulation, antitrust policy, and competition. No one knew, when new entry first was permitted, how to manage a competitive environment in which some of the players had exclusive franchises and were regulated as common carriers. The courts and the regulatory agencies addressed these tensions in an ad hoc, market-by-market fashion that (whatever the merits of the resulting rules) did not simplify the legal picture. Put as broadly and simply as possible, the changes in the industry presented the courts and regulators with two issues: first, how much competitive freedom to grant the established carriers (especially the local telephone companies); and second, how to regulate new, competitive telecommunications companies.

How Much Freedom for the Established Carriers?

As we noted, divestiture left AT&T with no significant limitations on the businesses it could enter (although it remained subject to regulation of its rates and earnings in the long-distance market). The local carriers formerly affiliated with AT&T, however, were expressly forbidden to enter particular markets; and both the Bell companies and the independent local carriers continued to operate under systems of regulation that denied them flexibility in pricing and required them to maintain artificially high rates for a number of their services.

Not surprisingly, the Bell companies argued that they should be permitted to enter the markets from which they were barred, and both AT&T and the local carriers (Bell and independents alike) asked for greater freedom to set cost-based prices for the markets in which they already faced competition. Courts and regulators balanced these arguments against two other policies: the desire to protect the competitive process from the perceived market power of the established carriers (especially the local companies, which continued to control the local network), and the continuing desire of many regulators (especially state regulators) to preserve as much as possible of the traditional subsidies. Each of these conflicts requires some explanation.

The first obstacle—the tension between the established carriers' demands for greater freedom and fear of their market power—was based on two concerns.

First, there was (and is) a persistent belief that when telcos [11] provide competitive services (such as cellular or information services) that competing companies must provide through interconnection with the local networks of those same telcos, the local carriers can impede their rivals by furnishing them with inferior forms of network access. (As we saw, allegations of such behavior helped bring about the divestiture of AT&T.)

Second, it was (and is) widely believed that local carriers have both the incentive and the ability to price their competitive services below cost and drive competitors from the markets in which this strategy is used. This practice is said to be risk-free for the local carriers, because they can manipulate the regulatory process to ensure that their rates for basic services will be set above the associated costs, thereby subsidizing the below-cost prices they charge for competitive services. (AT&T was accused of using this strategy, as well [12].)

Courts and regulators, in the exercise of their various jurisdictions over the industry, adopted a number of approaches to these perceived problems.

The solutions tended to be of three kinds: the *quarantine, structural separation,* and *nonstructural safeguards.*

The *quarantine* simply forbids carriers from entering markets in which their control of the local network might give them a competitive advantage. The most prominent example of the quarantine approach was the AT&T antitrust consent decree, which prohibited the Bell telephone companies from manufacturing telecommunications equipment or providing interexchange services. Another example was the statutory prohibition against local exchange carrier participation in the cable television business in their telephone service areas.

Structural separation requires carriers to offer competitive services only through separate subsidiaries. This approach reduced the telcos' ability to discriminate against competitors in the terms of network interconnection by requiring the subsidiaries to purchase access to the local network from the parent company on the same terms and conditions as competitors purchased them. And because the subsidiaries provided service through entirely separate personnel and facilities from those used by the parent company, there was thought to be less opportunity to shift any of the costs of competitive service to the regulated side of the house. Some examples of the structural approach were the requirement that Bell companies offer cellular service through separate subsidiaries, and the FCC's short-lived requirement that all so-called enhanced, or information, services of the local carriers be offered through separate subsidiaries.

Nonstructural safeguards were a means of permitting carriers to provide regulated and nonregulated services through the same facilities and personnel. They consisted of accounting and interconnection requirements that made discriminatory interconnection and below-cost pricing harder to achieve and easier to detect. The FCC adopted an extensive set of regulations intended to achieve this result in the enhanced services and customer-premises equipment markets (replacing an earlier, structural separation approach).

The second concern that inhibited regulators in granting greater freedom to the established carriers—the desire to preserve subsidies that support low rates for basic service—arose most starkly when a local carrier found that competitors were charging less for a service than the carrier was permitted to charge, or when the carrier wished to offer a service for which vigorous competition already existed. If the carrier's commission had a policy of requiring all nonbasic services to be priced well above the associated costs in order to support low rates for basic service, the carrier asked for relief from that requirement in order to compete. The most persuasive argument in these cases was that without pricing flexibility the carrier would have to withdraw the service altogether. So long as the carrier agreed to charge a price that contributed *something* to the support of basic service, the carrier could argue that a small subsidy is better than none.

The state commissions wrestled with this issue, especially as they designed and implemented "regulatory reform" plans that were intended (among other goals) to give greater pricing flexibility to carriers within their jurisdictions. The states tended to move cautiously in this area, balancing the legitimate needs of the local carriers for competitive breathing room with the desire to preserve as much of the subsidy structure as possible.

Regulation of New Entrants

Beginning with the limited competitive experiments of the 1960s and 1970s, the FCC and the state commissions had to decide whether and how new, competing service providers should be regulated. Because these companies did not control access to the local network and did not enjoy exclusive franchises, their ability to abuse customers and competitors seemed far less than that of the established carriers. On the other hand, these companies still provided a service "affected with a public interest," and at least some of the justifications for common carrier regulation still might apply to them.

This is another area in which ad hoc decision making was been the rule. As a general matter, new entrants were subjected to far less stringent regulation than AT&T and the local carriers; but the rules varied from service to service and from commission to commission, and even the legal basis for differential regulation could be murky. In fact, the United States Supreme Court found that the FCC lacked the authority to exempt some carriers from the requirement to publish their rates and services in tariffs.

Divided Authority and Differential Regulation

The complexity of telecommunications law was exacerbated by the overlapping jurisdictions of the courts and agencies that made the rules and the tendency of those authorities to impose different requirements on different service providers.

First, consider overlapping jurisdiction. Each court or commission that acted on telecommunications companies had some limitations on its jurisdiction: the court that enforced the AT&T decree could act against the Bell companies, but not against the independents; the FCC could prescribe rules for interstate services, but not intrastate services; the states could regulate intrastate services, but could not make rules for interstate communications. To know the rules under which it could provide a service, therefore, a telecommunications service provider first had to determine who had jurisdiction; and the provider might discover, in some cases, that the FCC, the state commission, and the courts all had something to say about the same service, and had developed arcane rules based on distinctions with no basis in the business or its technology.

Even when a service provider knew who had jurisdiction, it might have to locate its business in a scheme of differential regulation adopted by that authority. Many rules only applied to the larger (and presumably more powerful) carriers; each authority was entitled to draw the line where it thought appropriate. To know its obligations as a telecommunications company, therefore, the service provider might have to know (for any given service) whether it was a "dominant" carrier, or a "Tier 1" carrier, or where it fit in some other scheme of classification.

1996 to the Present: The Age of Competition

As the regulatory compromises of the Hybrid Age grew more unwieldy, many concluded that the system's tensions and contradictions could only be resolved when all telecommunications markets—including the local exchange network—were opened to full, effective competition. Once customers had alternatives to the Bells and the other incumbent carriers for local service, the local bottleneck that necessitated quarantines, separate affiliates, nonstructural safeguards, and price regulation should disappear. Eventually, all telecommunications markets might be deregulated.

The evolution in the climate of opinion was aided considerably by technology, which finally made it possible to think of the local exchange as something other than a natural monopoly. By the 1990s, wireless technologies, new local networks based on optical fiber, and cable television systems all seemed to offer potential alternatives to the existing local, wireline networks. In order for customers to use these alternatives, however, a number of regulatory and competitive obstacles had to be removed.

Most fundamentally, the ability of new competitors to enter the local service market was controlled by the licensing authority of the state regulatory commissions, which varied considerably in their willingness to certify new service providers. By the mid-1990s, the states generally were permitting intrastate toll competition and some states also had authorized operators of alternative fiber networks to provide local service to business customers. There was no reason to believe, however, that all states would permit competitive business and residential service in the local exchange in the near future.

Another equally important obstacle was the continuing control of the incumbents over access to their local customers and exchange facilities. New, local competitors could achieve only a niche presence unless the incumbents completed calls from the competitors' customers to the incumbents' customers, and delivered calls from the incumbents' customers to the competitors' networks. Realistically, incumbents would not offer the required level of cooperation without compulsion—that is, until they were legally required to interconnect with new entrants, and to charge their competitors no more than reasonable, cost-based rates to complete calls placed by competitors' customers to the incumbents' customers.

Finally, in order for local competition to flourish it would be necessary to reform the complex, entrenched system of subsidies that constituted the universal service system. Wherever the incumbents' local service rates—especially rates in rural and high-cost areas of the country—were kept artificially low by implicit and explicit subsidies, new entrants would be discouraged from attempting to compete with the incumbents for local service customers. Ideally, a competitive environment either would eliminate all universal service subsidies, or would provide those subsidies through explicit mechanisms that made support payments available on equal terms to incumbents and competitors alike.

The Telecommunications Act of 1996, which is discussed at length in this book, attempted to achieve all of these goals. The Act required the states to license competing local carriers, and imposed on the incumbents elaborate access and interconnection requirements that went well beyond the obligation to complete and deliver calls to and from their competitors. In fact, the 1996 Act directed the Bells and independents to make the switches, access lines, trunks, and other elements of their local networks available to competitors at cost-based rates, and to sell their local services at wholesale rates to resellers. The Act also ordered the FCC to reform the universal service system to make it explicit and procompetitive.

The Act also went beyond the opening of the local market to furnish a blueprint for the opening of all telecommunications markets to competition and the eventual deregulation of the telecommunications industry. Notably, the Act provided for the entry of telephone companies into the cable television business, and established a framework under which Bell operating companies could become eligible to manufacture telecommunications equipment and provide interexchange service within their local service regions—lines of business from which they were barred by the AT&T divestiture consent decree.

However, as this book discusses in some detail, the short-term effect of the 1996 Act has been to increase, rather than reduce, the complexity of the regulatory environment. The Act, as supplemented by decisions of the FCC, the state commissions, and the federal courts, has created a complex of legal controversies exceeding anything generated by the divestiture of AT&T. And those disputes have been complicated by business and technical innovations, such as the sudden rise of Internet telephony, that were unanticipated even in 1996. What we have chosen to call the Age of Competition, therefore, is so far anything but an age of deregulation.

Conclusion

With this brief, historical background, the succeeding chapters sketch the legal and regulatory environment in which each of the major categories of regulated telecommunications companies operates. From now on, theory and history will appear only in the context of particular issues, and in no more detail than the understanding of those issues requires. As we work through that process, the themes introduced in the last several pages will reappear in many forms.

Endnotes

[1] For example, the Telecommunications Act of 1996 defines "telecommunications" as "the transmission, between or among points specified by the user, of information of the user's choosing, without change in the form or content of the information as sent and received." 47 U.S.C. § 153(43).

[2] Technology has created some borderline cases that strain this—and all other—definitions. Cable television and direct-broadcast satellite service, for example, combine features of both telecommunications and broadcasting.

[3] There was also telegraph service, which predated the telephone by several decades and was, in the context of its time, the greater communications revolution. Western Union, the telegraph behemoth, refused to purchase Alexander Graham Bell's patent in 1876. In the following decades, the telephone slowly but inexorably eclipsed the older technology.

[4] When competitors proposed to sell telephones and other so-called customer-premises equipment, the Bells made the additional argument that such "foreign attachments" could harm the public by causing physical harm to the network.

[5] *Carterfone Service in Message Toll Telephone Service,* 13 FCC 2d 420, *recon. denied,* 14 FCC 2d 571 (1968).

[6] *MCI,* 18 FCC 2d 953 (1969), *recon. denied,* 21 FCC 2d 190 (1970).

[7] *MCI Telecommunications Corp. v. FCC,* 561 F.2d 365 (D.C. Cir. 1977), *cert. denied,* 434 U.S. 1040.

[8] Antitrust is intended to protect the competitive process from monopolistic and collusive interference, and so long as competition in the telecommunications industry was unlawful (that is, while there was no competitive process with which to interfere) antitrust had little role to play. Once limited competition in telecommunications was permitted, however, the size and economic power of the established telephone companies—and especially the Bell system—attracted new kinds of scrutiny.

[9] AT&T was temporarily restrained from entering the electronic publishing business, but that prohibition expired on February 8, 2000.

[10] The prohibitions on information services and nontelecommunications enterprises were removed by the district court, and the manufacturing and interexchange restrictions were supplanted by the Telecommunications Act of 1996, which established the conditions under which Bell operating companies could enter those markets.

[11] "Telco" is a common abbreviation for "telephone company."

[12] The fear is plausible only when two conditions exist: first, where some of the same inputs (e.g., facilities and personnel) are used to provide both regulated and competitive services; and second, where the regulated services are permitted to earn revenues sufficient to recover the associated costs. Under these circumstances, a carrier might allocate some of the costs of providing competitive services to the regulated services, where they can be recovered through rates set by a regulatory commission on a "cost-plus" basis. Incentive regulation plans, which are discussed later in this book, discourage such cost-shifting strategies.

Part I
Incumbent Local Exchange Carriers

Introduction to Part I

The local, land-line telephone company is the most familiar kind of telecommunications service provider. If you are like most residents of the United States, your local telephone company completes your local calls, delivers your long-distance calls to your interexchange carrier, and may provide the access line connecting your computer modem with your Internet service provider.

Before the Telecommunications Act of 1996 became law, almost all local telephone companies (known in the industry as local exchange carriers or LECs) enjoyed legally protected monopolies of local calling within their service areas. The 1996 Act, however, made possible a new class of LECs—known in the industry as competing local exchange carriers, or CLECs—that would offer exchange telephone service in competition with the existing LECs. In order to distinguish the CLECs from the carriers with which they would compete, the 1996 Act referred to the existing LECs as incumbent local exchange carriers, or ILECS [1]. Accordingly, if your local telephone company was providing local telephone service on Februrary 8, 1996—the date the 1996 Act became law—that company is an ILEC.

The distinction between ILECs and CLECs is fundamental. Because ILECs control extensive embedded networks and effectively enjoy a 100% share of the local market within their service areas, they are in a position to prevent competitors from entering or effectively serving local telephone markets. In order to control this power and promote competition, the 1996 Act subjects the ILECs to substantially greater legal obligations than the CLECs. The legal and regulatory environment of the ILEC, therefore, must be understood separately from that of the new carriers with which the ILECs must compete.

Before we describe the legal obligations of ILECs, we should more completely understand what an ILEC does. Imagine, therefore, that you own a company called Peninsula Telephone. Peninsula serves 20,000 telephones from two switches located in two central office buildings.

The physical layout of Peninsula is shown in Figure 1.1, which depicts Peninsula Telephone, along with representative customers and interconnected carriers. IXCs are interexchange carriers,; ISPs are Internet service providers; ILECs are incumbent local exchange carriers other than Peninsula Telephone; and CLECs are competing local exhange carriers. Each customer's telephone is connected to one of Peninsula's switches by a pair of copper wires called a subscriber line (or access line). If one Peninsula customer dials the number of another Peninsula customer, the switch will establish a temporary connection between the two customers' subscriber lines, creating a channel over which the customers can talk [2]. If a customer dials the number of someone outside Peninsula's calling area, that same switch will connect the calling customer's subscriber line to a trunk (a high-capacity line that carries many conversations at once) to transport the call from Peninsula's office to the switch of the calling customer's interexchange carrier. (As Figure 1.1 shows, the connection to the interexchange carrier passes from the customer's serving central office to a specialized switch, called an access tandem, that connects a number of interexchange carriers to Peninsula's local network.) The interexchange carrier then will transport the call to the switch of the called party's LEC (whether ILEC or CLEC), which in turn will deliver the call to its customer [3]. When your company completes the first call—the local call—it is said to provide exchange service. When your company delivers the second call—the long-distance call—to the calling party's interexchange carrier it is said to provide exchange access service [4]. The core business of an ILEC consists in providing these two services.

Your goal, as Peninsula's owner, is that of any businessperson: you want to earn enough revenue from your services to cover your costs and return a profit to the owners (i.e., you and any other investors). Your freedom to act in achieving these goals, however, is limited by two sets of obligations that weigh more heavily on ILECs than on most businesses: regulation (both state and federal) and the antitrust laws.

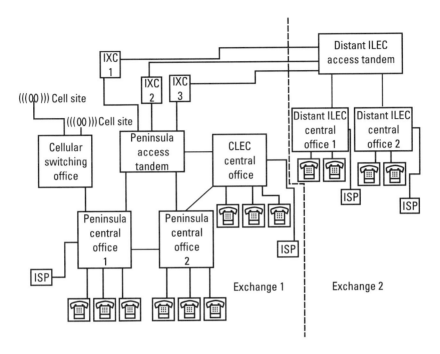

Figure 1.1 Peninsula Telphone.

What follows explains these constraints, not merely in the abstract, but as they affect particular business decisions. We look first at how much an ILEC may earn and how it must price its services, and then at how an ILEC must behave in its dealings with other businesses that seek access to its facilities. Finally, we discuss some additional obligations that apply only to ILECs formerly owned by AT&T.

Endnotes

[1] 47 U.S.C. § 251(c).

[2] When two ends of a conversation are connected by facilities that remain dedicated to the conversation throughout its duration, that connection is called a circuit-switched connection. More recent technologies, including those used by the Internet, transmit several conversations simultaneously over a single facility by breaking those conversations into "packets" that use common facilities and are reassembled into discrete conversations for delivery to their destinations.

[3] Physically, all ILECs may be characterized as combinations of switches, subscriber lines, and trunks. There are enormous differences in scale, of course: some rural ILECs are smaller than Peninsula, while other companies serve millions of subscribers from thousands of switches.

[4] Peninsula also receives incoming long-distance calls from the interexchange carriers and delivers those calls to Peninsula's customers. This function, too, is exchange access service.

1

How Much May Incumbent Local Exchange Carriers Charge for Local Exchange Services?

Because Peninsula is an ILEC rather than an unregulated business, your freedom to set rates for your core services is limited. Your state public utilities commission (PUC) will control your rates for intrastate services, including local exchange service, interconnection rates charged to CLECs, and access charges for intrastate calls carried by interexchange carriers (IXCs). The Federal Communications Commission (FCC) will control your rates for services within its jurisdiction, including access charges for interstate calling. This chapter describes how your state PUC will regulate your rates for local exchange services. Later, in Chapter 3, we describe how state and federal regulators will determine your rates for exchange access and interconnection with other types of service providers.

Regulatory control does not mean that you will wait passively for the PUCs to tell you what your rates will be. In fact, the initiative in ratemaking lies chiefly with the ILEC, which develops proposed rates, presents those rates to the appropriate PUC in a public document called a tariff, and performs the prescribed recordkeeping and reporting on which the PUC will rely in determining whether those rates are justified. The PUCs respond to the ILECs' rate proposals by permitting them to take effect without comment, or by investigating and possibly rejecting them [1]. If a PUC decides to investigate it can choose from a spectrum of procedural options, ranging from modest requests for additional data to extensive evidentiary hearings.

Each PUC has its own rules for determining whether an ILEC's proposed rates are justified, but they all use variants of two approaches—rate-of-return regulation and price-cap regulation.

The rate-of-return approach—once the dominant form of rate regulation—still is used by many state PUCs (especially in regulating intrastate rates of smaller ILECs) [2] and by the FCC in regulating the interstate access charges of some small and mid-size ILECs [3]. Under this regime, the PUC does not try to set individual rates on a service-by-service basis; instead, it decides how much revenue the ILEC may earn overall, and permits the ILEC to set rates calculated to produce that amount. So long as ILEC's earnings do not exceed the allowable maximum, the PUC will not disturb particular rates unless they are "unreasonable" or "unreasonably discriminatory."

The price-cap approach has been adopted by the FCC for regulation of the interstate access charges of the larger ILECs, and is used by most state PUCs as well. Under price caps the PUC sets maximum rates for particular services, but does not establish a maximum level of earnings. If the ILEC can earn more within the rate ceiling by cutting its costs, it is permitted to do so; but if its profits exceed a prescribed level, the carrier may be required to "share" part of the largess with ratepayers through refunds or rate reductions [4].

To show how rates and earnings are set under these two regulatory systems, we now return to Peninsula Telephone and work through some simplified examples of ratemaking.

I. Peninsula Telephone: Local Exchange Service Rates Under Rate-of-Return Regulation

We saw earlier that rate-of-return regulation is based on an overall cap on earnings (a number often called the ILEC's "revenue requirement") rather than scrutiny of individual rates. If Peninsula operates in a rate-of-return state, therefore, your principal regulatory goal is to persuade your state PUC to set a revenue requirement that is adequate to the needs of your business. Once you know your revenue requirement, ratemaking becomes a matter of setting prices for regulated service that are sufficient, in light of anticipated demand, to yield the permitted amount of revenue.

Peninsula's PUC will set the revenue requirement based on a more or less intensive inquiry into Peninsula's expenses and investment [5]. Regardless of the procedure used, the PUC will try to establish a level of earnings that permits Peninsula to do two things: first, recover all of its reasonably incurred operating expenses; and second, earn a reasonable return on the capital invested in the enterprise. The process may be summarized by the equation [E + r(RB)] = RR, where E represents the ILEC's anticipated expenses, r is the permitted rate of return on investment, RB is the rate base (i.e., capital invested in the enterprise), and RR is the company's revenue requirement.

Calculation of E—Peninsula's anticipated expenses—is straightforward. During the time in which the proposed rates will be charged, Peninsula will incur expenses—salaries, rent, fuel, and the like—associated with providing exchange service. Peninsula will tell the PUC what kinds and amounts of expense it anticipates for the years in which the revenue requirement will apply. The PUC may disallow some of Peninsula's expenses as unreasonable, but all allowed expenses will become part of the revenue requirement [6].

The second item—the rate-of-return calculation—requires more explanation.

Like all ILECs, Peninsula is a capital-intensive company: buildings, switches, trucks, and transmission facilities are expensive to acquire, upgrade, and replace; growing demand and advances in technology exert constant pressure to expand the network and replace obsolete equipment and facilities. Peninsula must raise the funds for all of this in the capital markets—by borrowing, selling bonds, and issuing preferred and common stock [7].

The constant need to raise capital creates two sets of claims on the revenue requirement. First, fairness to Peninsula's investors requires that they earn a reasonable return on the money they have contributed to the enterprise. (Bondholders have a contractual right to a specific rate of interest, and shareholders expect—although they are not guaranteed—a return comparable to that paid on securities offering similar risks [8].) Second, Peninsula must protect its creditworthiness and ability to raise additional capital [9].

The rate base/rate of return calculation is intended to satisfy these claims at minimum cost to ratepayers. To determine the appropriate level of earnings, the PUC must arrive at two numbers: the value of the capital invested in the enterprise (the "rate base"), and the percentage return on the rate base that will compensate investors fairly and attract new investment (the "rate of return"). Multiplying these two numbers yields a figure that will be added to the allowable operating expenses to produce the total revenue requirement [10].

To illustrate the setting of the revenue requirement, suppose that Peninsula is facing rising costs and advises its state PUC that it must increase its exchange service rates substantially. The PUC schedules a rate hearing and directs Peninsula to file data supporting all elements of the proposed, new revenue requirement. (We make the simplifying assumption that the PUC's inquiry will ignore all services except local exchange service, and that Peninsula offers only flat-rate local calling, at the same rate for all customers, with no optional features.)

A. Allocating Costs to Local Exchange Service

Before calculating Peninsula's revenue requirement, there is one refinement we must describe (i.e., the correct allocation of costs to the service for which the rates will be charged).

As we noted earlier, Peninsula's PUC is setting the revenue requirement for Peninsula's *local exchange service rates*. It will set a level of earnings sufficient to recover the costs associated with that service only.

But local exchange service is not the only service Peninsula provides. For example, we already saw that Peninsula provides interstate access service, the rates for which are regulated by the FCC [11]. Peninsula also provides fully competitive services (such as yellow pages) that are not regulated by either PUC. Fairness to Peninsula's ratepayers requires that the exchange service revenue requirement not include any of the costs of providing these other services.

Allocating these costs would be simple if Peninsula had three separate divisions—one providing interstate access service, another providing exchange service, and the third providing unregulated service—and if each division had its own plant, equipment, and staff. Then the costs of each division could be directly assigned to the service or services it provided. Like all ILECs, however, Peninsula uses many of its people and much of its plant to provide more than one service. How are these so-called common costs divided?

The FCC and the states have developed elaborate accounting systems to address this problem. One system, called jurisdictional separations, allocates costs between services regulated by the FCC and services regulated by the states [12]. Another system, usually called Part 64 accounting after the section of the FCC rules in which it is addressed, allocates costs between regulated and nonregulated services [13].

We may assume that Peninsula has complied with all cost allocation requirements, and therefore has appropriate cost data with which to support its request for an exchange service rate increase. We turn now to the calculations it will present to its PUC.

B. Calculating Peninsula's Revenue Requirement

To calculate Peninsula's operating expenses, you choose a recent year for which actual cost figures are available. (The year chosen for this purpose is called the "test year.") You may find, for example, that in the most recent calendar year Peninsula paid $200,000 in salaries, $60,000 in rent, and $15,000 in taxes; and that annual depreciation on Peninsula's tangible property totaled $40,000. You also know that your employees have a new union contract that will raise your salary expense to $240,000, that your new lease on the central office includes a rent increase to $70,000, and that your taxes have been raised to $18,000. You adjust the test year numbers upward accordingly, and present your PUC with the following operating expense numbers (see Table 1.1).

Your next task is to calculate the rate base portion of the revenue requirement. As we noted earlier, the rate base is the present value of the investment the bondholders and stockholders have made in the business, and on which those investors expect a reasonable return. The rate base typically consists of the depreciated value of the capital assets (buildings, plant, and equipment) used to provide exchange service [14], plus any working capital (cash, stockpiled supplies, and the like) needed by the enterprise [15].

We may suppose that Peninsula's tangible property consists of its building, constructed 10 years ago for $100,000; the switch, purchased 10 years ago for $200,000; and access lines and trunks constructed 10 years ago at a total cost of $500,000. We also assume that Peninsula's PUC values property at its original cost, and depreciates property on a straight-line basis

Table 1.1
Operating Expenses

Operating and Maintenance	$310,000
Depreciation	$40,000
Taxes	$18,000
Total Operating Expenses	$368,000

over a useful life of 20 years [16]. Our rate base will be calculated as follows (see Table 1.2).

All that remains is to multiply the rate base by the rate of return and add the result to the operating expenses.

The rate of return, as we mentioned earlier, is the percentage by which the rate base must be multiplied to determine how much money will be needed to pay the investors (and perhaps provide some retained earnings, as well). How is the rate of return calculated?

Most PUCs set the rate of return by determining the ILEC's *cost of capital*—in other words, the return an ILEC must offer in order to attract investors. At the time it set Peninsula's rate of return (probably during the most recent rate case), the PUC examined Peninsula's capital structure and arrived at a weighted average of the cost of Peninsula's capital [17].

Assume that your PUC decided during the last rate case that a rate of return of 12.5% will compensate Peninsula's investors adequately and assure the markets that your company treats capital well. You multiply this percentage by your $500,000 rate base, giving you a return of $62,500. Adding this to your operating expenses gives you a total revenue requirement of $430,500.

C. A Word About Depreciation

Peninsula's operating expenses and rate base each included a reference to "depreciation." The reader may have noticed that those items, which we left

Table 1.2
Intrastate Rate Base

Telecommunication plant in service	$800,000
Plant under construction	—
Property held for future use	—
Total plant	$800,000
Less: accumulated depreciation	$400,000
Net plant	$400,000
Working capital	$100,000
Rate base	$500,000

undefined, had a substantial impact on the revenue requirement: depreciation increased expenses by $40,000 and reduced the rate base by $400,000.

Because ILECs invest so heavily in tangible property, regulators' decisions as to how that property is valued and how rapidly it may be depreciated can have a substantial impact on an ILEC's earnings—especially where that ILEC's rates are regulated on a rate-of-return basis. For this reason we should not leave the subject of regulation of Peninsula's earnings without saying a few words about depreciation.

The *concept* of depreciation, at least, is simple to state: it is the accounting recognition of the fact that tangible goods decline in value over time. This loss of value may have a number of causes: the object may become physically exhausted from wear and tear, or may lack the capacity to meet growing demand, or may have been made obsolete by advancing technology.

Depreciation only applies to durable goods, rather than supplies and equipment that will be used up over the short term. (The FCC draws the line, for this purpose, at goods with an expected useful life of more than one year.) If Peninsula spends $5,000 on a month's supply of fuel for its trucks, for example, that entire $5,000 goes into the revenue requirement for that year (as an operating expense) and is recovered from current rates. If Peninsula buys a computer for $5,000, however, the computer (which has a useful life greater than one year) is treated as an asset rather than a current expense, and only part of the $5,000 is recovered from rates charged in the year the computer is purchased.

How much of an asset's cost will be recovered each year depends on the useful life assigned to the asset and the method of depreciation used. For example, if the computer has a useful life of five years and is depreciated on a straight-line basis, Peninsula will claim operating expenses of $1,000 each year for five years, and will be permitted to charge rates sufficient to recover $1,000 in each of those five years. At the end of five years, the owners of Peninsula will have fully recovered their investment in the computer.

During the five years that Peninsula recovers the cost of the computer from its ratepayers, it also will reflect the decline in the computer's value in its rate base calculation. So, for example, at the end of the first year Peninsula not only will recover $1,000 from its ratepayers to cover depreciation expense; it also will reduce its rate base by $1,000 to show that the computer now is worth only $4,000. If the rate of return allowed on the rate base for that year is 10%, then the effect of depreciation on the revenue requirement is easy to calculate: depreciation raises operating expenses by $1,000 and reduces the rate base by $100 (10% of $1,000), for a net increase in Peninsula's revenue requirement of $900.

In the real world, of course, depreciation accounting is a complicated thing to implement. ILECs buy thousands of durable goods, with widely different useful lives, at different times. In order to track all of this property and impose realistic accounting requirements, the FCC's Uniform System of Accounts has created 22 categories of goods, which are disaggregated further into smaller categories, which then are classified into groups with the same year of initial use and groups with roughly equal life expectancy. The FCC values all of these items at their original cost, and depreciates them on a straight-line basis.

Historically, it has been deeply important to rate-of-return ILECs that they be allowed to depreciate their durable property at a realistic rate—a matter that is in the hands of their regulatory PUCs [18]. The money recovered from ratepayers for depreciation expense, unlike money recovered to pay wages or buy fuel, can be used to invest in a new plant rather than spent to meet current requirements. Depreciation charges create a pool of cash reserves on which ILECs rely heavily to fund their construction budgets.

Regulation of depreciation is substantially less important, however, to ILECs subject to rate regulation on a price-cap basis, as described later in this chapter. Where a PUC's price-cap rules require carriers to "share" some portion of their profits with ratepayers, high depreciation rates will permit ILECs to earn more revenue before the sharing requirement is triggered. Where a PUC does not require sharing, however (as the FCC does not), prescription of depreciation rates by regulatory PUCs is largely anachronistic. In apparent recognition of this fact, the Telecommunications Act of 1996 makes FCC prescription of depreciation rates optiona [19].

D. Peninsula's Rate Structure

Now that you know how much Peninsula is permitted to earn, overall, from exchange services, your next task is to charge rates that, in light of anticipated demand, are reasonably calculated to produce that amount. This takes us from the question of Peninsula's overall rate *level* to the question of Peninsula's rate *structure*.

To simplify our discussion of rate structure, we assume that Peninsula will charge each customer a flat monthly rate for local exchange service, and will not charge rates that vary with usage. With this assumption, and supposing Peninsula to have only one kind of exchange service to offer, the question of rate structure seems to answer itself: divide the revenue requirement by the total number of customers and charge each customer the quotient. How can it possibly be more complicated than that?

To show how it can be more complicated, consider three Peninsula customers, all of whom purchase the same exchange telephone service. One customer lives on a farm several miles out of town, at the end of a mile-long access line; another is a young woman, living near the central office, who works all day and uses her telephone to make occasional, personal calls in the evenings and on weekends; and the other is a druggist, operating a pharmacy down the street from the central office, who uses his telephone to take dozens of prescription orders each day.

If you were designing Peninsula's rate structure on a clean slate, you might notice that the cost of serving each of these customers is not the same. The farmer has caused the company to make a substantial capital investment (the mile-long access line) for the sole purpose of providing him with service; if he pays the true cost of his service, he certainly will pay more than the customer who lives in town. Similarly, the druggist, who uses his telephone heavily during peak usage hours, is costlier to serve than the working woman who calls only at off-peak times [20].

If the rate structure reflected all of these cost differences, the working woman would pay least for her service, and the druggist and the farmer each would pay more. And to ensure that traffic-sensitive costs were faithfully recovered, all customers would pay for local service on a metered-use (by the minute or by the call) basis, with higher rates during peak calling periods.

Now assume that you are not writing on a clean slate, but are following the ratemaking policies of a typical state PUC. In the real world, the rates you charge each of your three customers will be based not on the costs you incur to serve them, but on regulatory policies designed to encourage universal service [21].

One such policy is geographic averaging (discussed in greater detail in Chapter 11). Under this policy, the farmer will not pay the full cost of his mile-long access line; instead, he will pay the same rate as the customer who lives across the street from the central office. The farmer's access line is subsidized by the urban ratepayer. Also, if an ILEC serves so many rural and high-cost customers as to raise its cost of providing access lines substantially above the national average, the company will recover explicit subsidy funding from the High-Cost Fund of the universal service system.

Another policy—often called value of service pricing—imposes above-cost rates on business customers and uses the surplus to subsidize basic, residential service [22]. (This subject also is discussed in Chapter 11.) The rationale is that business customers (such as our druggist) derive substantial economic benefit from their telephones, while residential subscribers do not. So the druggist pays more—not because he uses his telephone often, and during peak calling hours—but simply because he has been targeted to provide a subsidy.

Designing Peninsula's rate structure for basic exchange service, therefore, will be a bit more complicated than simply dividing the revenue requirement for the service among the number of subscribers. The departures from this model will be dictated primarily by regulatory policy rather than differences in cost of service.

E. Discriminatory Rates: The Regulator's View

As our discussion of rate structure shows, different customers may find themselves paying different prices for services that appear quite similar. Where customers feel aggrieved by these disparities, they may complain to the appropriate PUC that the rates are discriminatory. Since PUCs typically operate under statutes and regulations that forbid "undue" or "unreasonable" discrimination among telephone subscribers, these complaints must be heard.

Obviously, a claim of discrimination will not succeed merely because the complainant is paying more than the cost of his or her service, or is subsidizing some other ratepayer's service. Only if the subsidy or disparity is not among those sanctioned by regulatory policy does the claim have a chance of success.

In bringing and defending claims of undue discrimination it should be kept in mind that PUCs have broad latitude to improvise policy. Consider, for example, the decision of the District of Columbia's Public Service PUC, denying in part a requested increase in the rates for semipublic telephones (i.e., pay telephones located inside places of business) not because they are more "basic" than other services, but because they are less attractive to drug dealers than public telephones and should be encouraged [23].

F. Discriminatory Rates: The Economist's View

Economists may lack the policy perspective of regulators, but they do understand efficiency; and they generally agree that when the price of a service differs sharply from its cost, that price causes an inefficient allocation of society's scarce resources. This principle, in turn, yields an elegantly simple definition of price discrimination: for an economist, it is discriminatory to charge different rates to customers who cost the same to serve, or to charge the same rate to customers who do not cost the same to serve [24].

It is commonplace to observe that the economist's perspective is shown little deference in traditional ratemaking [25], but economic efficiency is becoming more influential in ratemaking for competitive services and in the development of nontraditional, or incentive, regulation. We shall have more to say about cost-based rates, therefore, when we consider these subjects.

II. Peninsula Telephone: Local Exchange Service Rates Under Price-Cap Regulation

Most state PUCs have grown dissatisfied with traditional regulation and have adopted versions of incentive, or price-cap, regulation.

The perceived problem with rate-of-return ratemaking is that it discourages productive efficiency [26]. So long as the only way for an ILEC to increase its earnings is to increase the size of its rate base, ILEC managers will have an incentive to overinvest. The result, of course, is higher rates for consumers of the ILECs' services [27].

Price-cap regulation tries to create incentives that favor cost reduction rather than cost padding. It does this by setting maximum rates, rather than a maximum level of permitted earnings, and letting the ILECs keep all or most of any profits achieved by cost cutting. In this sense the price-cap environment resembles competition: ILECs that fail to become efficient will earn less, while ILECs that cut costs aggressively will earn more [28].

Price-cap regulation is simple in concept, but somewhat more complex in practice. Notably, because the PUCs do not want ILECs to do *too* well under the new regime, many regulators have added two ongoing, corrective processes called *indexing* and *sharing*.

To illustrate how indexing and sharing work, assume that Peninsula's PUC has put in place a price-cap regime. At the time the new rules take effect, Peninsula is charging $30 a month, per subscriber, for local residential exchange service. Peninsula chose that rate to achieve the revenue requirement set at the conclusion of the most recent rate case, and Peninsula's PUC decides that the rate is presumptively valid and useful as a benchmark. The PUC tells Peninsula, therefore, that it may continue to charge $30 a month for one more year, at which time the rate will be indexed according to the following formula [29]:

$$R_t = [R_{t-1}(1 + I - X)]Z$$

R_t in this formula is the new rate Peninsula will be required to charge after the first year, and R_{t-1} is the old rate of $30.00 per month. The rest of the formula tells us how to get from the old rate to the new one by accounting for certain changes in Peninsula's costs that are expected to occur during the first year under price caps.

The first of these changes, represented by I, is the anticipated percentage increase in the prices of the goods and services Peninsula must buy in order to provide exchange service—in other words, the rate of inflation. We will assume that the PUC defines I according to the Gross National Product Price Index (GNP-PI), and finds a change in GNP-PI of 3.5% over the first year [30].

The second cost change, represented in the formula by X, assumes that the cost of providing service in the telecommunications industry will continue to decline. Specifically, X is the rate by which cost reductions (productivity gains) in the telecommunications industry are expected to exceed productivity gains in the economy as a whole [31].

This so-called productivity index is intended to force Peninsula to reduce its rates to match, not its own improvements in cost control, but those typical for its industry. In other words, Peninsula is strongly encouraged to be at least as productive as the average company in its business. We will assume that Peninsula's PUC adopts a productivity index of 4.5% [32].

The last cost change, represented by Z, is the increase or decline in so-called "exogenous" costs. Exogenous costs are those caused by governmental actions that affect LECs exclusively or disproportionately. So, for example, a tax increase that affected all corporate taxpayers equally would not be included in the Z factor; but a new tax levied exclusively on telephone-switching equipment, or a change in the separations formula, would be included.

To calculate Z, Peninsula must add the *change* in exogenous costs (up or down) to the revenues Peninsula expects to earn from exchange service in the coming year, and divide that sum by the same revenue figure. This will produce a percentage by which rates may be increased or decreased to account for exogenous costs. For purposes of our illustration, assume that a telephone-switch tax has been imposed and will cost Peninsula $500 in its first year under price caps, against revenues of $100,000.

So to index Peninsula's $30.00 rate and come up with a new one, we plug into the formula an economy-wide inflation rate of 3.5%, a productivity index of 4.5%, and exogenous costs of $500 on anticipated revenues of $100,000 [33]. This gives us the following calculation:

$$R_t = [R_{t-1}(1+I-X)Z]$$
$$R_t = [30(1+.035-.045)][(100{,}000+500)/100{,}000]$$
$$R_t = [30(.99)](1.005)$$
$$R_t = 29.70(1.005)$$
$$R_t = 29.8485$$

Peninsula's new monthly rate for basic exchange service, therefore, is $29.85.

If the indexing formula works as intended, and if the inflation and productivity numbers match Peninsula's actual experience over the first year, then Peninsula will recover its operating expenses and earn the same return on investment by charging $29.85 that it made by charging $30.00. If Peninsula cannot at least match the anticipated cost reductions of the average company in its industry, however, it will fare more poorly under price caps. If Peninsula can achieve cost reductions that exceed the industry average, it will earn profits that it could not have achieved under rate of return.

But not *too* much profit. Peninsula's PUC, like the public utility PUCs of many states, will require Peninsula to refund to its ratepayers all or part of any earnings that exceed a so-called benchmark rate of return.

Assume, for example, that Peninsula's PUC allows the company to keep 100% of earnings up to a return of 11.5%, but allows it to keep only 50% of any earnings above that level. If Peninsula is relentlessly efficient over its first year and reduces its operating expenses by $50,000, it will earn $110,347.50 over its costs (assuming that demand stays at the same level as in year t−1) even though its maximum rate has been lowered to $29.85. If Peninsula's depreciated investment (what used to be called its rate base) in the first year is still $500,000, an 11.5% return on that rate base would be only $57,500; so Peninsula must refund half the difference between $57,500 and $110,347.50, or $26,423.75.

Even with sharing, Peninsula does better under this price-cap formula than it would have done under rate of return. After refunding half the excess over its benchmark of 11.5%, it still earns a rate of return of almost 17% for its shareholders. Incentive regulation, therefore, remains an attractive option for ILECs that have confidence in their ability to cut costs.

Incentive regulation does raise one obvious question, however: since most subscribers have no alternative to their local ILEC as a source of telephone service, and if price caps make cost cutting the royal road to profits, what will prevent the ILECs from sacrificing quality of service to achieve greater cost reductions? The PUCs generally have instituted quality-of-service reporting requirements, and reserve the right to monitor quality more rigorously (or even to impose explicit service quality standards) if incentive regulation appears to have the unintended effect of degrading service [34].

Finally, we should point out some states not only have replaced rate-of-return regulation with incentive regulation, but are deregulating—or studying the possibility of deregulating—certain local exchange service rates altogether. Notably, as of September 1999, Idaho had deregulated all non-basic local exchange services and Nebraska had ended rate regulation for all nonresidential local exchange services [35]. It is reasonable to expect, however, that most states will deregulate ILEC local exchange service rates only after the ILECs' share of that market is significantly smaller than it is today.

Endnotes

[1] The PUC may investigate on its own initiative or in response to a complaint that the tariffed rate is unlawful.

[2] As of September 1999, price-cap regulation had replaced rate-of-return regulation for local exchange services in 70% of states and the District of Columbia. "Price-Caps Standard Form of Telco Regulation in 70% of States," *Communications Daily*, Sept. 8, 1999.

[3] In a 1997 order, the FCC noted that companies subject to rate-of-return regulation for their interstate services accounted for only about 10% of the nation's telephone access lines. In the Matter of Regulatory Reform for Local Exchange Carriers Subject to Rate-of-Return Regulation, 12 FCC Rcd. 2259 § 5 (1997): most of the companies still subject to rate-of-return regulation for their intestate services were small companies (defined, among other criteria, as carriers with fewer than 50,000 access lines) and mid-size companies (defined as carriers with between 50,000 and approximately 1,000,000 access lines). *Id.*; *see also* Regulatory Reform for Local Exchange Carriers Subject to Rate-of-Return Regulation, 7 FCC Rcd. 5023, 5024 (1992); Regulation of Small Telephone Companies, 2 FCC Rcd. 3811, 3812 (1987).

[4] The Federal Communications PUC eliminated the sharing requirement from its price-cap rules in 1997. Price-Cap Performance Review for Local Exchange Carriers, 12 FCC Rcd. 16642, 16700 (1997).

[5] Comprehensive, formal investigations into all of an ILEC's expenses and investment are infrequent. In the intervals between such "rate cases," adjustments to the revenue requirement may be made on the basis of more limited inquiries.

[6] Disallowed expenses are said to go "below the line" (i.e., they are absorbed by the ILEC's shareholders rather than its ratepayers).

[7] Not all carriers have access to all of these capital markets. Many of the smaller, rural ILECs obtain their financing primarily through loans from the Rural Utility Service (RUS) of the United States Department of Agriculture.

[8] This requirement has a constitutional dimension. A PUC that deliberately prevented investors from achieving a reasonable return would commit an impermissible, uncompensated "taking" of private property without just compensation. U.S. Const. amend. V; *Reagan v. Farmers' Loan & Trust Co.*, 154 U.S. 362 (1894); *Stone v. Farmers' Loan & Trust Co.*, 116 U.S. 307, 331 (1886). This does not mean that the PUC must guarantee the rate of return; but it must give the ILEC a fair opportunity to achieve it.

[9] The best way to satisfy the second of these claims, of course, is to satisfy the first. If existing investors are earning an adequate return, new capital will be attracted to the company at reasonable cost.

[10] The revenue requirement may be expressed as the formula $R = O + (V - D)r$, where R is the revenue requirement, O is the total of operating expenses, V is the value of the property used in the regulated business, D is the accumulated depreciation on tangible property, and r is the rate of return.

[11] Peninsula also provides interconnection and related services for CLECs and access service for intrastate toll calls carried by interexchange carriers. These rates also are regulated by Peninsula's state PUC.

[12] 47 C.F.R. § 36.1 *et seq.*

[13] *Id.* §§ 64.901–64.904.

[14] It often is said that tangible property must be both *used* and *useful* in providing regulated service. Investments that do not meet both of these criteria may be disallowed for ratemaking purposes.

[15] These are the principal items in the rate base, but others may be included where appropriate. Peninsula may be holding property for future use, or may have intangibles of an ascertainable value (such as water rights). Offsets also may be required: customer-service deposits, for example, are considered contributions to capital and must be excluded from the ratebase on the ground that investors are not entitled to a return on capital contributed by customers.

[16] We say more about depreciation at pages 10–12.

[17] The PUC will arrive at a separate cost of capital for each type of instrument the ILEC issues, by examining the returns earned on the issues of other companies that offer their investors comparable risks. The PUC then calculates a weighted average of those separate rates of return, reflecting the percentage of the ILEC's capital that each of those issues represents.

[18] The FCC regulates depreciation of tangible goods to the extent those goods are used to provide interstate service. The states may apply different rates and methods of depreciation for those costs of durable goods that are allocated, through the jurisdictional separations process, to intrastate service. *See* Louisiana Public Service *PUC v. FCC*, 476 U.S. 355 (1986).

[19] 47 U.S.C. § 220(b). For a discussion of the declining relevance of depreciation regulation, see Cincinnati Bell Telephone Company, Southwestern Bell Telephone Company, and US WEST Communications, Inc.; Prescription of Revised Depreciation Rates, 13 FCC Rcd. 6221 (1998)(separate statement of PUCer Furchtgott-Roth).

[20] The druggist, along with other peak-hour callers, causes *traffic-sensitive* investment (such as switching capacity) that would not otherwise be necessary. The farmer, along with other rural customers, imposes *nontraffic-sensitive* costs (such as the extended access line) that would not otherwise be necessary. We have more to say later about the impact of traffic-sensitive and nontraffic-sensitive costs on the setting of access charge rates.

[21] Although some dismiss universal service as a politically inspired anachronism, in fact many households still lack basic telephone service. The 1990 U.S. census counted five million housing units without telephone service: among African-American households the figure was 15%, and among Native Americans it was 20%. See, e.g., Doyle, "No-Phone Homes," *The Atlantic Monthly*, June 1993, p. 77.

[22] Optional residential features (e.g., touch-tone dialing, call waiting, and call forwarding) also are priced above cost and also subsidize basic service.

[23] Chesapeake and Potomac Telephone Company, 130 PUR 4th 310, 424 (District of Columbia Public Service PUC 1992).

[24] We say more about economically efficient pricing in Appendix A.

[25] "In this area (of rate-discrimination claims), the PUCs typically proceed only in response to particular complaints and all too often, from the economist's standpoint, the PUCs resolve such controversies on bases other than economic efficiency." Kahn, A., *The Economics of Regulation*, pp. 55–56, 1988.

[26] Productive efficiency is achieved when firms minimize their costs of production. See discussion at pages 223–228.

[27] Stated more carefully, the problem is that when the authorized rate of return exceeds the cost of capital, the ILEC is encouraged to overinvest. When the authorized rate of return is lower than the cost of capital, the ILEC is encouraged to underinvest. See, e.g., Averch and Johnson, "Behavior of the Firm under Regulatory Constraint," 52 *American Economic Review* 1052, 1962. Although the PUCs scrutinize operating expenses and investment to ensure that ILECs do not incur unnecessary costs, regulators are not parallel sets of managers; their ability to second-guess thousands of financial decisions is limited.

[28] We compare the incentives created by price caps with those typical of competitive markets at pages 235–237.

[29] The formula we use here is based on the one adopted by the California PUC for the intrastate services of Pacific Bell and GTE California, Inc. See Alternative Regulatory Frameworks for Local Exchange Carriers, 107 PUR 4th 1, 85 (California Public Utilities PUC 1989) (California Price-Cap Decision).

[30] It might seem more accurate to rely on the anticipated increases in the prices of the actual goods and service purchased by the ILEC, rather than a less-discriminating, economy-wide measure of inflation. The disadvantage of this more specific approach is that it would rely on company-supplied data, which the PUC would have to scrutinize. As the California Public Utilities PUC pointed out when confronted with this question, such an exercise would approach the complexity of a rate case and frustrate a secondary purpose of price-cap regulation (i.e., to simplify the administration of rate regulation). *Id.* at 89.

[31] The telecommunications industry has outperformed the productivity of the economy as a whole for decades. *See*, e.g., In the Matter of Policy and Rules Concerning Rates for Dominant Carriers, 3 FCC Rcd. 3195, 3401-02 (1988).

[32] As with the inflation index, the PUC will not accept actual, historic productivity gains by Peninsula as the basis for calculating X.

[33] Strictly speaking we would include, not only the exogenous costs themselves, but a benchmark rate of return, established by the PUC, on those costs—another survival of rate-of-return regulation.

[34] *See*, e.g., *California Price Cap Decision,* 107 PUR 4th at 151.

[35] "Price Caps Standard Form of Telco Regulation in 70% of States," *Communications Daily*, Sept. 8, 1999.

2

How Much May an ILEC Charge for Nonbasic Services?

As the previous chapter points out, the FCC and state public utilities commissions exercise considerable control over the rates ILECs charge for their core local exchange and exchange access services. Some commissions regulate those rates indirectly, by setting caps on the ILECs' overall earnings and permitting the ILECs to set rates at levels that will achieve those earnings (the rate-of-return approach). Others regulate those rates directly by setting caps on particular rates and letting ILECs retain some or all of the profits they achieve by reducing their costs of service (the incentive, or price-cap, approach). The commissions justify their continuing control over basic ILEC rates on the ground that ILECs still have a monopoly of local exchange telephone and exchange access services, and that so long as this monopoly persists ILECs will have strong incentives to charge excessive (i.e., above-cost) rates for those services [1]. Accordingly, the commissions control rates or earnings to ensure that ILECs do not exploit their monopoly power to the detriment of consumers.

But in addition to the basic telephone services for which they do not face effective competition, ILECs offer Internet access, wireless service, "Yellow Pages" directories, telephone equipment, and other products and services in competition with independent suppliers. Because the ILECs do not enjoy monopoly positions in these markets, the usual rationale for rate regulation should not apply to the rates charged for these services. Are the ILECs' rates for nonbasic services, therefore, entirely unregulated?

For now, and probably for some years in the future, this question must be answered in the negative. Regulators will continue to enforce rules that limit the freedom of ILECs to set rates, even where the services in question are subject to competition, for at least two reasons. First, some commissions still want rates for nonbasic service to subsidize local, basic, residential service [2]. Second, regulators want to ensure that ILECs do not set above-cost rates for monopoly services and use the earnings from those services to cross-subsidize their competitive service rates. Each of these policies, and the rules by which regulators seek to achieve them, requires some explanation.

I. Pricing Nonbasic Services Under Rate-of-Return Regulation

Under traditional rate-of-return regulation, the commissions dealt with the diversity of ILEC services by requiring ILECs to price all of their nonbasic and business offerings at a premium, creating politically popular subsidies that kept basic residential rates artificially low [3]. So long as no one was allowed to compete with ILECs in their core markets, and so long as ILECs did not enter markets in which competition already existed, this system of subsidies worked quite well.

To show how competition changes everything, let's return to your role as owner of Peninsula Telephone. We assume, for the purpose of this discussion, that your state commission still regulates Peninsula's rates on a rate-of-return basis.

For many years Peninsula has been offering its large business customers a service called off-premises switching, or OPS. With this service a business with many employee telephones can have station-to-station dialing, direct dialing to each station from outside the business premises, and a single directory number for the business—all without the need to route calls through an attendant [4]. At first this service faced no competition and was quite successful.

Over the last several years a number of your best customers have canceled OPS in favor of increasingly sophisticated switching equipment that can be installed on their premises. These private branch exchanges, or PBXs, offer the same capabilities as OPS at substantially lower cost. When you ask these customers how you might recapture their business, they all give you the same response: you must lower your rates dramatically, and you must offer your customers some measure of pricing stability (i.e., enter into long-term contracts that protect them from increases the state commission might require to the tariffed OPS rates).

Next, you make an appointment with a member of your state commission's tariff review staff. At the meeting, you point out to the staffer that the PBXs available to your customers duplicate every feature of OPS at a drastically lower price. Unless you are allowed to lower your rates and offer your customers some measure of price stability, OPS service will disappear entirely.

The staffer explains the commission's ratemaking philosophy. The commission's primary legislative mandate is to keep basic telephone service as inexpensive as possible. In order to accomplish this purpose, all business services and all household services classified as nonbasic (such as TouchTone dialing, custom calling, and the like) are priced at a steep premium. When the commission has squeezed all the revenue it can from the (presumably more affluent) ratepayers who purchase the luxury services, it recovers the rest of the ILECs' revenue requirements from basic services at rates that probably are below cost [5].

Warming to her subject, the staffer counsels you not to take this news too hard; after all, the telephone business is rife with subsidies. The jurisdictional separations process, under which ILECs recover a percentage of their costs from intrastate rates and a percentage of their costs from interstate rates, still assigns a disproportionate share of nontraffic-sensitive costs to the interstate jurisdiction, providing a subsidy to local service generally. Similarly, rural ratepayers, who cost more to serve than city dwellers, are subsidized by urban ratepayers. She explains that the process is politically driven and will not be changed by mere logic.

You refuse to be discouraged, however, and you take your case to the full commission. You tell the commissioners that while you have no quarrel with universal service and are happy to have OPS contribute to the reduction of basic rates, the ratepayers are better off with a modest subsidy than with none at all; and OPS will contribute nothing to the cost of basic service if no one buys it.

You also report that you have done some checking on the costs Peninsula actually incurs to provide OPS. Your accountants and engineers tell you that when Peninsula decided to offer OPS it made no new capital investment; it bought some switch software for an annual license fee of $1,000, and the company spends about $10,000 a year to promote the service, program the switch with customers' routing instructions, and perform maintenance. You propose to the commission that you charge a rate that recovers those costs, plus a modest contribution toward the costs of other services.

The commission agrees that the existing premium for OPS service cannot be sustained in the face of competition, but it disagrees with your proposal to recover only the additional costs incurred to provide OPS. The commissioners prefer that the service be priced at a rate that recovers, not just the costs specifically incurred to provide OPS, but a share of Peninsula's costs incurred to provide basic services, as well.

This suggestion strikes you as contrary to the whole idea of competitive pricing. You are confident that your competitors, the PBX manufacturers, do not price their products to recover the costs they incurred to make other products. Because the markets in which they operate are competitive, their prices for individual products are driven rigorously toward the cost of making each product. Regulators can afford to load prices with irrelevant costs, but competitors cannot.

At this point you and the commissioners have joined a familiar debate about how costs of a multiservice firm should be attributed to individual services. The alternatives suggested usually are variants of two approaches: the so-called incremental cost approach, and the fully distributed (or fully allocated) cost approach. Your suggestion falls into the first category; the commissioners' suggestion falls into the second. The FCC has explained the difference this way:

> A fully allocated system divides the entire cost of a group of products among those products through direct assignment or by some allocation factor. The allocation factor does not necessarily reflect the costs caused by any particular service. An incremental costing system measures the additional cost created by the production of one particular product, given a set of other products that are being produced [6].

As the PUCs abandon traditional "cost-plus" regulation in favor of price caps or deregulated pricing, this debate over incremental and fully allocated cost pricing becomes less critical [7]. In states where rate-of-return regulation still is applied, however, and where the PUCs still encourage ILECs to subsidize basic telephone services, ILECs may continue to be required to price nonbasic services at a level that reflects, either an outright allocation of the costs of other services to the competitive offerings, or the incremental cost of the new service plus a "contribution" to the cost of basic service.

In addition to their desire to preserve subsidies favoring basic ratepayers as much as possible, regulators have another concern when faced with requests for flexible regulation. They are concerned that ILECs not only will stop subsidizing basic service, but will reverse the process and use their revenues from basic service to subsidize the new, competitive services.

Imagine, for example, that Peninsula was allowed to set its rates for OPS at any level it chose, with no commission scrutiny of the rates or their relationship to the associated costs. Because Peninsula faces price competition from the PBX manufacturers, it has a strong incentive to lower its rates and to underprice its competitors, if possible. Such price reductions will not impair profits if the costs of OPS can be reduced proportionately.

Like all companies, Peninsula can try to reduce its costs by becoming more efficient. But unlike most companies, Peninsula has another option (in theory, at least): it can price OPS at a level that does not recover the costs of the service, knowing that the costs will be included in Peninsula's overall revenue requirement and recovered from basic rates. (Such strategies also may lead to predatory pricing claims under the antitrust laws, about which we shall have more to say in a moment.)

This so-called cross-subsidy problem has induced rate-of-return commissions to couple flexible regulation of competitive service rates with various cost-reporting requirements. The ILECs must ascertain the costs incurred to provide their various services or categories of services (either on an incremental or fully allocated basis), and follow reporting procedures meant to assure the commission that their competitive services are recovering those costs and making a contribution toward the costs of basic services.

II. Pricing Nonbasic Services Under Price Caps

As we discussed in the previous chapter, price-cap regulation is becoming the dominant method of controlling the rates that ILECs charge for their basic services. As we also discussed, the theory of price caps is that consumers will benefit from a bargain in which ILECs agree to reduce their rates and regulators allow the ILECs to keep any profits they earn by reducing their costs to even lower levels.

Under price caps, no less than under the rate-of-return approach, many regulators pursue the twin goals of subsidizing basic service and preventing the subsidization of competitive services. Both goals are served by rules that limit the flexibility of ILECs to raise their rates for basic services and lower their rates for competitive services.

The FCC's price-cap rules are a complex example of this approach. When the Commission created its price-cap regime for the interstate services of local exchange carriers, it grouped those services into four "baskets":

(1) common-line services, (2) traffic-sensitive services, (3) special access services, and (4) interexchange services [8]. Each of these baskets was subject to its own price cap, thereby limiting the extent to which rates for particular services could be raised or lowered in relation to rates for other services. The FCC also created "bands" of prices, which placed a 5% zone of flexibility above or below the price cap in each basket. The Commission presumed that rates within the bands were reasonable, but required cost support for changes that exceeded the bands. The upper band effectively preserved existing subsidies by preventing ILECs from sharply raising the prices of basic services, and the lower band was expressly intended to prevent ILECs from pricing their competitive services at below-cost levels [9].

State PUCs also have structured their price-cap regulations to preserve existing subsidies and prevent subsidization of more competitive services with earnings from less competitive services. Like the FCC, the states commonly group services in separate categories and limit pricing flexibility within each category.

Notably, the states have recognized that some services, while not fully competitive, do face some level of competition. (Centrex generally is placed in this category.) Many PUCs permit these services to be priced flexibly: usually the ILECs are free to charge any rate that exceeds a floor based on the costs of the service, and that is lower than a ceiling set by the most recent tariffed rate for the service. The floor ensures that basic ratepayers will not subsidize the services in this category, and the ceiling ensures that customers for these services will not be exploited [10]. Many PUCs also permit ILECs to offer some services in the "partially competitive" category (notably Centrex service for the largest users) under contract, at rates that also are subject to floors and ceilings. The ILECs may enter into contracts for partially competitive services that provide for "rate stability" for some period of time [11].

In addition to the partially competitive services, ILECs offer services that are fully competitive. These may include Yellow Pages, message storage services that compete with answering machines, non-ILEC answering services, and inside wire and information services of various kinds. Individual services within this category may be deregulated altogether or merely granted more latitude in pricing than services in the other categories. Many states have worked out criteria for deciding whether a new or existing service faces sufficient competition to be deregulated. Some of these criteria are set out in Appendix A [12].

Finally, and before we leave the question of pricing of competitive services under price caps, we should address an obvious question: Why are regulators concerned that ILECs subject to price-cap regulation will subsidize their rates for competitive services with earnings from monopoly services? Is it not true that under price caps, any padding of the costs incurred to provide any service will result directly in lower profits?

Applied to "pure" price-cap approaches, this observation is entirely correct. Cross-subsidization is an artifact of rate-of-return regulation, which allows ILECs to recover all of their reasonably incurred costs. Under that regime, if costs incurred to provide a competitive service are successfully misallocated to a monopoly service, the ILEC will be allowed to recover those costs from captive ratepayers—permitting the ILEC to charge below-cost rates for competitive services without suffering an economic penalty. Under price caps, however, ILECs are not guaranteed recovery of any costs and any "padding" of costs, therefore, will result in lower profits.

However, some price-cap regulations are not pure expressions of the price-cap theory. As we pointed out in the previous chapter, some price-cap regulators have kept a vestige of rate-of-return regulation, called sharing, that requires ILECs to return to ratepayers all earnings that exceed a specified percentage return on investment.

Where ILECs are subject to these sharing requirements, they have an incentive to misallocate costs incurred to provide high-earning services to less profitable services. By means of such a strategy, an ILEC can minimize the amount of its earnings that must be shared with ratepayers.

There is a general recognition that sharing requirements are an inefficient holdover of rate-of-return regulation, and the FCC, in fact, eliminated sharing in 1997 [13]. Sharing still is a feature, however, of some state price-cap systems [14].

III. Pricing Nonbasic Services Under the Antitrust Laws

While most of the rules of antitrust are intended to protect consumers from paying excessively high prices (set by monopolists acting alone or by competitors acting in collusion), some antitrust decisions have found prices to be unlawfully low. The theory of these decisions is that powerful firms can price their products or services below cost, drive weaker competitors from the market, then take advantage of the absence of competition to raise price above competitive levels. If the strategy works, the monopoly profits earned after rivals have disappeared are supposed to exceed any losses suffered during the monopolist's below-cost pricing campaign.

The courts have grown skeptical of most of these so-called predatory pricing claims, chiefly because economists and legal scholars have argued persuasively that true predatory pricing strategies are unlikely to work [15]. The courts are rightly reluctant to condemn vigorous price cutting that merely reflects the lower costs of an efficient firm and benefits consumers through lower prices [16].

Where the ILECs are concerned, however, courts and economists alike acknowledge that true predatory pricing is at least conceivable. The principal mechanism is one we have already discussed in this chapter (i.e., cross-subsidization).

Cross-subsidization would be easiest in the world in which ILECs were regulated on a rate-of-return basis and the commissions did not track the costs of individual services. In this environment an ILEC might price its competitive services below cost, knowing both that the overall revenue requirement will include those costs and that the rates for other services will be set high enough to recover them. If no adequate cost accounting and reporting mechanisms were in place, such abuses would be hard to detect.

A harder case is presented in the present environment, where the commissions require the ILECs to report the costs incurred to provide competitive services (either on an incremental or fully allocated basis) and to set the rates for those services at a level that recovers the associated costs. Even here, though, creative accounting by the ILECs and inadequate supervision by the commissions might cause some of the costs of competitive service to be reported on the regulated side of the ledger, in which case below-cost rates for competitive services might appear to be compensatory.

As noted earlier, price-cap regulation is the least hospitable environment for cross-subsidies because price caps deny ILECs the usual incentives to pad costs. Where the price-cap rules require sharing of profits above a certain level, however, below-cost pricing of competitive services still may be a temptation, because the losses incurred (or, put another way, the costs of competitive service that are shifted to the regulated side) will reduce the sharing revenues otherwise available to ratepayers. That is why California's price-cap plan, for example, places a cost floor under rates for partly competitive services, allocates a share of embedded costs to fully competitive services, and does not count the costs and revenues of the riskiest ventures as part of the overall sharing mechanism [17].

In addition to subsidization of one service by another, rate-of-return regulation may encourage ILECs that have both regulated and nonregulated lines of business to practice another form of predatory pricing. Specifically, where an unregulated subsidiary sells products or services to the ILEC's regulated businesses, it may be tempted to do so at monopoly prices. So long as the prices paid become part of the revenue requirement for the regulated services, the ILEC will recover its rate of return and the unregulated business will earn monopoly profits. Only the ratepayers will suffer [18].

Having explained why predatory pricing is theoretically possible for ILECs, how can ILECs avoid claims of predatory pricing?

The courts have not precisely defined predatory pricing by a company that offers both regulated and unregulated services. The Court of Appeals for the Seventh Circuit has found, however, that so long as the price of a service covers its long-run incremental costs, that price is not predatory [19]. Where ILECs comply with commission requirements that services recover their incremental costs, this standard is met; and where ILECs comply with regulatory rules that call for fully allocated cost pricing, they exceed the standard recognized by the Seventh Circuit Court [20].

Similarly, affiliate transactions should mirror the normal transactions that take place between the nonregulated vendors and unaffiliated purchasers.

As most ILEC managers know by now, the antitrust laws must be taken seriously. Failure to comply with the antitrust laws can lead to criminal prosecution, or to private lawsuits in which successful plaintiffs will recover three times their actual damages, plus their attorneys' fees.

Endnotes

[1] For an explanation of monopoly, see Appendix A.

[2] These so-called implicit subsidies are discussed in more detail in Chapter 10.

[3] We also referred briefly to this phenomenon when we discussed Peninsula's rate structure.

[4] As the reader may recognize, this is a rough description of Centrex service. Nothing we say here about the fictional OPS service, however, necessarily matches the real-world costs, service architecture, or history of Centrex.

[5] Some commissions, in fact, have referred to basic services as "residual." As one commission explains, "[t]he rates for nonresidual service customarily are set well above the cost of providing these services. Once the rates for the nonresidual services are set and the revenue for such rates is projected, the revenue projected from such services is subtracted from the company's revenue requirement, thereby establishing the amount of revenue that must be recovered from the residual services in order to meet the company's revenue requirement. Rates for the residual service are then designed, not to recover their costs, but merely to recover the balance of the company's revenue requirement. To the extent that nonresidual services are priced above cost, therefore, the residual services can be priced below cost and still produce the company's revenue

requirement." Northwestern Bell Telephone Company, 126 PUR 4th 526, 52829 (Minnesota Public Utilities Commission 1991); *See also* GTE Southwest, Inc., 126 PUR 4th 194, 282 (Texas Public Utilities Commission 1989).

[6] In the Matter of Separation of Costs of Regulated Telephone Service from Costs of Nonregulated Activities, 2 FCC Rcd. 1298, 1311 (1987).

[7] As we shall see, however, price-cap systems sometimes include "sharing" requirements and other vestiges of rate-of-return regulation that keep these cost-allocation issues alive.

[8] Policy and Rules Concerning Rates for Dominant Carriers, Second Report and Order, 5 FCC Rcd. 6786 para. 22526 (1990), (Second Report and Order). Later, the Commission created additional baskets for video dial-tone and marketing expenses.

[9] *Id.* para. 226.

[10] *See*, e.g., California Price Cap Decision, 107 PUR 4th at 99; *See also* Indiana Bell Telephone Company, Inc., 102 PUR 4th 181, 185 (Indiana Utility Regulatory Commission 1989).

[11] *See*, e.g., New Jersey Bell Telephone Company, 102 PUR 4th 69 (New Jersey Board of Public Utilities 1989); Indiana Bell Telephone Company, Inc., 102 PUR 4th 69 (New Jersey Board of Public Utilities 1989); Indiana Bell Telephone Company, Inc., 102 PUR 4th 181 (Indiana Utility Regulatory Commission 1989). Some commissions also have considered and rejected the argument that individual contracts, because they have the effect of producing different rates for similar services, are discriminatory. *See*, e.g., New Jersey Bell Telephone Company, supra, 102 PUR 4th at 74.

[12] *See* pages 245-246, infra.

[13] Price-Cap Performance Review for Local Exchange Carriers, Fourth Report and Order in CC Docket No. 941, 12 FCC Rcd. 16,642 (1997) para. 156.

[14] California, for example, "suspended" but did not eliminate the sharing requirement of its price-cap regulations for GTE and Pacific Bell. Rulemaking in Third Triennial Review of Regulatory Framework for GTE California, Inc. and Pacific Bell (CPUC Decision No. 9810026, Oct. 8, 1998) para. 5.2.1.

[15] *See* Bork, R., The Antitrust Paradox, 149155 (1978).

[16] *See*, e.g., *Matsushita Elec. Industrial Co. v. Zenith Radio Corp.*, 475 U.S. 574, 589 and 595598 (1986).

[17] California Price-Cap Decision, supra, 107 PUR 4th at 99.

[18] Both the FCC and the state commissions scrutinize such "affiliate transactions," not as a matter of antitrust enforcement, but to protect ratepayers from abuse.

[19] *MCI v. AT&T,* 708 F.2d 1081, 115, 1191123 (7th Cir. 1983).

[20] The FCC itself "do[es] not entirely disagree" with the notion that rates in excess of the long-run incremental cost of providing a service are sufficient to prevent cross-subsidy.

Separation of Costs of Regulated Telephone Service from Costs of Nonregulated Activities, 2 FCC Rcd. 1298, 109 at 1312 (1987). Where the FCC requires telcos to exceed fully allocated costs, in fact, it does so not so much to prevent cross-subsidy as to ensure that revenues earned in nonregulated enterprises contribute something to the telcos' embedded costs. *Id.*

3

When Must an ILEC Interconnect Its Facilities with Those of Other Service Providers?

Most people know their telephone companies only as providers of service to end users (i.e., as providers of telephone services to residential and business customers that do not resell those services to others). But in fact, a large part of any ILEC's business consists in selling facilities and services that will be used as components of the buyers' own services. We have seen, for example, that IXCs buy from ILECs the access services that connect the IXCs to their customers. We also have seen that CLECs purchase discounted ILEC services for resale and buy unbundled ILEC network elements for use as components of the CLECs' own networks. Similarly, information service providers buy various functionalities from the ILECs for use in offering Internet access, remote data processing, and other services.

If ILECs were ordinary businesses, they could choose to offer—or not offer—such "wholesale" products and services according to their best business judgment, at whatever prices and other terms they thought appropriate. But ILECs, of course, are not ordinary businesses. They own and control an overwhelming preponderance of the switches, access lines, and other facilities through which non-ILEC service providers must reach their customers, and the ILECs' decisions concerning when and on what terms they will make their networks available to others can determine the extent—or indeed the very existence—of competition in a wide range of markets [1]. For this reason Congress, the FCC, the state PUCs, and the antitrust courts have intervened extensively to dictate when, to whom, and on what terms the ILECs will offer wholesale services and facilities.

Over the years, these interventions have created elaborate rules that vary according to the kinds of entities to which the ILECs provide wholesale facilities and services. So, for example, ILECs have one set of obligations to IXCs, another to CLECs, another to mobile telephone companies, and still another set of obligations to information service providers. Although the lines separating these bodies of rules have tended to blur in recent years—along with the boundaries separating the lines of business in which the ILECs' wholesale customers are engaged—it is useful to consider these sets of regulations separately. In this chapter, therefore, we look first at the obligations of ILECs to CLECs, then at the obligations of ILECs to IXCs, commercial mobile radio service (CMRS) providers, and information service providers (ISPs).

I. ILEC Services Provided to CLECs

When Congress passed the Telecommunications Act of 1996, it announced a national policy in favor of local telephone service competition [2]. As a result of the 1996 Act, the FCC and state governments may no longer prevent new, local telephone companies from competing directly with the Bells and other established ILECs in the ILECs' traditional service areas [3].

Simply announcing a policy of local competition, however, did not give new telephone companies a realistic chance to compete. Most fundamentally, a new telephone company cannot attract many customers if those customers cannot place calls to, and receive calls from, the customers of the established telephone companies. At the very least, therefore, Congress had to require the ILECs to *interconnect* their networks with those of their new competitors so that CLEC customers could talk to ILEC customers.

In fact, Congress imposed obligations upon ILECs that go well beyond simple interconnection with their local competitors. It is no exaggeration to say that the 1996 Act requires ILECs to place their facilities and services at the direct disposal of their competitors at prices and other terms that will help competitors take business away from the ILECs. Specifically, ILECs must assist their local competitors in the following ways:

- ILECs must interconnect their facilities and equipment with the facilities and equipment of CLECs when asked to do so [4].

- ILECs must establish reciprocal compensation arrangements under which they will pay the CLECs for completing calls from ILEC customers and the CLECs will pay the ILECs for completing calls placed by CLEC customers [5].
- ILECs must provide their competitors with access to ILEC subscriber lines, switches, and other so-called "unbundled network elements" at cost-based rates [6].
- ILECs must offer all of their retail telecommunications services to CLECs at discounted prices for resale to the CLECs' customers [7].
- ILECs must provide number portability and dialing parity (i.e., they must ensure that customers can switch to CLECs for local telephone service without changing their telephone numbers and can place calls as CLEC customers without dialing extra digits) [8].
- ILECs must notify the public of any changes to their networks that affect the transmission and routing of service or the interoperability of the ILECs' facilities and networks with those of the CLECs [9].
- ILECs must permit CLECs to collocate the CLECs' facilities with those of the ILECs [10].
- ILECs must negotiate interconnection agreements with CLECs in good faith [11].

Each of these obligations has been implemented by the FCC in elaborate orders and regulations. In order to understand the impact of these orders and regulations in practical terms, it might be useful to follow a fictional interconnection negotiation. For this purpose we bring back Peninsula Telephone, our fictional ILEC, and introduce a new player, a fictional competitor called Peninsula CLEC, or PenCLEC.

A. Negotiating the Interconnection Agreement Between Peninsula Telephone and PenCLEC

As president of Peninsula, you first learn about PenCLEC when your network manager brings you a faxed letter from PenCLEC's president. The letter announces that PenCLEC has obtained authorization to operate as a local exchange carrier in the State of Davis. The letter also states that PenCLEC plans to offer local telephone and exchange access services in the cities of Mosby and Cantrell, but for the immediate future will have a switching facility only in Mosby. PenCLEC wants to buy facilities from Peninsula for use in connecting PenCLEC's Mosby customers to its switch, but has not decided how it will serve the city of Cantrell. Accordingly, PenCLEC wants to talk to you about buying a complete set of unbundled network elements in Cantrell or, in the alternative, simply buying and reselling Peninsula's local exchange services in that city.

You and your network manager now have a drink to mark the passing of the glad days of monopoly in the State of Davis. Then you place a call to your regulatory lawyer in the state capital.

Your lawyer, in his brisk and unsentimental way, tells you that you must negotiate an interconnection agreement with PenCLEC and submit the agreement for approval to the state public utilities commission. Your lawyer also warns you to negotiate the agreement with great care, because Peninsula will be required to offer other CLECs the same terms that are contained in the agreement with PenCLEC [12]. The following week you have your first meeting with Randolph Buford, the CEO of PenCLEC.

1. The First Meeting: What PenCLEC Needs

At your first meeting with Buford, you admit to some confusion about how the relationship between your company and his will work. As a foundation for the interconnection discussions, you ask Buford to explain how, in his view, PenCLEC will use Peninsula's facilities to help PenCLEC's customers place and receive telephone calls.

Buford begins with PenCLEC's situation in Mosby, where PenCLEC has installed its own switch. According to Buford, PenCLEC will contact all of Peninsula's higher value customers—chiefly including Mosby's larger business customers—in the very near future and urge those customers to switch their local service to PenCLEC. When a customer agrees to take service from PenCLEC, PenCLEC will ask Peninsula Telephone to lease the customer's subscriber line (i.e., the copper wire connecting the customer to Peninsula's central office) to PenCLEC for as long as the customer continues to take local service from PenCLEC.

PenCLEC will need more from Peninsula, however, than control over customers' subscriber lines. To illustrate these additional needs, Buford begins by describing a typical telephone call placed from a PenCLEC customer in Mosby to a Peninsula customer in Mosby. He takes himself—his first customer—as an example.

When Buford picks up his telephone to place a call, an "off-hook" signal will be transmitted over Buford's subscriber line to the Peninsula Telephone central office [13]. At the central office, Buford's subscriber line will be linked to a cross-connect facility (also provided by Peninsula) that terminates, in turn, on multiplexing equipment owned by PenCLEC and collocated at a Peninsula central office. The multiplexing equipment will combine Buford's off-hook signal with traffic from other PenCLEC subscriber lines and send the signal over a trunk (also provided by Peninsula) that connects the Peninsula central office to PenCLEC's switch.

When PenCLEC's switch receives the off-hook signal from Buford's telephone, it will return a dial tone over the trunk connecting the PenCLEC switch to the collocated PenCLEC equipment in Peninsula's central office. The dial tone will travel over the cross-connect facility to Buford's subscriber line and from there to Buford's telephone. When Buford receives the dial tone he will dial the number of the called party and the PenCLEC switch—not Peninsula's switch—will route the call to the called party. The Peninsula facilities that connect Buford's access line with PenCLEC's switch will remain part of the end-to-end connection throughout Buford's call.

Buford next explains that because the called party in his example is a Peninsula Telephone customer, Peninsula must do still more to ensure that the call is completed. Peninsula must help transport the call *back* to Peninsula's network and deliver the call to Peninsula's customer. Accordingly, when the PenCLEC switch recognizes the number dialed by Buford as a number assigned to a Peninsula Telephone customer, it will signal the Peninsula switch to establish a connection with the subscriber line of the called party. That signal will be sent over a transmission facility jointly maintained by Peninsula and PenCLEC. Peninsula will transmit PenCLEC's ringing signal to the Peninsula customer Buford is attempting to call, and return a busy signal or establish a voice connection between the parties' telephones.

Finally, Buford reminds you that many calls placed by PenCLEC's customers will be long-distance calls. PenCLEC does not plan to provide service outside the two cities of Mosby and Cantrell, and therefore will need to hand off long-distance calls to the various interexchange carriers from which its customers receive long-distance service. For this purpose, too, PenCLEC wants access to Peninsula's facilities. Specifically, PenCLEC prefers not to establish direct links between its switch and the facilities of all the IXCs that serve Mosby: instead, Buford wants to connect PenCLEC's switch with a Peninsula Telephone switch called an access tandem, which in turn connects with all the facilities-based IXCs in the area [14].

By now Buford has made it clear that even in Mosby, where PenCLEC will have a facilities-based presence, service to PenCLEC's customers will be provided almost exclusively over subscriber lines, trunks, and other facilities furnished by Peninsula Telephone. How, you ask Buford, will PenCLEC serve customers in Cantrell, where it lacks even a switch of its own? Buford confesses that he has not decided how he will serve Cantrell, but he has studied the 1996 Act and knows that he has two alternatives.

One alternative is for PenCLEC to purchase unbundled network elements in Cantrell, just as Buford proposes to do in Mosby; except that in Cantrell, PenCLEC not only will use Peninsula Telephone's subscriber lines, cross-connect facilities, and trunks to transport its customers' traffic, but also will use Peninsula's switch to route traffic to and from those customers. Buford's hypothetical telephone call will travel from its point of origin to its destination exclusively over facilities acquired by PenCLEC from Peninsula.

Another method of serving Cantrell is for PenCLEC simply to buy complete services, rather than facilities, from Peninsula. Instead of purchasing from Peninsula the physical access line, cross-connect facilities, and interoffice trunks needed to provide local exchange service to a customer, PenCLEC simply could buy local exchange service from Peninsula at a discount. PenCLEC then could sell that service to *its* customers under PenCLEC's name and at prices set by PenCLEC.

Now that Buford has explained in broad outline how his company will provide service in cooperation with Peninsula, he asks for a meeting to begin negotiating an interconnection and resale agreement that will establish these service arrangements in detail. You agree that within the next week, Peninsula's attorney and chief network engineer will sit down with PenCLEC's attorney and chief network engineer. For at least part of the meeting, you and Buford also will participate.

2. The Second Meeting: Negotiating the Interconnection and Resale Agreement

At the second meeting, Buford begins the discussion by proposing an agenda. He suggests that the two companies first work out how they will interconnect their networks and exchange traffic between their customers—that is, how each company will transport and terminate calls placed to its customers by customers of the other company. Buford then suggests that the companies decide which unbundled elements of Peninsula's network will be provided to PenCLEC, and upon what terms those elements will be provided. Finally, Buford wants to discuss some Peninsula services that PenCLEC may want to purchase at a discount for resale to PenCLECs' customers. He also proposes—and the Peninsula representatives agree—that the interconnection and resale agreement will be sufficiently broad to cover the

Peninsula-PenCLEC relationship in the entire State of Davis, including the cities of Mosby and Cantrell.

3. Interconnection and Reciprocal Compensation

As agreed, the two companies first explore how their facilities will be interconnected for the exchange of traffic between them, and how the two companies will share the cost of those facilities and the transport and termination services that will be provided over those facilities. As Buford points out, the challenge these arrangements pose in Mosby is very different from the situation in Cantrell. PenCLEC has a switch in Mosby, located about a mile from Peninsula's central office, and must establish physical connections between that switch and Peninsula's facilities. In Cantrell, by contrast, PenCLEC has no facilities of its own.

The parties decide to sort out the situation in Mosby first. PenCLEC must interconnect with Peninsula in Mosby for two purposes: first, to exchange local calls between Peninsula and PenCLEC customers; and second, to provide exchange access service for IXCs serving PenCLEC customers. Each of these services requires somewhat different physical arrangements between the two companies.

For exchange of local traffic in Mosby, the two companies agree to establish a jointly maintained SONET network [15]. Each party will place an optical line terminating multiplexer (OLTM) at its own premises, and each party will install and maintain one-half of a fiber optic ring connecting Peninsula's network with PenCLEC's switch.

Access traffic presents a different problem. PenCLEC, like any local telephone company, must connect its customers with their interexchange carriers as well as other telephone subscribers in the same local service area. In order to do this, PenCLEC can choose to establish facilities directly from its switch to the point of presence of a number of some or all of the interexchange carriers that serve customers in PenCLEC's service area. As noted earlier, Peninsula already has a special switch, called an access tandem, that connects Peninsula's customers with all of their interexchange carriers. Recognizing the greater efficiency of this special switch, PenCLEC requests interconnection with Peninsula's access tandem. Accordingly, the parties agree to establish an access toll connecting trunk—a high-capacity optical fiber connection between PenCLEC's switch and the PenTel access tandem serving the city of Mosby.

Now you must decide how to allocate the cost of these facilities and of the exchanges of local traffic that will take place over those facilities. The 1996 Act and the Commission's rules divide these costs into two categories:

transport and termination. The transport cost is the cost of the facilities that carry traffic between the two networks, and termination cost is the cost each carrier incurs, within its network, to deliver traffic originated by customers of the other carrier.

You agree that because you and PenCLEC have each agreed to build one-half of the SONET ring connecting the two carriers, including the multiplexer placed at the termination point on each carrier's network, there is no need for money to change hands for the cost of transport.

This leaves the question of compensation for termination costs. The 1996 Act and the FCC's regulations require termination costs to be recovered by an agreement for reciprocal compensation, under which the carrier whose customer initiates a call compensates the other carrier for the cost of delivering the call to the second carrier's customer. You want to set the rate at a level that will recover your total investment in your network, but your state commission has adopted a pricing standard requiring ILECs to set their rates for reciprocal compensation and unbundled network elements at something called total element long-run incremental cost (TELRIC). Under this standard, the rate you charge a CLEC for any facility or network functionality can only recover the associated forward-looking costs, assuming use of the most efficient technology. You may not set the rate based on historical costs, including the cost of any less-than-optimal technology you may have deployed in your network.

Now you are faced with two choices. One possible approach to reciprocal compensation is for each company to calculate the cost of terminating each unit of traffic it receives from the other, and for the two companies to keep track of the volumes of traffic they terminate and bill each other accordingly. Another, far simpler, approach is to assume that traffic flows between the two companies will be about the same in each direction and will impose roughly the same costs on each carrier. Under this second assumption, it makes sense for each company to bear its own costs of terminating traffic for the other. In the interest of avoiding delay, you agree with PenCLEC that this so-called "bill-and-keep" approach is preferable for the initial term of the interconnection agreement [16]. You also agree that before the interconnection agreement comes up for renewal, the two companies will conduct cost studies to determine whether the two companies' costs of terminating traffic are, in fact, the same. If it turns out that the two companies' costs are not the same, either party to the agreement may demand a renegotiation of the terms under which each party will terminate traffic for the other.

For the connection between PenCLEC's switch and Peninsula's access tandem, the parties agree to enter into a meet-point billing arrangement, under which each company will charge interexchange carriers an agreed percentage of the applicable access charges under each company's tariff. The interconnection agreement will refer to this service as jointly provided switched exchange access service.

4. Unbundled Network Elements

Now that you have resolved the questions of interconnection and reciprocal compensation, it is time to discuss PenCLEC's access to your unbundled network elements in Mosby. Buford tells you that his company will need access to his customers' existing subscriber lines. He also will need transport facilities between PenCLEC's switch and Peninsula Telephone's switch, and local switching services for switching of calls placed by PenCLEC's customers.

On the question of subscriber lines, you agree that for each customer PenCLEC acquires in Mosby, Peninsula Telephone will sell PenCLEC the right to use, as an element of PenCLEC's local service provided to that customer, the copper loop that already connects that customer to the serving Peninsula central office. You also agree to permit PenCLEC to install, at Peninsula's central office serving that customer, the equipment needed to connect that customer with PenCLEC's switch over a transport facility provided by Peninsula.

To satisfy PenCLEC's need for transport facilities, Peninsula Telephone agrees to provide PenCLEC with access to trunks connecting Peninsula's central office switch to PenCLEC's switch. Peninsula agrees that transport will be provided at any level of capability requested by PenCLEC, including DS1, DS3, and optical carrier levels.

For switching of local calls placed by PenCLEC's customers, Peninsula Telephone agrees to provide three types of switching facilities. First, Peninsula will provide line-related elements, which include connections of lines to lines, lines to trunks, and trunks to trunks, and associated features such as telephone numbers, directory listing, dial tone, and access to 911, operator services, and directory assistance. These line-related elements will be provided at a flat per-line rate.

Second, Peninsula will provide a trunkside local switching element assessed on a per-minute-of-use basis. This rate element includes the switching functions of connecting lines to lines, trunks to lines, lines to trunks, lines to switched features, and trunks to trunks.

Finally, Peninsula will provide switch facilities for PenCLEC's use in connection with customized routing, such as calls to directory assistance or operator services.

5. Sale of Peninsula Services for Resale

Now that Peninsula and PenCLEC have decided on the terms for interconnection, reciprocal compensation, and access to unbundled network elements, the parties still must decide the terms on which PenCLEC may, if it chooses, purchase Peninsula services for resale to PenCLEC's customers in Cantrell.

Your first instinct, when you learned that PenCLEC wanted to resell Peninsula's services in direct competition with you, was to offer PenCLEC a narrow range of basic services at the same tariffed rates that your retail customers pay. In this way, you hoped to limit PenCLEC's ability to compete. Your lawyer, however, had cautioned you against this approach. Specifically, your lawyer pointed out that under the 1996 Act, Peninsula must offer for resale "any telecommunications service that (Peninsula) provides at retail to subscribers who are not telecommunications carriers"[17]. Your lawyer also pointed out that Peninsula must sell services to CLEC resellers, not at retail rates, but at wholesale rates that recover only the cost of providing wholesale service [18]. Specifically, this means that Peninsula's rates may not be set at a level that recovers the cost of retail sales, advertising, customer support, and other activities that are not incurred when Peninsula provides service to other carriers.

In preparation for this meeting, therefore, you calculated the discount from your retail rates that would result if you excluded all "avoidable" costs of providing your services at retail. The discounts varied from service to service, ranging from a low of 18% to a high of 22%. Your lawyer reviewed the calculations and advised you that because the discounts fell within the FCC's "default range" of 17% to 25%, they probably would be approved by your state commission. Accordingly, you offer to sell PenCLEC all of your tariffed retail services at discounts ranging from 18% to 22% [19]. The offer is promptly accepted.

During the weeks following the second Peninsula-PenCLEC meeting, attorneys for the two companies draft and exchange written interconnection agreements. After some redrafting and further negotiation, you and Buford sign a final interconnection agreement and submit it jointly to the state public utilities commission for its review and approval.

B. Arbitration, Mediation, and Judicial Review

Although the 1996 Act makes the terms of ILEC/CLEC interconnection primarily a matter for negotiation between the parties, the Act also assigns substantial roles to the state commissions, the FCC, and the courts.

The Act gives the state commissions an especially vital role, by making them the supervisors and arbiters of interconnection arrangements between ILECs and CLECs. Specifically, the Act gives the state commissions four functions. First, state commissions must review—and may reject—negotiated interconnection agreements governing interconnections between carriers within their jurisdictions [20]. Second, state commissions may, upon request of any party, participate in interconnection negotiations as mediators [21]. Third, the state commissions may, upon request of any party, arbitrate and decide open issues that arise in the course of negotiations [22]. Finally, the state commissions are expressly empowered to set the rates at which ILECs will provide CLECs with transport, termination, unbundled network elements, and services for resale [23].

The Act does not give the FCC a hands-on role in supervising interconnection between ILECs and CLECs, but gives the FCC authority to enact general rules governing interconnection arrangements—an authority the Commission has exercised with considerable thoroughness [24]. The Act also provides that if a state commission fails to act on a request to review a negotiated interconnection agreement, the FCC shall issue an order preempting the state commission's authority within 90 days after the Commission receives notice of the state commission's failure to act [25]. After issuing its preemption order, the FCC is to complete the process of reviewing—and approving or rejecting—the negotiated interconnection agreement [26].

Finally, the Act provides that any person aggrieved by a state commission's decision concerning an interconnection agreement may bring a petition for judicial review in an appropriate United States District Court [27].

II. ILEC Services Provided to IXCs

There once was only one interexchange carrier—AT&T's Long Lines Division—and that carrier was affiliated with the large ILECs that served most of the telephone subscribers in the United States (the Bell telephone companies). In this environment Long Lines worked out the terms of its access to ILEC facilities with little controversy and little need for intervention by the commissions or the courts [28].

Two momentous developments turned interexchange carrier access into a source of contention. One was the emergence of specialized interexchange carriers that competed with AT&T's Long Lines Division; the other was the divestiture of the Bell system in 1984, which ended the affiliation between AT&T and the Bell telephone companies.

Even before divestiture, the FCC had coupled its decisions permitting limited interexchange competition with decisions requiring the ILECs to interconnect with the new carriers on reasonable terms and conditions [29]. After divestiture, the ILECs formerly affiliated with AT&T were required by the terms of the antitrust consent decree to provide all interexchange carriers with *equal* access—that is, access on terms and conditions functionally equivalent to those already enjoyed by AT&T [30]. The FCC then developed new rules, implementing both its own interconnection polices and the equal access requirements of the AT&T consent decree, and extending the equal access rules to all non-Bell ILECs [31]. More recently, the Telecommunications Act of 1996 confirmed that existing regulations governing interconnection between ILECs and IXCs would continue in effect pending further rulemaking by the Commission [32].

The rules that have emerged from this tangled history give the ILECs obligations to the IXCs, and to the IXCs' customers as well. The following examines these requirements, as developed before the Telecommunications Act of 1996 and as supplemented by rules promulgated in response to the 1996 Act [33].

A. Equal Access Defined

Before we can explain intelligently what *equal* access is, we should say a little more about what *access* is.

When an ILEC customer makes a long-distance call through an IXC, his or her local exchange carrier is far more than a passive conduit through which the conversation flows from the customer to the interexchange carrier. The ILECs at each end of the call provide a complex package of facilities and services that help to set up, route, take down, and bill the call.

First, the originating and terminating ILECs must provide links that connect their facilities with those of the IXC [34]. Those connecting facilities must have adequate transmission quality and sufficient capacity to carry the anticipated volume of traffic.

Second, the originating ILEC must determine which IXC is to receive the call and direct it accordingly. This means that either the ILEC switch must be programmed in advance with the customer's choice of IXC, so that each interexchange call placed by the customer will go to the preferred IXC automatically; or the ILEC must provide each IXC with an access code for the IXC's customers to dial each time they place an interexchange call.

The ILECs at both ends of the interexchange call must provide various sorts of signaling. Network control signaling, for example, includes such functions as sending the called party's number to the IXC so the IXC can determine how to route the call; another type of signaling, called answer supervision, informs the originating ILEC that the call has been answered, and informs the terminating ILEC when the caller has hung up.

The IXCs' customers also need to call "information" (directory assistance) in distant cities. This function, too, is performed by ILECs as part of the package of access services. Access also includes the maintenance and testing of the facilities used to provide access, and the provision of information the IXCs need in order to bill their customers. Chief among the latter services is the transmission to the IXC of the calling party's number—a service called automatic number identification, or ANI.

The AT&T antitrust consent decree defined access as including all of these services and required that access, so defined, be made available to all IXCs through arrangements that were "equal in type and quality to that provided for the interexchange telecommunications services of AT&T and its affiliates"[35].

While this seems clear enough, neither the AT&T consent decree, the consent decree applicable to the GTE companies [36], the judicial decisions entering the decrees nor the FCC's equal access decisions imposed a requirement of absolute technical equality. Judge Greene accepted a definition of "equal access" as access whose "overall quality in a particular area is equal within a reasonable range which is applicable to all carriers" [37]. Similarly, the FCC, while asserting its complete jurisdiction over the technical implementation of equal access, declined invitations to adopt detailed quality standards; instead, it encouraged the industry to resolve technical questions through consensus [38].

The interconnection quality requirements of equal access, as defined by the decree, by Judge Greene, and by the FCC, generally were satisfied by the service designated by the Bell telephone companies as "Feature Group D." This is the premium access package purchased by AT&T and most other IXCs, and includes trunk-side interconnection to the ILEC switch, dialing parity, automatic number identification, and a number of optional services [39].

B. How Equal Access Has Been Implemented

In the 1980s, when the equal access requirements first were imposed, the thousands of ILECs around the country varied widely in their technical ability to comply.

The biggest (although not the only) obstacle was dialing parity. During the many decades when AT&T Long Lines was the only long-distance company, customers reached AT&T simply by dialing a long-distance call.

Equal access meant that the ILECs had to make it as easy for a customer to reach the newer IXCs as it was for them to reach AT&T. The equal access rules contemplated that this would be done by presubscription (i.e., each customer would select an IXC, and the customer's ILEC would program its switch with the customer's selection). Each time the customer dialed a long-distance call to a destination outside the ILEC's service area, the switch would match the customer's telephone number with the carrier identification code (CIC) of the customer's presubscribed carrier and route the call accordingly [40].

This worked very well with software-directed switches that could be programmed, but not with the older, electromechanical equipment still used in many central offices. To require immediate, universal conversion to equal access would force the ILECs to scrap all of their electromechanical equipment at tremendous cost. Such a requirement would be burdensome for all ILECs, but especially for the smaller, independent companies.

A crash program of this kind was not only expensive, but unnecessary. Many of the electromechanical offices served small towns and rural areas that were economically unattractive to the new IXCs, who had no immediate intention of making the investment required to serve them.

The AT&T and GTE consent decrees, therefore, set out conversion timetables that recognized these realities. The Bell companies were to offer equal access from each of their switches for which an equal access request was made by September 1, 1986. (Upon request, the court would grant individual, *temporary* waivers of the deadline for electromechanical offices that were uneconomical to convert to equal access.) The GTE decree established a timetable that required equal access to be implemented by December, 1990 at all switches serving more than 10,000 access lines. (Waivers also could be requested under the GTE decree.) The AT&T and GTE deadlines all have passed, and effectively all of the nation's access lines have been converted to equal access.

The FCC adopted a separate set of requirements for the non-Bell and non-GTE ILECs. These companies were required to convert their electronic end offices to equal access within three years of the receipt of a reasonable request from any IXC. Where end offices were equipped with electromechanical switches, equal access was to be implemented as soon as practicable, but no timetable was imposed. Where equal access was not yet implemented under these rules, the independent carriers were to offer automatic number identification where possible, and access to IXCs using the shortest possible access codes, with access from rotary phones to the services of each IXC.

When Congress passed the Telecommunications Act of 1996, it mandated dialing parity for local as well as long-distance calling [41]. In adopting rules to implement this requirement, the Commission decided that customers should have the opportunity, not only to dial the same number of digits for a local call regardless of the LEC from which they obtained service, but also to presubscribe—if they desire—to separate carriers for their intraLATA toll calls and interLATA toll calls [42]. This so-called "2-PIC" rule imposed a new equal access obligation on LECs, and the Commission ordered that all LECs would "provide intraLATA and interLATA toll-dialing parity no later than February 8, 1999" [43].

Customers are always free to change their presubscribed interexchange carriers, and the ILECs are required to accept and process those changes. Unfortunately, customers have been subjected to ferocious IXC telemarketing campaigns, and many customers have complained to their ILECs that PIC changes processed on the customers' behalf were unauthorized. Complaints concerning these so-called "slamming" practices became so widespread that the Commission has had to adopt formal rules to govern verification of PIC change orders [44].

C. How ILECs Charge for Services Provided to IXCs

The system of charges for IXC access is older than, and still separate from, the rules under which ILECs are compensated for providing CLECs with transport, termination, and access to unbundled network elements [45]. In fact, the IXC access charge system, although recently modified and still in the process of reform, dates back to the predivestiture relationship between the Bell system and its early competitors in the long-distance business. Because of the importance to that industry of reasonable, cost-based interconnection rates, the Commission has prescribed both the structure and the pricing methodology of interstate access charges [46].

The FCC's rules recognize that a substantial part of the traffic carried by ILECs is access traffic, and a substantial part of the ILECs' investment in plant and equipment is therefore incurred to support the provision of access service. Subscriber lines are used in part to originate and terminate interexchange traffic; other lines connect the ILEC switch to the IXC switch; and ILECs must purchase the switching capacity and software needed to recognize the calling party's IXC and switch the call accordingly. ILECs are entitled to compensation for this investment and for the associated operating and maintenance expense. The FCC permits ILECs to recover the costs of access for interstate calling through a complex set of charges, some of which are paid by the IXCs and some of which are paid by telephone subscribers.

The IXCs and their customers purchase more than one kind of access service from the ILECs. *Switched access* is used by residential customers and smaller business subscribers; as the name implies, switched access customers reach their IXCs by dialing through the ILEC central office switch. *Special access*, used by larger business customers with a heavy volume of interexchange calling, connects customers to their IXCs over private lines (i.e., nonswitched facilities devoted exclusively to the customer's use).

The subscriber's contribution to recovery of the cost of switched access is in the form of a fixed, monthly fee called the end-user common line (EUCL) charge, also known as the subscriber line charge (SLC). This charge recognizes that the subscriber line, connecting the customer's telephone to the ILEC central office, is used not only for local calling but also to originate and terminate long-distance calls carried by the IXCs. The subscriber line is therefore part of the plant used to provide access service; and the SLC recovers part of the ILEC's investment in the subscriber line that is attributed to interstate access service.

For political reasons, however, the SLC never has been set at a level sufficiently high to recover all of the costs attributable to use of the subscriber line for interstate calling. Accordingly, some portion of that cost historically has been recovered through access charges imposed upon IXCs—specifically, through a per-minute (rather than flat) rate element called the carrier common line charge (CCL).

The time-sensitive CCL charge has been a source of controversy ever since it was adopted. In order to appreciate the basis for this controversy, it is necessary to understand the difference between traffic-sensitive (TS) and nontraffic-sensitive (NTS) costs.

The best example of a traffic-sensitive ILEC cost is the cost of a switch. The cost of a switch depends on how much it will be used: an ILEC needs a bigger, more expensive switch to provide exchange and exchange access service together than it needs to provide exchange service alone; and it needs a bigger switch to carry a lot of access calls than it needs to carry fewer access calls. This is what is meant when we say that the cost of a switch (and of certain ILEC facilities used for access, such as common trunks) is traffic-sensitive.

The best example of NTS plant used for access is the subscriber line. The typical subscriber line is a twisted pair of copper wires, designed to carry one conversation at a time. Its cost does not vary with the kind and amount of traffic it is expected to carry. Even if the ILEC could be assured that the subscriber never would make a long-distance call, it still would give the subscriber the same line, at the same cost, as it would give to a subscriber who intended to call across the continent fifty times a day. This is the sense in which the cost of the subscriber line is said to be nontraffic sensitive.

In its original access charge scheme, the FCC proposed to recover all traffic-sensitive access costs from usage-sensitive charges on the IXCs, and all NTS access costs from flat charges on the subscribers. This idea makes economic sense. Because all of the costs of NTS facilities are incurred to give the subscriber a dedicated connection to the local exchange—whether the subscriber ever uses that connection or not—the subscriber should pay the entire cost of the plant dedicated to his or her use. Because the cost of that plant does not vary with usage, the charge to the subscriber for that plant also should not vary with usage.

Unfortunately, the FCC's original plan met with considerable opposition. It turned out that if subscribers were to pay for all the NTS costs they caused, the SLC would have to be set at about $6.00 per month. Many considered this charge a threat to the universal availability of telephone service, and the FCC was persuaded to reduce the SLC for residential customers.

The balance of the NTS costs did not simply disappear, however; they had to be recovered somewhere. The result was the CCL—a usage-sensitive rate imposed on the IXCs to recover that part of the interstate NTS costs not included in the SLC. The IXCs, in turn, have passed the CCL along as part of the usage-sensitive rates they charge to their long-distance customers.

This scheme, while politically acceptable, had a number of vices. It inflated the access charges paid by IXCs and their customers, and it penalized heavy users of interexchange services, who paid far more than their share of the NTS costs of access. These distortions were aggravated by another problem (i.e., the long-standing overallocation of costs to the interstate jurisdiction through the separations process). Although the FCC has reduced this subsidy, it persists, making the recovery of NTS costs through usage-sensitive access charges even more troubling.

The present access charge regulations reflect the FCC's effort to reform the system by squeezing out subsidies and moving toward recovery of NTS costs through NTS charges. Specifically, in an order entered in 1997, the Commission stated that to the extent the SLC charge does not recover the interstate portion of ILECs' local loop costs, the shortfall should be recovered through a flat-rated—rather than usage-sensitive—charge imposed on

IXCs [47]. The flat-rated charge, called a presubscribed interexchange carrier charge (PICC), should be recovered from the IXC to which the customer has presubscribed for interstate calling.

The Commission's adoption of the flat-rated PICC charge is only part of the FCC's program to reform the access charge structure. Notably, although the Commission's 1997 order did not immediately raise the ceiling on the SLC charge (then set at $3.50), the order directed that the SLC ceiling for primary residential and single-line business lines would increase gradually until it reached a level that permitted "full recovery of the common-line revenues from flat charges assessed to both end users and IXCs" [48]. At the same time, the Commission raised the SLC ceiling immediately for nonprimary residential and multiline business lines [49]. The Commission also determined that to the extent rising SLC charges and the new PICC charge, together, did not recover the interstate cost of the subscriber line, ILECs could continue to charge the usage-sensitive CCL. The Commission contemplates, however, that the CCL eventually will be eliminated [50].

In addition to the charges intended to recover the cost of the subscriber line attributable to interstate calling, the access charge regime includes charges for local switching, transport, information services, and other facilities and services provided to IXCs. These charges combine flat-rate and usage-sensitive elements, as appropriate [51].

Special access services are different from, and less complex than, switched access charges. The special access customer does not use an ordinary subscriber line to connect his or her telephone to the central office; instead, the customer buys a private line, which connects to the central office but is not switched there. The private line connects to a transmission facility that carries the customer's calls to the switch of the customer's IXC (often after passing through a second ILEC central office).

For special access, ILECs generally impose a flat charge for transmission between the subscriber's telephone and the ILEC central office, a distance-sensitive charge for transport from one central office to another (if required), and another flat charge for transmission from the final ILEC central office to the IXC switch. The segments connecting central offices to the subscriber and the IXC are called channel terminations; the segments connecting one central office to another are called channel mileage. These charges vary according to the grade of facility provided (i.e., the carrying capacity of the facility and whether it is designed to transmit voice or data).

Although ILEC access charges still are closely regulated, the FCC's goal is to deregulate those charges as CLECs come to provide an increasing competitive alternative to ILEC access services. In the years that have passed since enactment of the 1996 Act, however, the slow pace of local competition has not permitted the Commission to take that step.

III. ILEC Services Provided to CMRS Providers

CMRS providers offer point-to-point wireless services, interconnected with the public switched telephone network, to the public. Services classified as CMRS include cellular telephone, satellite telephony, interconnected specialized mobile radio, and personal communication service (PCS). (Chapter 9 gives a more complete description of the technology, business, and regulatory status of CMRS and private mobile radio service (PMRS).)

The FCC requires all ILECs to provide these carriers with interconnections of reasonable quality at reasonable rates. Specifically, ILECs must: (1) establish reciprocal compensation arrangements under which ILECs compensate CMRS providers for terminating calls placed by ILEC customers, and CMRS providers compensate the ILECs for terminating calls placed by CMRS customers; (2) establish "reasonable charges for interstate interconnection provided to CMRS licensees"; and (3) furnish a CMRS provider with "any form of interconnection arrangement the (ILEC) makes available to any other carrier or customer, unless the (ILEC) meets its burden of demonstrating that the provision of such interconnection is either not technically feasible or economically reasonable" [52].

Although the Commission has asserted plenary jurisdiction over the physical terms of interconnection between ILECs and CMRS providers, the FCC has not preempted the authority of the states to regulate the *intrastate* rates charged for those interconnections. The respective spheres of state and federal jurisdiction over CMRS interconnection rates, in fact, is somewhat confused and merits a brief explanation.

When cellular telephone companies began operation in the 1980s, the FCC concluded that it had the power to preempt state regulation of the *physical* terms of interconnection between LECs and cellular telephone companies, but lacked the authority to preempt the *pricing* of such interconnection [53]. Accordingly, the Commission imposed upon LECs the obligation to negotiate with cellular telephone companies and provide "reasonable interconnections for the services (the cellular telephone companies) provide" [54], including efficient, trunk-side interconnections with LEC central offices. As to interconnection pricing, however, the Commission determined that its authority extended only to charges designed to recover the interstate costs of interconnection (i.e., access costs incurred in connection with interstate calls). Because relatively few calls to and from mobile units are interstate, this left the states with the lion's share of the regulatory burden.

Later, Congress passed the Omnibus Budget Reconciliation Act of 1993, which expressly preempted state regulation of market entry by, or rates charged by, providers of commercial or private mobile radio services [55]. Although rates charged to CMRS providers for intrastate interconnection affect CMRS providers' charges to their customers and arguably are within the scope of the 1993 statute's preemption language, the FCC declared that it would not preempt state regulation over the rates for intrastate interconnection "unless the charge for the intrastate component of interconnection was so high that the price effectively precluded interconnection" [56].

The FCC's authority over ILEC/CMRS interconnection rates grew still more confused when Congress passed the Telecommunications Act of 1996. As we have seen, that Act gives the states a substantial role in ascertaining the rates for interconnection between ILECs and other telecommunications carriers. Because CMRS providers are telecommunications carriers, the 1996 Act appeared to conflict with the preemption authority granted to the Commission in the Omnibus Budget Reconciliation Act of 1993. The FCC's solution to this dilemma was to declare that ILEC/CMRS interconnection arrangements—including interconnection rates—would be controlled by the 1996 Act. The Commission also found, however, that the Omnibus Budget Reconciliation Act's preemption provisions continued in effect and would be used to assert FCC jurisdiction if CMRS providers did not obtain satisfactory interconnection arrangements under the process defined in the 1996 Telecommunications Act [57].

Until the Commission makes a contrary determination, therefore, the ILECs' interconnection obligations owed to CMRS providers are essentially the same as those owed to CLECs under Section 251 of the Telecommunications Act of 1996.

IV. ILEC Services Provided to ISPS and Other Providers of Information Services

One of the most complex subjects in telecommunications law is the Commission's effort to regulate the enhanced, or information, services provided by the Bell operating companies and AT&T; and to ensure that those companies provide their competitors in this market with efficient, nondiscriminatory interconnections and related services. That subject is covered in

Chapter 4, which deals with legal issues peculiar to the Bell operating companies.

For non-Bell ILECs, the picture is somewhat simpler. Because Internet service providers and other companies offering information services are not classified as telecommunications carriers, ILECs are not required to provide those companies with unbundled network elements, discounted services for resale, reciprocal compensation arrangements, or the other facilities and services mandated by Section 251 of the 1996 Telecommunications Act. Instead, the non-Bell ILECs deal with information service providers as end users rather than carriers, and must provide them with business lines, dial-up service, private lines, and other retail services at reasonable and nondiscriminatory rates and conditions. Any attempt to provide inferior service to information service providers, as a means of conferring a competitive advantage on information services provided by the ILEC, would run afoul of the common carrier obligations of Title II of the Communications Act and potentially would attract antitrust scrutiny, as well.

V. The Future of Interconnection and Intercarrier Compensation

This chapter has described a set of rules for interconnection and compensation, as between ILECs and different categories of service providers, that is more the product of history than of any consistent policy. Under the present regime, customers and carriers seeking access to the same ILEC facilities may be forced to pay dramatically different rates. For example, a CLEC can obtain ILEC access lines, switch ports, and interoffice trunks at unbundled network element rates. An IXC or competitive access provider that wants those same facilities, however, must purchase switched or special access at a much higher price.

The FCC is well aware of the inconsistencies in its interconnection and access regime, and has issued policy papers that propose moving toward an integrated regime in which similar ILEC facilities and services will be available to other carriers on comparable terms. Translating those proposals into rules will occupy the Commission for some years to come.

Endnotes

[1] An ILEC's monopoly of local exchange network facilities can harm competition in two ways. First, a monopoly supplier can charge above-cost prices for its network-based

products and services, reducing the consumer welfare of the ILEC's customers and the society generally. Second, where network-based services or facilities are used by purchasers as inputs to services that compete with those of the ILEC, an ILEC has a strong incentive to deny those inputs to their competitors or provide them at excessive cost or inferior quality. These practices inhibit competition and may result in ILEC monopolies over all products and services that rely on access to the ILECs' network. For a more thorough discussion of the effects of monopoly power, see Appendix A, "The Economic Background of Telecommunications Law."

[2] The Conference Report on the 1996 Telecommunications Act declares the policy of Congress to provide for "a procompetitive, deregulatory national policy framework by opening all telecommunications markets to competition." 104th Congress, 2nd Session, Report 104458 to accompany S. 652 (Jan. 31, 1996).

[3] *See* 47 U.S.C. § 253.

[4] *Id.* § 251(c)(2). Such interconnection must be provided for the transmission and routing of telephone exchange service and exchange access at "any technically feasible point within the carrier's network." *Id.* § 251(c)(2)(A)-(B). Such interconnection also must be provided "on rates, terms, and conditions that are just, reasonable, and nondiscriminatory," in accordance with the 1996 Telecommunications Act and the terms of the interconnection agreement negotiated between the ILEC and CLEC. *Id.* § 251(c)(2)(D). The ILEC will not comply with the Act if the interconnection arrangements it offers to CLECs are not at least equal to the interconnections it provides to itself, its affiliates or any other carrier. *Id.* § 251(c)(2)(C). All telecommunications carriers—including those that are neither ILECs nor CLECs—have a more general obligation to establish direct or indirect connections with the facilities and equipment of other telecommunications carriers. *Id.* § 251(a)(1).

[5] *Id.* § 251(b)(5). All LECs—both ILECs and CLECs—are required to establish these reciprocal compensation arrangements with other local exchange carriers. *Id.*

[6] *Id.* § 251(c)(3).

[7] *Id.* § 251(c)(4). All LECs are prohibited from restricting resale of their services, but only ILECs have the additional obligation to provide their services to resellers at discounted rates. *Id.* § 251(b)(1).

[8] *Id.* §§ 251(b)(2)-(3). All LECs have these obligations under the 1996 Act. Number portability is required "to the extent technically feasible," and the dialing parity obligation includes "the duty to permit (other LECs) to have nondiscriminatory access to telephone numbers, operator services, directory assistance, and directory listing, with no unreasonable dialing delays." *Id.*

[9] *Id.* § 251(c)(5).

[10] *Id.* § 251(c)(6).

[11] *Id.* § 251(c)(1).

[12] The 1996 Telecommunications Act states that a local exchange carrier "shall make available any interconnection service, or network element provided under an agreement

approved under this section to which it is a party to any other requesting telecommunications carrier upon the same terms and conditions as those provided in the agreement." *Id.* § 252(I).

[13] When a conventional telephone handset is placed in its cradle, no electrical current flows to the central office. The telephone then is said to be in on-hook status. When the handset is removed from the cradle, the switchhook contacts of the telephone close, generating direct current that alerts the local carrier's switch to generate and send a dial-tone signal to the customer.

[14] As noted earlier, telephone companies with many central offices in a local service area establish access tandems that are interconnected with several central offices, so that IXCs serving the area are not required to establish a direct connection with each central office.

[15] SONET stands for synchronous optical network. SONET is a high-speed transport service based on fiber optic cabling, especially well-suited for carrying traffic of multiple customers or carriers. In order to ensure reliability, SONET typically is deployed in a ring architecture in which one set of strands is a spare set that can be used to reroute traffic if one of the primary sets of fiber strands is cut.

[16] The 1996 Act expressly provides that "bill-and-keep" arrangements are an acceptable form of reciprocal compensation. *Id.* § 252(c)(2)(B)(i).

[17] 47 U.S.C. § 251(c)(4)(A).

[18] *Id.* § 252(d)(3); Implementation of the Local Competition Provisions in the Telecommunications Act of 1996, CC Docket No. 96-98 (First Report and Order, 1996) § 911.

[19] *Id.* § 910.

[20] 47 U.S.C. § 252(e).

[21] *Id.* § 252(a)(2).

[22] *Id.* § 252(b)-(c).

[23] *Id.* §§ 252(c)(2), 252(d).

[24] *Id.* § 251(d)(1).

[25] *Id.* § 252(e)(5).

[26] *Id.*

[27] *Id.* § 252(e)(6).

[28] To the extent the commissions were concerned about this relationship, they were chiefly interested in ensuring that local exchange carriers (both Bell and independents) received adequate compensation for carrying Long Lines' traffic.

[29] *See*, e.g., Specialized Common Carrier Service, First Report and Order, 29 FCC 2d 870, 940 (1971), *aff'd sub nom. Washington Utilities & Transport Commission v. FCC,*

[29] 513 F.2d 1142 (9th Cir.), *cert. denied sub nom. National Association of Regulatory Utilities Commissioners v. FCC,* 432 U.S. 836 (1975).

[30] GTE also entered into a consent decree, containing equal access provisions similar to those of the AT&T decree. Compliance with both decrees was supervised by Judge Harold Greene of the United States District Court in Washington, D.C.

[31] *See,* e.g., Investigation into the Quality of Equal Access Services, 60 Rad. Reg. (P&F) 417 (1986).

[32] 47 U.S.C. § 251(g).

[33] CLECs also connect IXCs with their customers, and are required to do so as part of their general obligations to interconnect with other carriers and permit customers to select two presubscribed IXCs—one for intraLATA toll calling and another for interLATA toll calling. *See* 47 U.S.C. §§ 251(a), 251(b)(3); *see also* Implementation of the Local Competition Provisions in the Telecommunications Act of 1996, CC Docket No. 6-98, Second Report and Order and Memorandum Opinion and Order (rel. Aug. 8, 1996). These requirements of the Telecommunications Act of 1996 apply to both ILECs and CLECs. However, the specific equal access requirements developed prior to enactment of the 1996 Act apply only to ILECs.

[34] Interconnection takes place at a facility designated by the IXC. Some IXCs interconnect directly with ILEC end offices, while others interconnect with an ILEC facility called an access tandem. The access tandem can connect several IXCs to a number of ILEC central offices.

[35] Modified Final Judgment (AT&T Consent Decree) (hereinafter "MFJ"), Appendix B, A.1; *see United States v. American Telephone and Telegraph Co.,* 552 F. Supp. 226 (D.D.C. 1982). The GTE Consent Decree required access "that is equal in type, quality, and price for all interexchange carriers and information service providers, including any information services of a (GTE telephone company)." Final Judgment (GTE Consent Decree) V. A; *see United States v. GTE Corp.,* 1985-1 Trade Cas. (CCH) 66,355 (1984).

[36] The GTE decree imposed a number of restrictions on GTE but did not require the outright divestiture of any of its operations—which at that time included manufacturing and long-distance service.

[37] *United States v. Western Elec. Co.,* 569 F. Supp. 990, 1063 (D.D.C. 1983).

[38] "It would be unrealistic for us to attempt to delve further into these complexities and to set standards governing each aspect of (equal access interconnection). Such standards would not only be complicated and difficult to enforce, they would also retard the ability of exchange carriers to improve their service quality when advances in technology allow them to do so." *Investigation into the Quality of Equal Access Services,* 60 Rad. Reg. 2d (P&F) 417, § 133 at 447 (1986).

[39] In addition to Feature Group D, the ILECs have offered less sophisticated (and less expensive) arrangements to IXCs that want them. Feature Group A, for example, provides only a line-side connection and requires the IXC's customers to dial a seven-digit

number to reach the carrier, then dial the called number after receiving a second dial tone.

[40] Under the Telecommunications Act of 1996 and the Commission's rules adopted pursuant to the Act, customers must be permitted to presubscribe to an interLATA toll carrier and—if the customer desires—a separate intraLATA toll carrier. 47 U.S.C. § 251(b)(3); Implementation of the Local Competition Provisions of the Telecommunications Act of 1996, CC Docket No. 96-98, Second Report and Order and Memorandum Opinion and Order § 37 (1996)(Second Local Competition Order).

[41] 47 U.S.C. § 251(b)(3). This requirement applied to both ILECs and CLECs.

[42] Second Local Competition Order § 37.

[43] *Id.* § 59.

[44] Policies and Rules Concerning Changing Long Distance Carriers, 7 FCC Rcd. 1038 (1992); *see also* 47 U.S.C. § 258.

[45] The Telecommunications Act of 1996 expressly preserves the FCC's existing rules governing interconnection between ILECs and IXCs. 47 U.S.C. § 251(g).

[46] All of the larger, and most smaller, ILECs set their interstate exchange access rates on a price-cap basis. The states continue to regulate access charge rates for intrastate toll calls.

[47] Access Charge Reform, CC Docket No. 96-262 (First Report and Order rel. May 16, 1997) § 55.

[48] *Id.* § 38.

[49] *Id.* § 39.

[50] *Id.* § 102.

[51] For a list of the access charge elements, *see* 47 C.F.R. § 69.4.

[52] Interconnection between Local Exchange Carriers and Commercial Mobile Radio Service Providers, 11 FCC Rcd. 5020 (1996) § 21.

[53] The Commission arrived at this conclusion by applying the preemption analysis set out by the U.S. Supreme Court in *Louisiana Public Service Commission v. FCC*, 476 U.S. 355 (1986).

[54] In the Matter of the Need to Promote Competition and Efficient Use of Spectrum for Radio Common Carrier Services, 59 Rad. Reg 2d (P&F) 1275, § 12 at 1278 (1986). Under that decision, a federal agency may not preempt state regulation unless the Congress expressly has authorized preemption or enforcement of the state regulation would make enforcement of a valid federal regulation impossible.

[55] Omnibus Budget Reconciliation Act of 1993, Pub. L. No. 103-66, Title VI, § 6002, 107 Stat. 312 (1993). The preemption provision is codified at 47 U.S.C. § 332(c)(3).

[56] Interconnection between Local Exchange Carriers and Commercial Mobile Radio Service Providers, 11 FCC Rcd. 5020 § 20 (1996).

[57] Implementation of the Local Competition Provisions in the Telecommunications Act of 1996, CC Docket No. 96-98 (First Report and Order rel. Aug. 8, 1996) § 1025.

4

Special Cases: The Bell Operating Companies

So far we have described those regulatory and antitrust obligations that apply to *all* ILECs. We found that all ILECs must provide exchange and exchange access service at reasonable and nondiscriminatory rates, and that their rates for basic services are regulated either directly (on a price-cap basis) or indirectly (through a cap on earnings determined under a rate-of-return formula). We also saw that ILECs are required to interconnect with other carriers, including CLECs, on reasonable terms and conditions.

We now turn to additional obligations and restrictions that apply only to some of the largest and most visible ILECs (i.e., the local exchange carriers that once were owned by AT&T as part of the unitary Bell system). These companies—known in the industry as Bell operating companies or BOCs—have been subjected to unique obligations in two principal areas. First, as part of the antitrust decree (also known as the Modification of Final Judgment or MFJ) entered at the divestiture of the Bell system, the BOCs were entirely prohibited from engaging in certain lines of business, including interexchange service and manufacturing of telecommunications equipment [1]. More recently, those decree obligations were perpetuated, in part, in the Telecommunications Act of 1996, which also defines a process by which BOCs may be relieved of those obligations [2]. Second, as a result of a lengthy and complex FCC rulemaking process, the BOCs were subjected to certain constraints in their provision of Internet access and other information services. The following describes these requirements, and their background, in detail.

I. The BOC Restrictions in the Telecommunications Act of 1996

By historical accident and the fortunes of litigation, most of the largest local exchange carriers in the United States are severely limited in the kinds of services they may provide. Because they once were under common management and were accused of sundry violations of the antitrust laws, they now are bound by statutory restrictions that limit their ability to provide interexchange service or manufacture telecommunications equipment.

A complete understanding of the BOC restrictions requires considerable background in the history and legal theories attending the divestiture of the Bell system, as well as the post-divestiture litigation through which the original decree provisions have been interpreted and modified. In keeping with the practical approach of this book, we forego a preliminary discussion of that theory and history and go directly to the specific statutory restrictions as they apply to the operations of the Bell telephone companies today. In the course of that discussion, we will fill in the theoretical and historical background where necessary.

A. The InterLATA Restriction

The first and most significant restriction on BOC activities is the so-called "interLATA restriction." According to Section 271(a) of the 1996 Act, no BOC or BOC affiliate may provide telecommunications services that originate in a LATA within the BOC's service area, and terminate outside that LATA, unless that BOC or affiliate has first complied with an extensive "competitive checklist" intended to show that the BOC faces competition in its local service area [3]. In order to understand this restriction, it is important to know what a LATA is and when a BOC may be said to provide "interLATA service." Our search for an understanding of these questions, in turn, will take us back to the AT&T divestiture consent decree.

1. LATAs Defined

One of the chief purposes of the AT&T divestiture was to divorce AT&T's long-distance services from the local exchange services provided by the AT&T-owned BOCs. The Justice Department believed that once their relationship with AT&T was severed, the BOCs would have no incentive to furnish AT&T's long-distance services with more efficient interconnections, on more favorable terms, than those they provided to the competing interexchange carriers.

The Justice Department lawyers did not believe, however, that severing the existing affiliation between AT&T and the BOCs would be enough. If the post-divestiture BOCs remained free to offer their own interexchange services, they might use their continuing dominance of the local networks in their service areas to impose competitive disabilities on nonaffiliated interexchange carriers. Instead of favoring AT&T, they now would favor their own long-distance affiliates. The consent decree, therefore, prohibited the BOCs from providing any "interexchange telecommunications services" [4].

This term, however, is somewhat misleading. The term "interexchange service" in the decree did not refer to services between local telephone exchanges (i.e., between areas served by any of the 7000 BOC central offices). The BOCs were not confined to service within these relatively small areas.

Nor did the interexchange restriction prevent BOCs from offering all long-distance service. As commonly understood, long distance refers simply to calls for which toll charges, rather than local calling charges, apply. The BOCs offer substantial long-distance service in this sense, and the decree court made it clear that the demarcation point between local and permitted toll service was a matter for state regulators to decide [5].

Instead, interexchange service, within the meaning of the MFJ, was defined in reference to a kind of service area created especially to implement the consent decree—the local access and transport area, or LATA. A LATA was defined in Section IV(G) of the decree as "one or more contiguous local exchange areas serving common social, economic, and other purposes, even where such configuration transcends municipal or other local governmental boundaries" [6]. LATAs can be as large as a state or as small as an area with 10,000 subscribers; their size is based on an estimate of the minimum area necessary to make it economical for competing IXCs to establish a point of presence and make the investment necessary to provide service.

AT&T, the Justice Department, and Judge Greene all were involved in the definition of these new entities. When the process was complete the territories served by the BOCs were divided into 163 LATAs. (Even independent company territories were assigned to BOCs as "shadow" LATAs, so that the independents could connect to IXCs within the adjoining BOC territories to which they were assigned.) Almost all LATAs lie within the territory of a single state, but a few cross state boundaries.

When the decree prohibited interexchange telecommunications services, therefore, it really prohibited *interLATA* telecommunications services. And when the 1996 Act preserved the interexchange restriction in Section 271, it adopted the more precise term *interLATA service* to define the restriction [7].

2. The InterLATA Restriction Explained

The 1996 Act does not simply adopt and perpetuate the interLATA restriction of the MFJ. In fact, the 1996 Act expressly preempts the MFJ and replaces the interLATA restriction of the antitrust decree with a somewhat narrower restriction that may be removed, on a state-by-state basis, through the process described in the next section of this chapter.

Notably, although the MFJ prohibited BOCs from providing any communication between a point inside a LATA and another point outside that LATA, the 1996 Act restricts only "in-region" interLATA services (i.e., communication between a point in a LATA located in the BOC's local service area and a point outside that LATA). With this change, Congress signaled its encouragement to the BOCs to compete immediately in the long-distance markets outside their monopoly service areas (i.e., to bring new competition to the long-distance markets in areas in which they did not control the local exchange)[8].

The 1996 Act also permits the BOCs to provide so-called "incidental" interLATA services. These include the transport across LATA boundaries of audio and video programming, alarm monitoring services, video or Internet services provided over dedicated facilities to elementary and secondary schools, commercial mobile radio service, and certain other services listed in subsection g of Section 271 of the statute [9]. The Act expressly provides that this list of incidental interLATA services must be "narrowly construed" so that these services will not become the pretext for premature BOC entry into mainstream interLATA telecommunications services [10].

Finally, the 1996 Act permits BOCs to terminate interLATA services (i.e., deliver interLATA communications to customers) in their regions.

With these exceptions, the 1996 Act prohibits any BOC from providing interLATA services—including both telecommunications and information services—originating in any state within the BOC's service area until the BOC has obtained the FCC's permission to do so [11]. As the next section shows, the FCC may not grant such relief from the interLATA restriction until the BOC has satisfied an extensive series of market-opening requirements set out in Section 271(c) of the 1996 Act.

3. Escaping the InterLATA Restriction

The 1996 Act reflects the same concern that prompted the interexchange restriction of the MFJ (i.e., the concern that if BOCs simultaneously monopolize local exchange service and provide interexchange service, they will use their control of the local exchange to discriminate against competing IXCs). The MFJ neutralized this threat by quarantining the BOCs within their local exchanges. The 1996 Act takes a different approach—permitting the BOCs to offer in-region interLATA service, but only after those BOCs have demonstrated that they have opened their local exchange markets to competition.

Specifically, under Section 271 of the 1996 Act, a BOC may provide in-region interLATA service in any state in which it: (1) has entered into an approved access and interconnection agreement with one or more facilities-based local exchange competitors[12]; or (2) has not received a request for access and interconnection from a competitor, but has obtained approval from the state commission of a statement of terms and conditions under which the BOC will provide access and interconnection to local competitors [13]. The BOC's access and interconnection agreements or statements of terms and conditions must meet the requirements of a "competitive checklist" consisting of the following elements.

a. Interconnection with Local Competitors

As we discussed in Chapter 3, Section 251 of the 1996 Act requires ILECs to interconnect their networks with those of all telecommunications carriers that request such interconnection. Without such internetwork connections, CLEC customers will be unable to call ILEC customers and ILEC customers will be unable to call CLEC customers. Accordingly, interconnection arrangements between ILECs and other telecommunications carriers must provide for the transmission and routing of telephone exchange service and exchange access at any technically feasible point within the carrier's network, at a level of quality that is at least equal to that which the ILEC provides to its own affiliates or other parties, on rates, terms, and conditions that are just, reasonable, and nondiscriminatory [14]. Charges for access and interconnection also must conform to the pricing standard set out in Section 252(d)(2) of the 1996 Act [15].

b. Access to Network Elements

As we also discussed in Chapter 3, ILECs are required, upon request from competing telecommunications carriers, to provide subscriber lines, central office switch ports, interoffice trunks, signaling systems, and other elements of the ILECs' own networks for use by the requesting carriers in providing competing services. In order to obtain interLATA authority, a BOC must show that its interconnection agreements or statements of terms and conditions make unbundled network elements available "on an unbundled basis at any technically feasible point on rates, terms, and conditions that are just, reasonable, and nondiscriminatory" [16]. Like interconnection for the transport and termination of traffic, access to unbundled network elements must be priced according to the standard set out in Section 252(d)(2) of the 1996 Act.

c. Access to Poles, Ducts, Conduits, and Rights-of-Way

In addition to interconnecting with local competitors and providing access to unbundled network elements, BOCs must—if they want to provide in-region, interLATA service—permit other carriers to place facilities on the poles and in the ducts, conduits, and rights-of-way that the BOCs own or control [17]. This obligation is part of the more general requirement, set out in Section 224 of the Act, that any utility must grant "a cable television system or any telecommunications carrier with nondiscriminatory access to any pole, duct, conduit, or right-of-way owned or controlled by it" [18].

d. Local Loop Transmission

Before they may be granted authority to offer in-region, interLATA service, BOCs also must demonstrate that they are providing or are prepared to provide "local loop transmission from the central office to the customer's premises, unbundled from local switching or other services" [19]. This item of the competitive checklist, which the FCC also identified as one of the required, ILEC unbundled network elements in its order implementing the local competition provisions of the 1996 Act, is defined by the Commission as "a transmission facility between a distribution frame, or its equivalent, in an incumbent LEC central office, and the network interface device at the customer's premises" [20]. In order to satisfy this requirement, the ILEC must show that it has "a specific and concrete legal obligation to furnish loops and that it is currently doing so in the quantities that competitors reasonably demand and at an acceptable level of quality" [21].

e. Unbundled Local Transport

BOCs seeking interLATA authority also must show that they are providing or will provide "local transport from the trunkside of a wireline local exchange carrier switch unbundled from switching or other services" [22]. As defined by the FCC, this local transport obligation consists of two services. One service is dedicated transport—a BOC transport facility devoted to a particular customer or carrier that provides telecommunications between wire centers owned by BOCs or requesting carriers [23]. The other service is shared transport, consisting of transmission facilities used by more than one carrier between central office switches, between central office switches and tandem switches, and between tandem switches in the BOC's network [24].

f. Unbundled Local Switching

Approval of in-region BOC interLATA service also requires that the BOC switch local calls for other telecommunications carriers [25]. Unbundled local switching consists not only of the ability to connect dial-up calls for delivery to their destinations, but also of access to all capabilities of the switch that are available to the BOC's customers [26].

g. Access to 911, Directory Assistance, and Operator Services

BOCs seeking in-region interLATA authority must provide other telecommunications carriers with access to the BOCs' emergency 911 services, directory assistance services, and operator call completion services [27]. BOCs also must provide their competitors' customers with access to the BOCs' white pages directory listings [28].

h. Access to Telephone Numbers

At the time of enactment of the 1996 Act, telephone numbers under the North American Numbering Plan still were assigned by the BOCs. The 1996 Act directed the FCC to "create or designate one or more impartial entities to administer telecommunications' numbering and to make such numbers available on an equitable basis" [29]. The 1996 Act also directed that until the new entity and its rules or guidelines were established, BOCs seeking in-region interLATA authority were required to provide "nondiscriminatory access to telephone numbers for assignment to the other carrier's telephone exchange service customers" [30]. After the establishment of the new entity and its rules or guidelines (which was, in fact, accomplished in 1997), BOCs seeking in-region interLATA authority must comply with the guidelines, plan, or rules of the independent numbering administrator [31].

i. Access to Databases and Signaling

BOCs may not be granted in-region, interLATA authority unless they provide their local competitors with nondiscriminatory access to BOC databases and signaling necessary for call routing and completion [32]. In order to satisfy this checklist item, a BOC must show that it provides its competitors with "the same access to these call-related databases and associated signaling that it provides itself" [33].

j. Number Portability and Dialing Parity

The competitive checklist also requires BOCs seeking in-region, interLATA authority to demonstrate that they have implemented, and are in compliance with, all FCC regulations concerning number portability and local dialing parity [34]. The former requirement—number portability—means that consumers must be able to keep the same telephone number when they transfer their local service from the BOC to a competitor. The latter requirement—local dialing parity—means that customers of CLECs can make local telephone numbers by dialing the same number of digits that BOC customers dial.

k. Reciprocal Compensation Arrangements

Reciprocal compensation is an agreement between a BOC (or other CLEC) and local competitors under which the BOC compensates the competing carrier for terminating calls placed by the BOC's customers and the competing carrier compensates the BOC for completing calls that flow in the opposite direction. In order to obtain in-region, interLATA authorization, a BOC must demonstrate that it has such "arrangements in place, and that it is making all required (reciprocal compensation) payments in a timely fashion" [35].

l. Resale

BOCs seeking in-region, interLATA authorization also must show that they are offering "all of (their) retail services at wholesale rates without unreasonable or discriminatory conditions or limitations such that other carriers may resell those services to an end user" [36].

When a BOC has satisfied all elements of this competitive checklist, the FCC may grant the BOC in-region, interLATA authority for calls originating in the state for which the BOC's application was made. Before making this determination, however, the Commission is required to consult with the Attorney General of the United States and the state PUC for the state in which the proposed interLATA service will originate [37]. The Commission is not required to follow the recommendations of the Attorney General and the state PUC, but is required to give "substantial weight" to the Attorney General's evaluation of the BOC's request [38].

4. The Manufacturing Restriction

The 1996 Act expressly perpetuates the MFJ's restriction on BOC manufacturing and sales of telecommunications equipment, and manufacturing of customer-premises equipment, until such time as a BOC or its affiliate obtains authority to provide interLATA service under Section 271 of the Act [39]. The 1996 Act also requires the BOCs to conduct their manufacturing operations through separate affiliates for three years after in-region, interLATA authority is obtained, subject to extension of the three-year period by the FCC [40].

Because the 1996 Act expressly adopts the manufacturing restriction as defined in the MFJ, rather than—as in the case of the interLATA restriction—replacing the MFJ's definition with a new, somewhat different definition, an understanding of the manufacturing restriction of the Act requires an understanding of the manufacturing restriction of the MFJ [41].

Equipment Covered by the MFJ Restriction

In a nutshell, the decree (as interpreted by Judge Greene and the Department of Justice) prohibited the BOCs from *manufacturing* or *providing* any *telecommunications products*, and from *manufacturing* any *customer-premises equipment (CPE)*. (The BOCs were permitted to *provide* customer-premises equipment that they did not manufacture.)

Telecommunications equipment (the stuff that BOCs may not manufacture *or* provide) [42] was defined in the decree as "equipment, other than customer-premises equipment, used by a carrier to provide telecommunications services" [43]. This definition certainly included the switches, wires, and other equipment used to transport, switch, and process telecommunications signals. How much else it included was unclear, although the Department took the view that it does not include "test equipment that is not integrated with, or incorporated in switching and transmission facilities or CPE " [44].

Customer-premises equipment (the stuff the BOCs could not manufacture, but could provide) included any equipment used on the premises of a customer (but not another carrier) to "originate, route, or terminate telecommunications" [45]. It certainly included telephone handsets, PBXs, and key telephones. Whether it also included transmission facilities is unclear.

While BOCs were permitted to provide (lease or sell) CPE, they could do so only to customers and not to other carriers. A BOC could sell CPE to carriers or resellers, however, when that equipment would be used only for the carrier's or reseller's internal use, rather than in the provision of the carrier's services.

The Meaning of "Manufacturing"

Because nothing was simple in the MFJ world, even the definition of "manufacturing" had to be litigated.

The BOCs understood manufacturing, as that word was used in the decree, to refer to the physical fabrication of equipment. They did not take it as extending to the design and development of products to be fabricated by others, and they certainly did not believe they were prohibited from writing software that might be used in telecommunications equipment or CPE. Judge Greene read the decree differently, however, and the Court of Appeals affirmed his determination that manufacturing included the design and development of equipment, as well as the writing of software that is "integral" to the operation of hardware [46].

The BOCs *were permitted* to engage in network engineering essential to provide local service, develop network software, and modify software that was intended by its vendor to be modified by the user. They also were permitted to test, and adopt standards for, equipment that they would not be allowed to design, develop, or fabricate [47].

The 1996 Act expressly provides that for purposes of its manufacturing restriction, "the term 'manufacturing' has the same meaning as such term has under the AT&T Consent Decree" [48]. Accordingly, all of the history and lore of the MFJ's manufacturing restriction, as briefly described here, is relevant to an understanding of the restriction contained in the 1996 Act.

5. The Electronic Publishing Restriction

The 1996 Act also restricted the BOCs' involvement in electronic publishing, which it defined as "the dissemination, provision, publication, or sale of news; entertainment; editorials, columns, or features; advertising, photos, or images; or other like or similar information" [49]. Specifically, the Act provided that a BOC could not engage in electronic publishing except through an entity, described as a "separated affiliate," that maintained its own financing, management, books and records, and observed other separation requirements described in the Act [50]. The electronic publishing restrictions also included a "sunset date," providing for their expiration four years after the enactment of the 1996 Act; accordingly, the BOCs no longer were subject to those restrictions after February, 2000 [51].

6. Alarm Monitoring and Payphones

The 1996 Act also imposes some requirements concerning BOC participation in the alarm monitoring and pay telephone markets.

Concerning the alarm-monitoring market, the Act does not permit BOCs to provide alarm-monitoring service until five years after the date of enactment of the 1996 Act [52]. The Act does, however, permit BOCs to continue to provide alarm-monitoring services that they were providing on

November 30, 1995; but prohibits those BOCs from acquiring an equity stake in, or control of, any other unaffiliated alarm-monitoring entity until five years after enactment of the 1996 Act [53].

Concerning the payphone market, the 1996 Act requires BOCs to make fundamental changes in the arrangements under which they provide those services [54]. These changes are fully described in Chapter 7 of this book.

7. Separate Affiliates and Other Safeguards

Some of the most complex provisions of the 1996 Act, and the FCC rules and orders implementing the Act, describe the structural and transactional terms on which BOCs may provide in-region, interLATA services (including interLATA information services other than electronic publishing and alarm monitoring) and manufacture equipment after they have received the FCC's permission to enter those markets. Most importantly, the BOCs may provide these services only through separate affiliates (i.e., entities that have separate books, officers, directors, and employees from those of the affiliated BOC, deal with the BOCs at arms' length in all transactions, and comply with a number of nondiscrimination requirements set out in Section 272 of the Act) [55].

The separate affiliate requirements of the 1996 Act are not perpetual. As applied to manufacturing and interLATA telecommunications services, those requirements cease to apply (sunset) three years after the date a BOC is authorized to provide in-region interLATA service, unless the FCC extends the sunset date by rule or order [56]. As applied to interLATA information services, those provisions expired in February, 2000 [57].

B. Interconnection with Data Networks and Other Enhanced Service Providers: The *Computer III* Rules

So far we have described ILEC services as nothing more than the transmission of human speech from one end user to another. We have portrayed local exchange service as something that lets two people within the ILEC service area talk to each other; and we have portrayed exchange access service as something that helps a person within the ILEC service area talk to someone outside the ILEC service area. As described, these services fit the traditional understanding of basic telephone service.

As most people are aware, this picture no longer exhausts the uses to which telephone facilities are put. Most notably, computers have been talking to each other over the telephone network for many decades. These conversations (and similar exchanges between fax machines and other devices) take place in electronic streams that represent bits of data. They may be carried by special data lines, or they can take place over the same ILEC facilities that carry voice traffic.

So long as the ILEC is only *transmitting* information (whether voice or data), and delivering it instantaneously to the receiving party in the same form as it was sent by the originating party, it is doing something that courts and regulators have found familiar and comforting. But human ingenuity, of course, has learned to do more with electronic data than merely transmit it. Information (including human speech) can be stored, retrieved, and manipulated. It can be converted from one medium or format to another. And telephone lines can carry information to and from the databases and processors that do all of this.

We all are familiar with services that act on information transmitted over telephone lines. Voice messaging services, for example, record calls and play them back when the customer is ready to hear them. On-line research databases, responding to user commands sent over phone lines, search for information and transmit it to the customer's computer. And the Internet gives users access to a wealth of information and services, ranging from e-mail to streaming video.

The courts and the regulators are comfortable when these services are provided by ISPs and other entities not affiliated with any ILEC. When an ISP buys ILEC services to connect its customers to the Internet, nothing is amiss. But when LECs seek to offer these services themselves, questions arise. Specifically, the courts and regulators are concerned that when the ILECs—and BOCs, in particular—offer information services in competition with others, they may use their control of the local exchange to place their competitors at a disadvantage.

These concerns led to the prohibition, in the AT&T divestiture decree, against BOC provision of information services. That restriction remained in effect, except for some services offered under waivers, until it was removed by Judge Greene in 1991. Since the information services prohibition was removed, the principal constraint on BOC provision of these services has been the system of regulation adopted by the FCC.

The Commission became concerned about information services (which it placed in the roughly equivalent category of "enhanced services") long before the divestiture of the Bell system. Its earliest regulations on this subject included a separate subsidiary requirement [58] under which AT&T's affiliates could offer enhanced services only through entities having their own personnel, facilities, and books of account [59]. (After divestiture, these regulations were applied to the divested Bell operating companies.) The BOCs themselves could offer only "basic" services that did not store or manipulate the information transmitted in any way, and they were required to interconnect with both affiliated and unaffiliated enhanced service providers according to published tariffs.

The FCC later abandoned the separate subsidiary approach and imposed a system of nonstructural safeguards under which the Bell operating companies provide enhanced services using the same facilities and personnel they use to provide basic services. The accompanying safeguards are complex—so complex, in fact, that the Commission declined to impose them on the independents—and proceedings spanning several years were needed before the BOCs had finished even the preparatory work needed to comply with them. In the following paragraphs, we look at each element of this so-called *Computer III* regulatory scheme in turn.

1. The ONA Model

The BOCs' first task was to structure a menu of facilities and services through which enhanced service providers (including the BOCs' own enhanced services) would obtain flexible access to the local networks.

In 1986 the Commission outlined a general approach called the Open Network Architecture (ONA) model, under which enhanced service providers (conventionally called ESPs) would obtain access to defined services (such as switching, signaling, and billing) through different types of access links. The services themselves would be called basic service elements, or BSEs; and the access links over which the services would be provided were to be called basic serving arrangements, or BSAs [60]. So, for example, an ESP might order switching service (a BSE) over a trunkside BSA (equivalent to the Feature Group D access provided to IXCs) or over a lineside BSA (equivalent to Feature Group A). Available BSAs also include private lines, packet switching [61], and other arrangements ESPs might use to connect their facilities with the local network.

The BOCs then developed, and submitted for the Commission's approval, specific ONA plans that varied in detail from region to region. These plans were approved after some Commission-ordered amendments, and the BOCs filed federal tariffs implementing their ONA plans in November of 1991.

The implementation of ONA is another example of shared federal/state jurisdiction. The FCC required the BOCs to file federal tariffs setting out the ONA services and facilities, and to file state tariffs for intrastate ONA offerings. (A joint federal/state conference was formed to coordinate this effort.)

As a result of the dual jurisdiction over ONA services, ESPs and customers often have a choice of obtaining facilities from state or federal tariffs. (The federal/state conference has developed relative usage criteria to determine when a facility used for both intrastate and interstate traffic must be ordered from the federal, rather than state, tariff.)

The ONA rules tried to ensure parity of access for BOC and non-BOC ESPs. The BOCs were required to buy access for their own enhanced services from the same tariffs as the non-BOC ESPs. If a BOC intended to offer any new service that required a new BSE, it was required to amend its ONA plan to include the new BSE at least 90 days before the new enhanced service would be offered. If an ESP submitted a written request to a BOC for a new BSE, BSA, or other ONA capability, the BOC was required to respond within 120 days, stating whether and when it will meet the request, and at what price.

2. Comparably Efficient Interconnection

It was not enough for BOCs simply to offer a flexible menu of access services to ESPs, and provide their own services through the same tariffed access arrangements; BOCs also were required to ensure that the access provided to the non-BOC ESPs is actually comparable, in quality and cost, to the access provided to the BOCs' own ESP services.

The question of comparability is especially important where the BOCs will provide their own enhanced services at the same premises from which they provide basic services, or will integrate those services in the same software and hardware that they use to provide basic services. Under these circumstances, even if the BOC ESP services ordered access from the same tariffs as the competing ESPs, collocation at the BOC premises might give the BOCs' ESP services advantages in cost and transmission quality over the services of competitors. The Commission did not respond to this perceived problem by requiring collocation, but it did impose a number of comparably efficient interconnection, or CEI, requirements intended to minimize the BOCs' advantages.

First, the Commission encouraged the BOCs to provide precisely the same interconnection facilities to competitors that they provided to their own ESP services. Where identical facilities were not provided, the BOCs still were to furnish competitors with facilities having comparable (if not precisely equal) performance characteristics (including transmission speed, error rates, and availability of various options). As with equal access, the FCC imposed no detailed specifications for technical equality, but looked to customer perception and the effect of different arrangements on the ability of ESPs to compete.

Second, the Commission adopted a "two-mile rule" under which all ESPs located within two miles of the serving central office (including BOC services within the central office) must be charged the same rate for transport. This gives the non-BOC ESPs some incentive to place their facilities at an efficient distance from the serving central office, but eliminates the advantage that otherwise would be enjoyed by BOC services and ESPs that collocate their facilities with those of the BOCs.

Third, the Commission directed that the same "common interconnection charge" be imposed on all ESPs (including BOCs) regardless of the form of interconnection used. This rule is intended to minimize any cost advantage the BOCs otherwise might gain by offering more efficient interfaces to their own enhanced services [62].

3. Other Safeguards

In addition to the ONA and CEI rules, the Commission imposed an assortment of additional safeguards intended to prevent the BOCs from using their control of the network to the advantage of their own enhanced services, or from imposing disabilities on competing ESPs.

One such safeguard was the requirement that the BOCs disclose changes in their networks that may affect the interconnection arrangements of competing ESPs. Relevant technical information concerning such changes was required to be made available well in advance of their implementation.

Another set of rules dealt with the disclosure and use of so-called customer proprietary network information, or CPNI. This is customer information the BOCs maintain as part of their provision of basic services: it includes such data as the design of customers' telecommunications services and the extent of usage of those services.

Both competitors and end users expressed concerns, during the *Computer III* proceedings, about the uses to which CPNI might be put. Competing providers of enhanced services complained that the BOCs could use CPNI to determine which customers were using the services of non-BOC ESPs, and to target those customers for BOC marketing efforts. Users were concerned that CPNI should be protected from disclosure because it "may reveal information concerning businesses' manufacturing or marketing processes, new services, plans for territorial expansion, and other information that businesses would want to be kept from any disclosure" [63]. The Commission was urged to limit the access of ESPs (BOC and non-BOC alike) to this information.

The FCC decided that the BOCs would be allowed to use CPNI to market their enhanced services unless a customer requested that it not be so used [64]; and that the BOCs need not release this information to third parties unless the customer requests that the BOCs do so.

The Commission took a different approach, however, to so-called aggregated CPNI. This is noncustomer-specific information, generated from the BOCs' databases, that shows "usage levels and traffic patterns for the network services in (the BOCs') service areas" [65]. Some parties had argued that this information "would be of substantial value in the technical and economic design of enhanced services," and the Commission agreed that aggregated CPNI should be made available to competitors and other parties on the same terms as it is provided to the BOCs' enhanced service providers [66].

Finally, the BOCs were required to adopt and follow procedures that ensured that when competing ESPs order services from the BOCs, they would receive the same quality of installation and maintenance service as the BOCs' enhanced services receive. In addition to describing their installation and maintenance procedures to the Commission in writing, the BOCs were required to submit periodic performance reports [67].

4. Accounting Safeguards

The *Computer III* safeguards we have described so far were designed to prevent discrimination against non-BOC ESPs in the terms and conditions of interconnection. But what about the other, perpetual concern where ILECs provide both regulated and unregulated services (i.e., cross-subsidization)? To prevent basic services from subsidizing the BOCs' enhanced services, the Commission relied on its Part 64 accounting requirements for regulated and nonregulated services, and on incentive regulation.

The Part 64 rules require the BOCs (and all dominant carriers) to separate the costs incurred to provide regulated services from the costs incurred to provide nonregulated services (including enhanced services). Each BOC has adopted an FCC-approved cost allocation manual to govern this process: the costs are separated on a fully allocated (rather than incremental) basis, and the allocation is done before the separations process divides the regulated costs further between the federal and state jurisdictions. Also, as we have discussed, the new FCC regime of incentive, or price-cap, regulation deprives the BOCs of much of their ability to benefit from cross-subsidization of competitive services.

Between them, incentive regulation and the cost allocation rules leave little room for the BOCs to subsidize enhanced services from the revenues earned in providing basic services.

5. State Regulation of Enhanced Services

As we have explained previously, the FCC is empowered to regulate interstate telecommunications, and has interpreted that mandate broadly. At the same time, Section 2(b) of the Communications Act preserves the right of the states to regulate intrastate telecommunications, and the courts have not hesitated to restrain the Commission when it appears to encroach on that right.

In developing its *Computer III* rules, the Commission took a broad view (as it had in the earlier Computer II proceedings) of its authority to preempt inconsistent state regulation. The Commission found both that enhanced services were not covered by the Act's reservation of power to the states, and that in any case it was impossible for the Commission to implement its policies if the states were permitted to regulate enhanced services [68]. Accordingly, the Commission preempted all state tariffing requirements and other state common carrier regulation of enhanced services. The Commission also preempted: "(1) all state structural separation requirements applicable to the provision of enhanced services by AT&T, the BOCs, and the independents; (2) all state nonstructural safeguards applicable to AT&T and the BOCs inconsistent with federal standards; and (3) all state nonstructural safeguards applicable to the independents more stringent than those applied to AT&T and the BOCs" [69].

The Court of Appeals for the Ninth Circuit disagreed with the FCC's view of its authority to preempt the states. The court found in particular that Section 2(b) of the Act does apply to enhanced services, and that the Commission had not shown that complete preemption was necessary to avoid frustrating its goals. The court ordered the Commission to revisit its preemption decision under the appropriate standard [70].

In accordance with the decision of the Court of Appeals, the Commission made a more discriminating review of particular areas of potential state regulation of enhanced services. The Commission did not attempt to reinstate its prohibition on state tariffing and other common carrier requirements. The Commission decided to preempt any state requirements for separation of facilities and personnel used to provide "jurisdictionally mixed" enhanced services (i.e., any services not provided on a purely intrastate basis); but ruled that states could require structural separation for the provision of intrastate enhanced services, and could require a separate corporate entity, with separate books of account, for the intrastate portion of jurisdictionally mixed enhanced services [71]. (The same rule was applied to state structural separation requirements imposed on AT&T and the independents.) The

Commission also preempted any state CPNI rules applicable to the BOCs, AT&T, or the independents that require prior authorization where such prior authorization is not called for by the FCC's rules. Finally, the Commission preempted state network disclosure rules that require disclosure at a time different from the time specified in the federal rules [72].

So long as the states do not violate the Commission's limited preemption policy, they are free to establish their own rules to govern intrastate enhanced services and enhanced services interconnection [73].

6. Post-1996 Reform of the *Computer III* Rules

The 1996 Act directed the FCC to undertake periodic reviews of its regulations and eliminate those that no longer served the public interest [74]. Following a review of its *Computer III* rules pursuant to this authority, in 1999 the Commission decided to eliminate two of its requirements under those rules. Specifically, the Commission eliminated the requirement that BOCs file or obtain preapproval of CEI plans and amendments to those plans "before initiating or altering an intraLATA information service" [75]. Similarly, the Commission in that order eliminated certain of its network disclosure rules [76]. As competition in information and telecommunications services improves, the Commission can be expected to relax its complex *Computer III* regime still further.

Endnotes

[1] *United States v. AT&T,* 552 F. Supp. 131, 226-234 (D.D.C. 1982), *cert. denied,* 460 U.S. 1001, 103 S.Ct. 1240, 75 L. Ed. 2d 472 (1983). The complete text of the MFJ appears at 552 F. Supp. 226-234.

[2] Telecommunications Act of 1996, Pub. L. 104-104, 110 Stat. 56 (1996).

[3] 47 U.S.C. § 271.

[4] MFJ § II(D)1.

[5] *United States v. Western Elec. Co.,* 569 F. Supp. 990, 995 (D.D.C. 1983).

[6] MFJ § IV(G).

[7] 47 U.S.C. § 151(21). The 1996 Act defines interLATA service as "telecommunications between a point located in a local access and transport area and a point located outside such area." *Id.*

[8] Not surprisingly, many members of Congress were disappointed when BOCs chose to merge with, rather than compete with, other BOCs.

[9] 47 U.S.C. § 271(g).

[10] *Id.* § 271(h).

[11] The Commission has found that under the 1996 Act, "BOCs may not provide interLATA information services, except for those designated as incidental interLATA services under Section 271(g), in any of their in-region states prior to obtaining Section 271 authorization (to provide interLATA services in those in-region states)." Implementation of the Nonaccounting Safeguards of Sections 271 and 272 of the Communications Act of 1934, as amended, 12 FCC Rcd. 2297 (1997).

[12] 47 U.S.C. § 271(c)(1)(A). The BOC must be providing access and interconnection to an unaffiliated competitor that is offering local exchange service to business and residential subscribers. In order to qualify as "facilities-based," the interconnecting carrier must be offering service exclusively over its own telephone exchange facilities, or predominantly over its own facilities in combination with the resale of the telecommunications services of another carrier. *Id.* The Commission has taken the position that a carrier offering service predominantly through unbundled network elements obtained from an ILEC qualifies as facilities-based under this statutory definition.

[13] *Id.* § 271(c)(1)(B). A BOC may rely upon a statement of terms and conditions, rather than actual interconnection with a competitor, if no competing carrier has requested interconnection within three months of the date the ILEC applies for in-region, interLATA authority. *Id.*

[14] *Id.* §§ 271(c)(2)(B)(i), 251(c)(2).

[15] *Id.* § 251(d)(2).

[16] *Id.* §§ 271(c)(2)(B)(ii), 251(c)(3).

[17] *Id.* § 271(c)(2)(B)(iii).

[18] *Id.* § 224(f)(1).

[19] *Id.* § 271(c)(2)(B)(iv).

[20] Local Competition First Report and Order, 11 FCC Rcd. at 15691. Because the Commission has identified the local loop as one of the required unbundled network elements, BOCs seeking interLATA authority are required to provide this item by both Sections 271(c)(2)(B)(ii) and 271(c)(2)(B)(iv) of the competitive checklist. The next two items on the checklist—local transport and local switching—also are required unbundled network elements under the Commission's Local Competition Order.

[21] BellSouth Louisiana Order, 13 FCC Rcd. at 20637.

[22] 47 U.S.C. § 27(c)(2)(B)(v).

[23] Wire centers are facilities containing local exchange service switches (e.g., BOC central offices).

[24] *See* Application by Bell Atlantic New York for Authorization under Section 271 of the Communications Act to Provide In-Region, InterLATA Service in the State of New

York, CC Docket No. 99-295 (Memorandum Opinion and Order rel. Dec. 22, 1999) § 343 (Bell Atlantic New York Order). Tandem switches aggregate traffic from more than one central office. A specialized tandem switch, called an access tandem, connects IXCs to two or more central offices of a local exchange carrier for originating or terminating access service.

[25] 47 U.S.C. § 271(c)(2)(B)(vi).

[26] Bell Atlantic New York Order § 343.

[27] 47 U.S.C. § 271(c)(2)(B)(vii).

[28] *Id.* § 271 (c)(2)(B)(viii).

[29] *Id.* § 251(e).

[30] *Id.* § 271(c)(2)(B)(ix).

[31] *Id.*

[32] *Id.* § 271(c)(2)(B)(x).

[33] Application of BellSouth Corporation, BellSouth Telecommunications, Inc., and BellSouth Long Distance, Inc., for Provision of In-region, InterLATA Services in Louisiana, 13 FCC Rcd. 20599, 20610 (1999).

[34] 47 U.S.C. §§ 271(c)(2)(B)(xi)-(xii).

[35] *Id.* § 271 (c)(2)(B)(xii); BellSouth Louisiana Order, 13 FCC Rcd. at 20611.

[36] BellSouth Louisiana Order, 13 FCC Rcd. at 20611.

[37] 47 U.S.C. § 271(d)(2).

[38] *Id.*

[39] *Id.* § 273(a).

[40] *Id.* § 272(f)(1).

[41] *Id.* § 273(h).

[42] The decree speaks of telecommunications products as the forbidden category, but the term is not defined in the decree. "Telecommunications equipment" is defined in the decree, however, and the Judge and the Department have used that term as a stand-in for "telecommunications products."

[43] MFJ Section IV(N).

[44] Reply of the United States Concerning Southwestern Bell Corporation's Restructured Proposal to Acquire Communications Test Design, Inc. at 7, *United States v. Western Elec. Co.,* No. 82-0192 (D.D.C. May 20, 1988).

[45] The BOCs also may supply equipment used by customers to terminate, maintain or multiplex their access lines. (Multiplexing is the process of concentrating traffic, usually

[46] Both Judge Greene and the Court of Appeals were profoundly opaque on the distinction between permitted and forbidden software, saying only that the BOCs are restrained from writing "firmware" that is "integral" to the hardware device in which it resides. *United States v. Western Elec. Co.,* 894 F.2d 1387, 1394 (D.C. Cir. 1990); *United States v. Western Elec. Co.,* 675 F. Supp. 655, n. 54 at 667 (D.D.C. 1987).

[47] Indeed, the development of standards for both local networks and equipment is a principal reason for the creation, as part of the divestiture, of Bell Communications Research (Bellcore).

[48] 47 U.S.C. § 273(h).

[49] *Id.* § 274(h).

[50] *Id.* § 274(b).

[51] *Id.* § 274(g)(2).

[52] *Id.* § 275(1).

[53] *Id.* § 275(2).

[54] *Id.* § 276.

[55] *Id.* § 272. A complete discussion of Section 272 requirements and the Commission's implementing rules is beyond the scope of this introductory book. For the FCC's rules enacted under Section 272 and the rationale for those rules, *see,* e.g., Implementation of the Non-Accounting Safeguards of Sections 271 and 272 of the Communications Act of 1934, as amended, 11 FCC Rcd. 21905 (1996).

[56] 47 U.S.C. § 272(f)(1).

[57] *Id.* § 272(f)(2).

[58] We noted in the introduction that courts and regulators tend to deal with ILEC participation in competitive markets by imposing quarantines, separate subsidiary requirements, or nonstructural safeguards. The information services/enhanced services world is one area in which all three of these approaches have been tried: the quarantine contained in the consent decree, the FCC's structural separations approach, and finally the Commission's present regime of nonstructural safeguards.

[59] *See* Amendment of Section 64.702 of the Commission's Rules and Regulations, Second Computer Inquiry, 77 FCC 2d 384, *modified on recon.,* 84 FCC 2d 50 (1980), *further modified on recon.,* 88 FCC 2d 512 (1981), *aff'd sub nom. Computer and Communications Industry Ass'n v. FCC,* 693 F.2d 198 (D.C. Cir. 1982), *cert. denied,* 461 U.S. 938 (1983), *aff'd on second further recon.,* CC 84-190 (released May 4, 1984). This so-called *Computer II* inquiry succeeded an earlier scheme of separate subsidiary regulation, generally referred to as the *Computer I* regime.

[60] In the Matter of Filing and Review of Open Network Architecture Plans, 5 FCC Rcd. 3103, para. 4 at 3104 (Memorandum Opinion and Order released May 8, 1990). The ONA model also calls for Complementary Network Services (CNSs), which are services purchased by end users in conjunction with enhanced services, and Ancillary Network Services (ANSs), which include deregulated services (such as billing and collection) available to ESPs. *Id.*

[61] Packetizing is a process of sending data in separate, addressed "bursts." When one packet has been transmitted, the channel is freed for the transmission of another packet to the same or a different destination.

[62] The ESPs also enjoy substantially lower interconnection charges than the usage-sensitive charges paid by the IXCs. They are allowed to use cheaper, state-tariffed exchange services instead of interstate access services for their interstate access lines. If they use local lines, they pay the end-user common line (EUCL) charge, as other end users do; if they order from access tariffs, they pay the carrier common line charge.

[63] Policy and Rules Concerning Rates for Competitive Common Carrier Services and Facilities Authorizations Therefor, 104 FCC 2d 958, 1087 (Report and Order released June 16, 1986) (Phase I Order).

[64] The Commission later changed this rule as it applies to larger business customers. If BOC enhanced services personnel want to access the CPNI of customers with more than twenty lines, they first must obtain the customer's permission.

[65] Phase I Order, *supra,* 104 FCC 2d at 1087.

[66] *Id.*

[67] *Id.* at 1055-56.

[68] Where the Act has not expressly created an exception to Section 2(b) that plainly applies to the service at issue, the Commission may preempt state regulation of the service only where necessary to avoid frustrating some valid goal that is within the Commission's jurisdiction under the Act. And even where this requirement is met, the Commission may preempt only those aspects of state regulation that cannot be separated and into intrastate and interstate components. *Louisiana Public Service Commission v. Federal Communications Commission,* 476 US 355 (1986).

[69] The quotation is from the Commission's Report and Order in the subsequent *Computer III Remand* proceedings (In the Matter of *Computer III* Remand Proceedings: Bell Operating Company Safeguards and Tier 1 Local Exchange Company Safeguards, 6 FCC Rcd. 7571, 7625 (Report and Order released Dec. 20, 1991)(*Remand Order*)); for the preemption decision itself, *see* Amendment of Sections 64.702 of the Commission's Rules and Regulations (Third Computer Inquiry), 104 FCC 2d 958, 1128 (Report and Order released June 16, 1986).

[70] *California v. FCC,* 905 F.2d 1217 (9th Cir. 1990).

[71] As a practical matter, of course, the interexchange restriction of the MFJ—and later the Telecommunications Act of 1996—severely limits the ability of the BOCs to provide interstate enhanced services.

[72] Remand Order, *supra,* 6 FCC Rcd. at 7632-37.

[73] *See,* e.g., the description of Oregon's intrastate ONA regulations in the *Communications Daily,* July 13, 1993, p. 8.

[74] 47 U.S.C. § 161.

[75] *Computer III* Further Remand Proceedings: Bell Operating Company Provision of Enhanced Services; 1998 Biennial Regulatory Review—Review of *Computer III* and ONA Safeguards and Requirements, 14 FCC Rcd. 4289 § 4 (1999).

[76] *Id.*

5

Competing Local Exchange Carriers

As we noted in the introduction to this book, CLECs are creations of the Telecommunications Act of 1996. Before the 1996 Act became law, the states generally refused to license new telephone companies to compete with the BOCs and other incumbents in the market for local exchange service [1]. The 1996 Act brought an abrupt end to this exclusionary practice. Today, state and local governments may not prohibit "the ability of any entity to provide any interstate or intrastate telecommunications service" [2].

Although CLECs may provide the same local exchange services and exchange access services as the BOCs and other incumbents, they are subject to different—and substantially less rigorous—regulatory constraints than the ILECs. Notably, once a CLEC has satisfied the certification requirements of its state commission and obtained its authorization to provide service, it may conduct its business without regulatory interference so long as it contributes to universal service and other funds, files state tariffs, and makes informational reports to regulators where required. Most importantly, CLECs are effectively free from direct regulation of their rates and earnings, either on a rate-of-return or price-cap basis, and lack the detailed interconnection obligations of ILECs.

This relative freedom from regulatory constraint, however, does not mean that CLECs can afford to ignore regulatory developments. In fact, the rules and decisions of the FCC and the state PUCs critically affect the ability of CLECs to obtain the ILEC facilities and services—including interconnection, access to unbundled network elements, and services offered for resale—without which local exchange competition would be largely illusory. Those commissions also have attempted to define the terms on which CLECs may obtain access to poles, conduits, and other rights-of-way, including access to shopping centers, office buildings, and other multiple tenant properties. Accordingly, the CLECs' interest in their regulatory environment has been no less intense than that of their incumbent competitors.

This chapter describes the principal concerns that state and federal regulation presents for CLECs. We begin with the requirements the states impose for CLEC certification and the conduct of CLECs' operations; after which we describe the evolving rules under which CLECs obtain needed access to ILEC facilities, ILEC services, and poles, conduits, and other rights-of-way.

I. State Certification of CLECs

Although the states may not prohibit CLECs from operating, they may impose reasonable requirements intended to "preserve and advance universal service, protect the public safety and welfare, ensure the continued quality of telecommunications services, and safeguard the rights of consumers" [3]. Pursuant to this authority, all of the states require CLECs to satisfy certification requirements before providing service. (Some states hold formal hearings on some or all CLEC applications for certification, while other states conduct only "paper" proceedings.) Although certification requirements vary from state to state, they all include some combination of the following elements.

A. Election of Facilities-Based or Resale Authority

All states require applicants for CLEC authority to state whether they will operate as facilities-based carriers or resellers. These categories are often ill-defined, but a general account of the differences between them is possible.

Facilities-based CLECs, as the name suggests, provide service wholly or partly through equipment and facilities that the CLECs own or control. In most states, ownership or control of a switch will qualify a CLEC as facilities-based. Similarly, a CLEC that will use unbundled network elements (UNEs) obtained from ILECs will be classified as facilities-based and may, in fact, need a facilities-based certification before an ILEC will permit it to obtain UNEs.

CLECs doing business as resellers purchase services from ILECs at a discount and sell those services to the CLECs' customers. Resellers typically "brand" the service, market the service, provide customer support, and set the rates at which the service is sold. Resellers are to be distinguished from sales agents, which sell telephone services on a commission basis and do not brand the service, set the rates for the service or provide customer support after the sale is made. Sales agents are not CLECs, do not require certification, and do not file tariffs or comply with other common carrier obligations.

B. Specification of Services to Be Offered

The states generally require applicants for CLEC certification to specify the services they will offer. If the CLECs later propose to add or eliminate services, they generally will be required to amend their applications accordingly.

This modest requirement raises a larger issue. After passage of the 1996 Telecommunications Act, state public utilities commissions received requests for certification from CLECs that had no intention of offering the full range of traditional telephone services. In fact, the commissions received requests from companies that planned to provide "data-only" services to business customers and avoid the voice and residential markets altogether. In all states, these proposals departed from past practice and the common carrier principle to which traditional carriers were expected to adhere. In some states, these proposals also violated regulations that expressly required carriers to provide basic telephone service to the public at large [4].

In spite of its novelty and inconsistency with past practice, the "data CLEC" concept did not meet with outright rejection by the state commissions. Some state commissions accommodated data CLEC applications by waiving the usual service requirements. Other commissions granted "conditional" data CLEC authority pending further review of the public-interest implications of permitting CLECs to offer a limited menu of business-oriented services [5]. As data CLECs become more entrenched and businesses become dependent on their services, states are unlikely to require CLECs to provide a full range of telecommunications services as a condition of certification [6].

C. Specification of Area to Be Served

A number of states certify all applicants for CLEC authority to serve the entire state, without necessarily requiring CLECs to exercise that authority in all areas of the state. Other state commissions require CLEC applicants to specify each local exchange they propose to serve, and grant authorization only for the exchanges so specified. Where a certification is for a specific service arca, the CLEC's certificate may include a "build-out" requirement, providing that the carrier's service must be available throughout the service area within a specified time.

D. Financial Ability to Serve

Most states require some showing of the CLEC applicant's financial ability to serve. California, for example, requires an applicant to show that it has $100,000 cash, or cash equivalent, available at the time the application is made. Texas imposes a complex formula, requiring the applicant to show that it possesses either:

1. The greater of $100,000 cash or cash equivalent *or* enough cash or equivalent to meet startup expenses, working capital requirements, and capital expenditures for the first two years of operations in Texas; or

2. That it is an established business entity and has shown a profit for two years preceding the date of application, as demonstrated by: (a) a long-term debt-to-capitalization ratio of less than 60%, (b) a return-on-assets ratio of at least 10%, and (c) $50,000 cash or cash equivalent.

Missouri also imposes alternative tests. Specifically, an applicant in Missouri must show that it has either:

1. A total debt-to-capital ratio no greater than 62% and a pretax interest coverage of at least 2.3×; or

2. A cash balance of four months operating expenses inclusive of interest expense and taxes.

Most states also require applicants to file financial statements with the state commission. Pennsylvania, for example, requires a number of filings including a tentative operating balance sheet and projected income statement for the first year of operations in Pennsylvania. Georgia requires extensive submissions, including the most recent certified report on examination of the applicant's financial statements, a current year operating budget and proposed budget for the next year, and current and next-year gross revenues and employment for the applicant's Georgia operations.

E. Construction Plans and Environmental Impact Statements

The principal difference between applying for certification as a facilities-based CLEC and applying as a resale CLEC is that facilities-based applicants may be required to submit detailed construction plans and environmental impact statements to their state commissions. These requirements are not very exacting when the applicant proposes only to install switches within existing buildings, but can be quite extensive when the applicant will install optical fiber or other facilities that require excavation or construction of new structures [7].

F. Tariff Filings

As we discuss in the next section, the states continue to require local exchange carriers to publish tariffs that set out the rates and other conditions of their intrastate services. Depending upon the state's regulations, tariff filings may be due as part of the application for certification, within a specified time after the application is granted, or before commencing service to any exchange area.

II. Ongoing CLEC Regulatory Obligations

Most, but not all, of the CLECs' regulatory obligations are imposed by state PUCs rather than the FCC, which has effectively deregulated CLECs' access charges and has taken no role in certification of CLECs to provide any service, whether interstate or intrastate [8]. In fact, the principal involvement of federal regulators in CLEC affairs—besides defining the interconnection obligations owed to CLECs by ILECs—is in the administration of certain "universal service" and other funds to which CLECs contribute and from which some CLECs obtain benefits.

The following identifies and summarizes the principal, ongoing regulatory obligations of CLECs under state and federal regulations.

A. Contributions to Universal Service and Other Funds

As we discuss at length in Chapter 10, providers of telecommunications services in the United States contribute to an array of state and federal funds that finance or subsidize services and activities related to the telecommunications industry and its customers. Some of these funds are used to subsidize particular services or service providers, including basic, residential service for low-income customers, providers of service to high-cost areas and persons with hearing and speech disabilities. Other funds underwrite the entire cost of numbering plan administration and the number of portability databases that permit customers to change service providers without changing their telephone numbers. CLECs contribute to all of these funds.

The most significant of the federal funds to which CLECs contribute is the high-cost fund of the FCC's universal service system. This fund, which subsidizes telephone companies that serve high-cost (chiefly rural and remote) areas of the country, is supported by contributions from every "telecommunications carrier that provides interstate telecommunications service" [9]. Specifically, a CLEC that provides interstate access or other interstate services must pay to the fund a percentage of its total interstate and international end-user telecommunications revenue [10]. The contribution factor (i.e., percentage of revenues that the CLEC must contribute) is calculated quarterly by the Administrator of the Universal Service Administrative Company (USAC). The contribution factor tends to be set in the range of 5% to 6% of end-user interstate telecommunications revenues [11].

All carriers that provide interstate telecommunications service also must contribute to the federal fund for Telecommunications Relay Service (TRS), an operator-assisted telephone service provided to the speech and hearing impaired. At this writing, the contribution factor for the TRS fund is .00038 of each contributor's end-user, interstate telecommunications revenues.

The FCC also oversees funds that support the administration of the North American Numbering Plan (NANP) and the regional databases that permit customers to transfer their telephone numbers from one local carrier to another. Contributions to the NANP fund are based on the contributor's end-user telecommunications revenues from both interstate and intrastate services, with a contribution factor at this writing of .0000577. Each carrier's contribution to the number portability fund depends upon the costs of the regional database from which that carrier is served. For each region, the FCC's Common Carrier Bureau sets a contribution factor that is assessed against carriers' end-user telecommunications revenues, both interstate and intrastate. Like the NANP fund contributions, payments to the number portability fund tend to be quite modest as a percentage of revenue.

The states also subsidize local residential service and service provided to high-cost customers. Some states rely for this purpose on the traditional system of so-called "implicit subsidies," in which certain services are required to be priced above cost in order to support below-cost rates for other services. An increasing number of states, however, have instituted universal service funds similar to the federal fund. These funds tend to be supported by contributions collected through a surcharge on end-users' bills for intrastate services.

The contribution factors for these state funds vary. Texas, for example, imposes a contribution factor of 3.6% of an end-user's bill for intrastate service. Similarly, Utah imposes a contribution factor of 1%; Nebraska collects 7%; and Kansas takes 7.9%.

States also may maintain separate funds for TRS service, service to low-income customers and other purposes. A California Public Utilities Commission order entered in February, 2000, for example, requires a newly certified CLEC to make the following contributions:

1. A .5% surcharge on all intrastate services for Universal Lifeline Telephone Service (a subsidy program for low-income customers);

2. A 1.92% surcharge for the California Relay Service and Communications Devices Fund (intrastate TRS);

3. A 2.6% surcharge on all intrastate services to support the California High-Cost Fund;

4. A .05% surcharge on a defined set of intrastate services for the California Teleconnect Fund (providing discounted service to qualifying libraries, schools, hospitals, health clinics, and community organizations) [12].

As noted earlier, the various state and federal funds are assessed only against revenues earned from telecommunications services. Accordingly, CLECs will not contribute to those funds on the basis of Internet access and other enhanced, or information, services.

B. Tariffs and Informational Filings

The states generally apply tariffing obligations to CLECs for their intrastate telecommunications services. However, some states permit CLECs to obtain relief from those obligations upon a showing that such relief will not harm the public interest.

The states also generally require CLECs to file reports with the state commission at prescribed intervals. Connecticut, for example, requires each

CLEC to file annual reports that include, at a minimum, the following information:

1. The number of customers for each service offered under the CLEC's certificate;
2. The number of lines subscribed to the CLEC's services;
3. Total intrastate revenues;
4. The intrastate minutes of use on a total service basis;
5. A description of physical changes in or additions to existing facilities expected for the next fiscal year and any expected change in use of those facilities;
6. Any changes in the information filed with the state commission at the time of the CLEC's certification [13].

C. Interconnection and Related Obligations

Because CLECs were not incumbent monopolists at the time of passage of the 1996 Telecommunications Act, they are not subject to the pervasive obligations imposed upon ILECs by Section 251(c) of the Act. Notably, CLECs are not required to make unbundled elements of their networks available to competitors, or to permit collocation of other carriers' facilities with theirs, or to sell their services to other carriers at a discount. The CLECs are, however, subject to the less exacting requirements of Sections 251(a) and 251(b) of the 1996 Act, which apply, respectively, to all telecommunications carriers and all local exchange carriers.

The obligations imposed by Section 251(a), which apply to all telecommunications carriers (including ILECs, CLECs, and mobile telephone companies) are straightforward. A telecommunications carrier must interconnect, either directly or indirectly, with the facilities of other telecommunications carriers; and a telecommunications carrier must not install network features or capabilities that frustrate use by persons with disabilities or interoperability among networks. The first of these requirements is effectively met by any carrier that connects with the public switched telephone network; the second requirement is further defined by Sections 255 and 256 of the Act and the Commission's rules adopted pursuant to those sections.

Section 251(b), which applies to all local exchange carriers (including both ILECs and CLECs), imposes five obligations.

The first of the 251(b) obligation is that of *resale*. Specifically, a LEC must not prohibit the resale of its telecommunications services or impose unreasonable or discriminatory conditions or limitations upon such resale [14]. This requirement does not mean, however, that a CLEC (as opposed to an ILEC) must offer its services at a discount to those who intend to resell those services [15]. It means only that LECs must sell their services at their customary retail rates to end users and resellers alike, without imposing conditions designed to frustrate the development of a resale market for the LECs' services.

The second of the 251(b) obligations is that of *number portability*. As the name suggests, LECs are required, to the extent feasible, to permit their customers to take their telephone numbers with them when they switch to other carriers [16]. The FCC has adopted rules to govern number portability, and regional databases have been established to facilitate the porting of numbers among carriers.

The third of the 251(b) obligations is that of *dialing parity*. This requirement means that a LEC must permit its customers to reach the customers of other carriers without dialing extra digits [17]. LECs also must permit all providers of telephone exchange and toll service to "have nondiscriminatory access to telephone numbers, operator services, directory assistance, and directory listing, with no unreasonable dialing delays" [18].

The fourth requirement is that all LECs must afford competing telecommunications service providers with access to their poles, ducts, conduits, and rights-of-way [19]. This requirement, which affects only facilities-based carriers with their own transmission facilities, is more fully defined in Section 224 of the Act and the FCC's rules adopted pursuant to that section. These requirements are discussed more fully in the following section and, as they affect ILECs, in Chapter 3.

Finally, Section 251(b) requires all LECs to establish "reciprocal compensation arrangements" with other carriers for the "transport and termination of telecommunications" [20]. This language refers to the terms upon which local carriers compensate each other for transporting and delivering the calls of each carriers' customers to customers of the other carrier. This obligation, also, is discussed at greater length in Chapter 3 and Section III of this chapter.

III. CLEC Access to ILEC Facilities, ILEC Services, and Rights-of-Way

For a CLEC, the most important regulatory obligations are not its own, but are those of the ILECs that still control the local telephone exchange everywhere in the United States. As we discussed at greater length in Chapter 3, if ILECs refused to interconnect with CLECs' networks, the CLEC industry would be nothing more than an archipelago of closed networks, offering their customers only the ability to communicate with each other. As we also discussed, if ILECs did not give their CLEC competitors access to unbundled network elements and sell services to CLECs at a discount, local competition would take much longer to develop. For these reasons, the 1996 Telecommunications Act imposed detailed obligations on ILECs in each of these areas, and made the opening of their networks to competition the condition for entry of the Bell operating companies into the interexchange marketplace.

The following discussion will not repeat the descriptions, in Chapters 3 and 4, of the obligations owed to CLECs by BOCs and other ILECs. It is worth noting here, however, some issues and requirements that have been of particular concern to CLECs as they have attempted to secure the cooperation of their (often reluctant) incumbent competitors.

A. Interconnection and Reciprocal Compensation

As we discussed in Chapter 3, the 1996 Act requires all ILECs and their CLEC competitors to negotiate arrangements under which ILECs pay CLECs for completing calls from ILEC customers to CLEC customers, and CLECs pay ILECs for completing calls from CLEC customers to ILEC customers. The Commission has determined that this reciprocal compensation obligation applies only to completion of local calls, and the interconnection agreements negotiated between ILECs and CLECs generally state that the parties will compensate each other for completion of calls between points within the same local calling area.

When the first interconnection agreements were negotiated, the parties generally did not address the question whether seven-digit calls placed to ISPs were local calls [21]. When CLECs acquired ISPs as local service customers, the CLECs took the position that calls placed to ISPs by ILEC customers should result in payment of reciprocal compensation to the CLECs.

The ILECs discovered, however, that ISPs were accounting for a significant share of the CLECs' customer base. Because ISP traffic tends to flow from end users to the ISP and not in the other direction, CLECs with a large number of ISP customers relative to non-ISP customers were entitled to receive more compensation money from ILECs than they were paying to the ILECs. The ILECs' response to this phenomenon was to refuse to pay reciprocal compensation for calls placed by their customers to ISPs served by

CLECs. As a rationale for this refusal, the ILECs argued that when end users place seven-digit calls to ISPs they are not making local calls but instead are initiating communications that extend from the customers' premises to Internet sites located outside the exchange—and even outside the state or country—in which the calling party is located. Accordingly, the ILECs simply declared that calls to ISPs are not local and should be classified as interstate traffic under the interconnection agreements.

The ILECs' refusal to pay reciprocal compensation for ISP traffic resulted in a petition to the FCC by the Association for Local Telecommunications Services (ALTS), requesting a declaration that ISPs still are classified as end users under the Commission's decisions and that seven-digit calls to ISPs therefore are local rather than interstate. CLECs requested similar rulings from state regulatory commissions, which are empowered to interpret and enforce interconnection agreements between ILECs and CLECs. The FCC did not act on the ALTS petition, but over 20 state regulatory commissions ruled that local calls placed to ISPs are eligible for payment of reciprocal compensation under existing interconnection agreements.

In March of 1999, the FCC entered a decision finding that seven-digit calls to ISPs are jurisdictionally interstate and should not, therefore, be classified as local traffic for purposes of eligibility for reciprocal compensation [22]. (The Commission did not, however, attempt to overturn those state decisions that had reached a contrary result.) The Commission also pursued a parallel proceeding to determine whether it should develop a separate system of compensation for termination of ISP-bound traffic. At this writing, that proceeding is still pending. A federal appellate court then added to the confusion by vacating the FCC's reciprocal compensation decision and remanding the issue (i.e., sending it back to the FCC) for further consideration [23].

The issue of reciprocal compensation for ISP-bound traffic continues to be contentious and confused. Besides the ongoing FCC proceedings, the reciprocal compensation question continues to be the subject of negotiation between individual ILECs and CLECs, as well as arbitration proceedings before state regulatory commissions.

B. Access to Unbundled Network Elements

The 1996 Act requires ILECs to provide their local competitors with access to "unbundled network elements" to be used "for the provision of a telecommunications service," but does not specify the particular network elements that must be provided [24].

The Commission addressed this question in 1996, when it directed the ILECs to provide access to the following elements of their networks to competing carriers:

1. Loops, including loops used to provide high-capacity and advanced telecommunications services;
2. Network interface devices;
3. Local circuit switching (except for larger customers in major urban markets);
4. Dedicated and shared transport;
5. Signaling and call-related databases;
6. Operations support systems;
7. Operator and directory assistance services [25].

The Supreme Court, however, determined that the FCC had applied an improper legal standard in generating this list and remanded the matter to the Commission for further proceedings [26]. Accordingly, the Commission undertook a new proceeding and determined that all of the elements on its list except the seventh (i.e., operator and directory assistance service) were necessary in order for CLECs to provide competitive local service and must be provided by ILECs as unbundled network elements [27].

Subsequent to its identification of this basic list of unbundled network elements, the Commission was presented with an additional problem by the rapid deployment of digital subscriber line (DSL) technology, which permits telephone customers to combine voice and high-speed data communications on a single subscriber line. Although CLECs were entitled to obtain end-users' subscriber lines as unbundled network elements, it was not clear that they were entitled to obtain access to just the upper frequencies of ILEC local loops conditioned for DSL. Without such access, customers that wanted to obtain DSL from CLECs would have to combine that service with the CLEC's voice service, or (if they wished to purchase CLEC DSL service without discontinuing their ILEC voice service) purchase a second line.

Fortunately for the CLECs, the FCC determined that ILECs must provide the upper frequency portions of their DSL-conditioned loops to CLECs as unbundled network elements [28]. This "line sharing" decision did not, however, resolve the question of how much ILECs could charge for shared lines provided to CLECs. Several state commissions have received petitions from CLECs, arguing that upper frequency portions of loops should be provided without charge because ILECs incur no incremental cost to transmit data on a line that already carries voice. (At least one Verizon employee apparently conceded this point with an incautious statement that the incremental cost of providing DSL over a voice line is zero.) CLECs have negotiated rates for shared lines that range from just over $5.00 per-line-per-month to over $8.00 per-line-per-month, and at least one state commission (Minnesota) has established a line-sharing rate of $6.05 per month.

C. Poles, Conduits, and Rights-of-Way

Section 224 of the 1996 Act provides that utilities, including local telephone companies, must "provide a cable television system or any telecommunications carrier with nondiscriminatory access to any pole, duct, conduit, or right-of-way owned or controlled by it" [29].

Under the authority granted in Section 224, the Commission has adopted extensive rules that amplify the meaning of "nondiscriminatory" access and establish principles for setting prices that telecommunications carriers (including CLECs) and cable television companies must pay for access to these facilities.

Some of the most troubling questions under Section 224 have involved access to multitenant environments such as apartment buildings, office buildings, office parks, and shopping centers. In most of these environments, telephone service is provided by the incumbent carrier. In order to reach individual tenants within these multitenant premises, the ILEC uses a distribution network that runs through conduit, risers, and other facilities controlled by the ILEC directly or by the owner of the premises.

The distribution facilities within these multitenant environments presented, potentially, a formidable obstacle to local competition. Without mandated access to these facilities, CLECs either could not offer their service to tenants or could not do so without paying exorbitant fees to premises owners.

Responding to these concerns, in 1999 the FCC asked for comment on a proposed rule that CLECs must be granted access to rights-of-way and riser conduit that public utilities own or control in multiple tenant environments [30]. In that same notice, the Commission asked for comment on a more difficult question (i.e., whether it has the jurisdiction to require building owners to grant CLECs access to building areas that are controlled, not by a public utility, but by the building owner himself) [31]. Although an affirmative answer to this question will pose obvious benefits for CLECs, adoption of an FCC rule that effectively imposes access obligations on property owners, rather than telecommunications service providers already subject to the Communications Act, might raise constitutional issues and certainly would start a long campaign of appellate litigation.

Conclusion

When Congress opened the local telephone exchange to competition, it gambled that new entry into local markets could be more than a niche business serving the specialized needs of business customers in urban areas. To the extent regulators can affect the outcome of this gamble, they generally have done so by regulating CLECs lightly and requiring ILECs to open their networks to competition to an extent that has no precedent in any other industry. The gamble's ultimate outcome will depend, in large part, on the vigor with which regulators can enforce the market-opening commitments of the ILECs, and especially of those BOCs that already have obtained entry into interexchange markets and have reduced incentives to demonstrate their cooperativeness.

Endnotes

[1] Before the 1996 Act became law, a number of states had certified competitive access providers (CAPs) to offer exchange access and specialized services primarily to business customers. The states continued, however, to protect the incumbents' monopoly over ordinary local exchange telephone service.

[2] 47 U.S.C. § 253(a).

[3] *Id.* § 253(6).

[4] Providers of nonvoice services also do not provide services, such as emergency 911 calling, that are required of all telephone companies. Unless the state commission has adopted rules that eliminate these requirements for data CLECs, applicants must request and obtain waivers before providing service.

[5] *See* Application of Northpoint Communications, Inc. for a Certificate of Public Convenience and Necessity Docket No. 990504 (Csm. OPUC Sep. 30, 1999); Application of DSLnet Communications, LLC for a Certificate of Public Convenience and Necessity, Docket No. 990119 (Conn. OPUC June 16, 1999).

[6] State commissions, however, limit eligibility for state universal service supports to those CLECs that provide basic telephone service to all who request it.

[7] *See,* e.g., Application of Intermedia Communications, Inc. for Reinstatement of Its Lapsed Certificate of Public Convenience and Necessity to Operate as a Facilities-Based

Competitive Local Exchange Carrier and to Offer Resale of Local Exchange Services in California, Decision No. 00-02-013 (Cal. PUC 2000).

[8] Although the FCC has jurisdiction over the CLECs' interstate access services, the Commission has exercised its forbearance authority to relieve CLECs of the obligation to file tariffs for their access services provided to IXCs in connection with interstate calls. However, in response to IXC complaints of price-gouging by CLECs, the Commission is considering whether some scrutiny of the relationship between CLEC access charges and the CLECs' cost of providing interstate access might be appropriate.

[9] 47 U.S.C. § 254(d).

[10] Still more precisely, the individual company's contribution is a product of: (1) the contributor's individual interstate and international end-user telecommunications revenues; and (2) a contribution factor that is equal to the ratio of total projected quarterly expenses of the fund to the total interstate and international end-user telecommunications revenues of all contributors. CLECs do not contribute to the FCC universal service fund on the basis of their interstate access charges to IXCs, but do contribute to the fund to the extent they impose access charges on their local service customers.

[11] CLECs and other telecommunications carriers also contribute, on the basis of revenues from interstate end-user telecommunications services, to a fund that subsidizes advanced telecommunications services to qualifying schools, libraries, and health care providers. 47 U.S.C. § 254(b)(6).

[12] Application of Intermedia Communications, Inc. for Reinstatement of Its Lapsed Certificate of Public Convenience and Necessity to Operate as a Facilities-Based Competitive Local Exchange Carrier and to Offer Resale of Local Exchange Services in California, Decision No. 00-02-013 (CPUC 2000).

[13] Application of DSLnet Communications, LLC for a Certificate of Public Convenience and Necessity, Docket No. 99-01-19 (CDPUC 1999).

[14] 47 U.S.C. § 251(b)(1).

[15] ILECs, of course, are required by Section 251(c) of the Act to provide telecommunications services to resellers at wholesale rates. *Id.* § 251(c)(4).

[16] *Id.* § 251(b)(2).

[17] *Id.* § 251(b)(3).

[18] *Id.*

[19] *Id.* § 251(b)(4).

[20] *Id.* § 251(b)(5).

[21] MFS Communications Company, however, took the precaution of filing a Petition for Partial Reconsideration and Clarification of the FCC's Local Competition Order, requesting that the Commission clarify that traffic destined for ISPs is eligible for reciprocal compensation under the 1996 Act. The FCC has not acted on the MFS petition, which was filed in September of 1996. *See* Ex Parte Procedures Regarding Requests for

Clarification of the Commission's Rules Regarding Reciprocal Compensation for Information Service Provider Traffic, CC Docket No. 96-98, Public Notice released Aug. 17, 1998.

[22] Implementation of the Local Competition Provisions in the Telecommunications Act of 1996, 14 FCC Rcd. 3689 (1999).

[23] *Bell Atlantic Telephone Companies v. Federal Communications Commission,* 206 F.3d 1 (D.C. Cir. 2000).

[24] 47 U.S.C. § 251(c)(3).

[25] Implementation of the Local Competition Provisions in the Telecommunications Act of 1996, 11 FCC Rcd. 15499.

[26] *AT&T Corp., et al. v. Iowa Utilities Board, et al.,* 525 U.S. 366 (1999).

[27] "FCC Promotes Local Telecommunications Competition; Adopts Rules on Unbundling of Network Elements," FCC Report No. CC9941 (Sept. 15, 1999).

[28] Deployment of Wireline Services Offering Advanced Telecommunications Capability and Implementation of the Local Competition Provisions of the Telecommunications Act of 1996, 14 FCC Rcd. 20912 (1999).

[29] 47 U.S.C. § 224(f)(1).

[30] Promotion of Competitive Networks in Local Telecommunications Markets, 14 FCC Rcd. 12673 (1999).

[31] *Id.*

Part II
Non-ILECS: The Competitive Telecommunications Industry

Introduction to Part II

In keeping with the pervasive regulatory regime to which ILECs are subject, Part I of this book concerned itself exclusively with the problems of these incumbent—and still dominant—carriers. Part II of this book discusses the problems of service providers that face widely varying degrees of legal and regulatory constraint. Some of those constraints—such as those that apply to interexchange carriers and international carriers—include less restrictive versions of the kinds of economic regulation to which ILECs are subject. Others—such as the kinds of legal liability for harmful content with which Internet service providers must be concerned—have no parallel in the universe of ILECs, CLECs, IXCs, and other common carriers. Part II summarizes the legal and regulatory environment of these and other service providers, including mobile telephone companies, payphone providers, and international carriers. Part II also takes a more detailed look at the universal service support system, which we introduced in Chapter 4.

6

Interexchange Carriers

The IXC—defined loosely as a telephone company that carries long-distance communications as its primary or exclusive line of business—is a product of the era when regulators permitted competition in long-distance markets but preserved the monopolies of incumbent carriers in local markets. This quasi-competitive interlude, which lasted from the early 1970s until 1996, can be divided into two phases. During the first phase, from the early 1970s until 1984, AT&T and the Bell system companies continued to offer an integrated package of local and long-distance service, but also were required to interconnect with, and permit their local customers to obtain service from, competing long-distance companies [1]. In the second phase, from 1984 until 1996, all IXCs—including AT&T—were separate from the Bell operating companies and telephone subscribers in the United States had no choice but to obtain their local and long-distance services from separate carriers.

There is little doubt that most telephone subscribers prefer to purchase their local and long-distance service from the same provider. For this reason, the IXC has always been an artificial artifact of regulation; destined to be absorbed, as soon as local competition became widespread, by telephone companies that bundled local service, domestic long-distance, and international calling in a single package. With the passage of the Telecommunications Act of 1996 and its mandate for local competition, that process could begin.

For now, however, the distinct long-distance market and the rules developed to regulate that market have not disappeared. In order to understand the regulatory environment of the IXCs, we should examine two subjects. First, we sketch briefly the historical background of the IXC industry and the circumstances that gave rise to IXC regulation. Second, we discuss the regulations to which IXCs are subject [2].

I. Origins and Growth of the IXC Industry

As some readers of this book are old enough to recall, there was a time when the separate long-distance telephone company did not exist. From the first days of the telephone industry until the 1970s, long-distance service was bundled with local service and provided—for most U.S. consumers—by the Bell operating companies and the Long Lines department of AT&T. Although customers were billed differently for local and toll calls, the two types of service were obtained from the same carrier.

The separate long-distance industry was a product of technology and regulation. On the technology side, the change was the advent of microwave radio, which provided a cheaper means of transmitting telephone conversations over long distances. On the regulatory side, the change was the willingness of the FCC to allow new, microwave-based companies to offer certain kinds of long-haul service in competition with AT&T.

Credit for transforming these developments into marketplace reality must go to Microwave Communications, Inc. (later known as MCI), which applied in 1963 for permission to build microwave radio facilities between St. Louis and Chicago. The new company's stated intention was to provide point-to-point, private-line communications to business customers with high volumes of telephone traffic between locations in the two cities. (MCI disclaimed any intention to compete in the switched long-distance market.) After several years of contentious proceedings, in which the established telephone companies argued that introducing competition into the high-margin private-line business would erode the subsidies that supported basic telephone service, the Commission granted MCI's applications [3].

MCI's success spawned many would-be imitators. In 1971, after consolidating requests from dozens of applicants, the Commission announced a policy of open entry into the intercity, private-line telephone marketplace, and ruled that the established local carriers must interconnect with the new IXCs for the origination and termination of private-line calls over the local exchange [4]. In 1976, the Commission also ruled that incumbent carriers must permit competing IXCs to resell the incumbents' long-distance services—a decision that made it possible for IXCs to offer nationwide service over a combination of their own facilities and resold AT&T service [5].

The Commission's policy decision to permit new entry into long-distance markets was expressly limited to competition for services other than ordinary, switched interexchange service. In maintaining this limitation, the Commission gave serious deference to AT&T's claim that more extensive competition would undermine the Bell system's ability to subsidize basic, local service with above-cost rates for long-distance and other nonbasic services.

The Commission's cautious approach was overtaken, however, by a radical decision of the U.S. Court of Appeals for the D.C. Circuit [6]. That decision had its origins in MCI's filing of a tariff for a service called Execunet, which would permit MCI's customers to make switched long-distance calls to any city served by MCI. The Commission rejected the Execunet tariff as inconsistent with its limited procompetitive policy. Unfortunately for AT&T, the D.C. Circuit ruled, upon judicial review of the FCC's decision, that MCI had the right to use its microwave facilities for any purpose not expressly prohibited. Because the FCC's policy in favor of limited competition had not been accompanied by a corresponding prohibition in the authorizations granted to MCI, the court found that MCI was permitted to provide ordinary, switched interexchange service over its licensed microwave facilities.

The Execunet decision marked the beginning of real competition in the interexchange marketplace. As that competition grew, the FCC confronted the question of appropriate regulation of the new IXCs. In the Commission's view, there was no reason for carriers that lack market power to file tariffs, seek FCC permission to construct interstate facilities, and discontinue interstate services, or comply with the other common carrier requirements of Title II of the Communications Act. Accordingly, beginning in 1980, the Commission entered a series of decisions that relieved various categories of competing carriers, including the new IXCs, of those obligations [7].

Both the Court of Appeals for the D.C. Circuit and the Supreme Court, however, rejected the Commission's so-called "forbearance" decisions as beyond the authority granted by Congress in the Communications Act [8]. Pursuant to these judicial rulings, the Commission maintained tariffing and related requirements for all IXCs until Congress, in the Telecommunications Act of 1996, authorized it to forbear from any common carrier requirements where it found such forbearance to be in the public interest [9].

As the FCC was trying to deregulate MCI and the other new IXCs, it maintained its common carrier regulation of AT&T, which clearly had market power in the local and interexchange marketplace and could, in the absence of control over its rates and interconnection practices, harm interexchange competition. After the judicial separation of AT&T from its Bell operating companies, however, the Commission undertook two initiatives that resulted in substantial deregulation of AT&T as well as its interexchange competitors. The first of these initiatives was the replacement of rate-of-return regulation of AT&T with price-cap regulation. The second was the classification of AT&T as nondominant in the interstate, interexchange marketplace and the resultant deregulation of AT&T.

A. Price-Cap Regulation of AT&T's Rates

In 1989 the FCC abandoned its rate-of-return regulation of AT&T's earnings in favor of price-cap regulation of AT&T's rates. The FCC's purpose was to reduce rates by giving the company incentives to become efficient. The Commission therefore adopted initial rates based on the most recent prices set under rate-of-return regulation and provided for an "indexing" mechanism that would moved the caps downward, paced by industry productivity gains offset by inflation.

Overall efficiency was not the only goal of the AT&T price-cap model, however; the Commission also wanted to protect customers from excessive or volatile rates, protect competitors from predatory pricing, and give AT&T some ability to change its rates on a streamlined basis in response to the market. These concerns led the Commission to adopt its complex "baskets and bands" version of price-cap regulation. To understand how this system worked and why it was adopted, it might be helpful to review the alternatives the Commission had before it [10].

One version of price-cap regulation would have set a separate cap, or ceiling, on each of AT&T's services or rate elements [11]. Under this approach, AT&T would have no ability to change its rates for any service in response to market conditions. The rates would change only by application of the indexing formula; AT&T could not depart from that formula or make changes on a "streamlined" basis. At the same time, there would be no danger of one service subsidizing another or of frequent and confusing rate changes.

At the other extreme, the Commission could have established an overall cap on the average of AT&T's rates for all of its services. Under that approach, AT&T could change its rates for individual services on a streamlined basis, so long as those changes did not result in an average rate in excess of the price cap.

The trouble with the first approach, as the FCC saw it, was that it gave AT&T too little flexibility. The trouble with the second approach was that it subjected ratepayers to excessive pricing volatility, and gave AT&T the means to subsidize its more competitive services by charging higher rates for residential services.

The Commission's solution lay between these extremes. The FCC separated AT&T's services into three groups, or baskets. The first basket included the services used most by residential customers and small businesses (domestic and international MTS, operator-assisted calls, credit card calls, and Reach Out America) [12]; the second basket consisted of toll-free 800 services [13]; and the third consisted of services for larger business customers, including Outward WATS [14], software-defined networks [15], and wideband private lines [16].

The services grouped within each basket were chosen because they were considered substitutable for each other. Each basket had its own price cap, and AT&T was required to keep the weighted average of actual prices within each basket (called the basket's actual price index, or API) below the price cap applicable to the basket. So long as the average of the prices did not exceed the cap for the basket, individual services could be priced flexibly.

But not too flexibly. In order to prevent customer confusion from excessive rate changing within baskets, the services or rate elements within the baskets were "banded"—that is, each service or rate element within each basket was assigned a maximum percentage of annual change in rates (up or down) that AT&T could make without forfeiting streamlined review of its tariffs.

To understand the functions served by establishing baskets of services, consider the first basket, which consisted of services offered to residential and small-business customers. Those services were placed in separate baskets so that residential customers, in particular, would not be subjected to price increases designed to let AT&T lower its prices for other, more competitive services [17]. Within the basket the Commission defined six service categories: domestic day, domestic evening, domestic night/weekend, international MTS, operator/credit card services, and Reach Out America. AT&T would lose its streamlined treatment if it "raise[d] the rates of domestic evening or domestic night/weekend by more than 4%, or raise[d] the rates of any other service category (in the basket) by more than 5% per year" [18]. Other limitations also were set: AT&T could lower its rates in all these categories no more than 5% per year, or it would forfeit streamlined review; and AT&T

was required to ensure that "the average residential rate per minute [did] not increase by more than 1% per year above the [price cap]" [19]. Later, the FCC deregulated the AT&T services in baskets 2 and 3; and ultimately, as discussed further below, the Commission declared AT&T nondominant and ended all direct regulation of AT&T's rates.

B. Classification of AT&T as Nondominant

In 1993, AT&T petitioned the FCC to be classified as a nondominant carrier in both domestic interexchange and international telecommunications markets [20]. Under the analysis then used by the Commission to determine whether particular carriers and types of carriers might be deregulated, classification as nondominant required a demonstration that the petitioning carrier lacked market power (i.e., the ability to raise prices or restrict output) in the market for which nondominant classification was sought. If the petitioner demonstrated that it lacked market power, the Commission would relieve it of certain common carrier requirements under Title II of the Communications Act.

In the early stages of its deregulatory initiatives, the Commission had made it clear that AT&T was a dominant carrier by any measure. Specifically, as the owner both of the AT&T long-distance network and the 23 Bell operating companies, the pre-divestiture AT&T controlled local access facilities for over 80% of the nation's telephones, had an overwhelming share of the MTS and WATS markets, and had private-line revenues that were thirteen times as great as the combined private-line revenues of all of its competitors [21]. Under these circumstances, the Commission properly imposed the full weight of common carrier regulation on AT&T.

After divestiture took place in 1984, however, AT&T lost its presence in the local loop and suffered a steady decline in market share in the interexchange marketplace. By 1994, "AT&T's market share, in terms of both revenues and minutes, fell from approximately 90% to 55.2% and 58.6% in terms of revenues and minutes respectively" [22]. For this and other reasons, the Commission found in 1995 that AT&T no longer threatened competition in the domestic, interexchange marketplace [23]. Accordingly, it declared AT&T nondominant and established new rules for AT&T that included the following:

1. AT&T would be freed from price-cap regulation for its residential, operator, 800, directory assistance, and analog private-line services;

2. AT&T could file tariffs for its domestic interexchange services on one-day's notice;

3. AT&T's domestic, interexchange tariffs would be presumed lawful;

4. AT&T could enter into contracts with other carriers without reporting those contracts or filing them with the FCC;

5. AT&T would not need the Commission's permission to construct, acquire, or operate transmission lines or extend service to any domestic point;

6. AT&T would be required to report additional circuits to the Commission only semiannually [24].

With the reclassification of AT&T as nondominant, the entire domestic, interexchange telecommunications industry was relieved of most of the requirements of Title II of the Communications Act.

II. Regulation of IXCs Today

As we have seen, by the time of enactment of the Telecommunications Act of 1996, the Commission had relieved IXCs—including AT&T—of rate regulation and many of the other, traditional elements of common carrier regulation. One of the Commission's most cherished initiatives, however—the elimination of tariff filing requirements for the IXCs and all other nondominant carriers—was rejected by the courts on the ground that the Commission lacked authority, under the Communications Act, to relieve any common carrier of the statutory obligation to "file with the Commission and print and keep open for public inspection schedules showing all charges for itself and its connecting carriers for interstate and foreign wire and radio communication" [25].

Fortunately, the 1996 Telecommunications Act gave the FCC the statutory authority it needed to detariff any interstate or international telecommunications service, so long as the Commission found that action to be in the public interest [26]. With that congressional mandate in hand, the Commission determined in 1996 that the interstate, domestic interexchange services of nondominant interexchange carriers must be detariffed [27]. The Commission's order was stayed for over three years, however, while the process of judicial review ran its course.

Finally, on April 28, 2000, the United States Court of Appeals for the D.C. Circuit upheld the Commission's order [28]. Pursuant to that decision, the following rules were established:

1. Not later than January 31, 2001 (later extended to April 30, 2001 for mass-market services), all nondominant carriers were to cancel their tariffs for domestic interexchange services;
2. Nondominant carriers could continue to tariff their dial-around service using the carriers' access codes;
3. Nondominant carriers must continue, in the post-tariff environment, to make the current rates, terms, and conditions of their detariffed services available to the public in at least one location during business hours, in an easy-to-understand format and in a timely manner;
4. When responding to an inquiry from the public concerning rates, terms, and conditions, carriers must note that such information is publicly available and explain how it can be obtained;
5. If a carrier has a Web site, it must make the rate, term, and condition information available on-line in a timely and easily accessible manner and must update this information regularly [29].

With the success of its detariffing initiative, the Commission largely completed its program to deregulate IXCs. In fact, in the present environment, IXCs have only two, principal regulatory concerns: first, to comply with state commission regulation of their intrastate services; and second, to comply with the general obligations to engage in reasonable and nondiscriminatory practices, as required by Title II of the Communications Act, and contribute to universal service and other funds to which telecommunications carriers contribute [30].

Endnotes

[1] The non-Bell local telephone companies also were required to implement "equal access" to IXCs, on a schedule established by the FCC.

[2] In Chapter 3, we described the interconnection obligations that incumbent local exchange carriers owe to IXCs.

[3] Application of Microwave Communications, Inc. for Construction Permits to Establish New Facilities in the Domestic Public Point-to-Point Microwave Radio Service at

Chicago, Ill., St. Louis, Mo., and Intermediate Points, 18 FCC 2d 953 (1969), recon. denied, 21 FCC 2d 190 (1970). The Commission previously had permitted private entities to build point-to-point microwave facilities for their own use. Allocation of Frequencies in the Bands Above 890 MHz, 27 FCC 359 (1959), recon. denied, 29 FCC 825 (1960).

[4] Establishment of Policies and Procedures for Consideration of Applications to Provide Specialized Common Carrier Services in the Domestic Public Point-to-Point Microwave Radio Service, 29 FCC 2d 870 (1971), aff'd, 31 FCC 2d 1106 (1971), aff'd sub nom. *Washington Util. & Transp. Comm. v. FCC,* 513 F.2d 1142 (9th Cir.), cert. denied, 423 U.S. 836 (1975).

[5] Regulatory Policies Concerning Resale and Shared Use of Common Carrier Services and Facilities, 60 FCC 2d 261 (1976), modified on recon., 62 FCC 2d 588 (1977), aff'd, *AT&T v. FCC,* 572 F.2d 17 (2d Cir.), cert. denied, 439 U.S. 875 (1978).

[6] *MCI Telecommunications Corp. v. FCC,* 561 F.2d 365, 182 U.S. App. D.C. 367 (D.C. Cir. 1977), cert. denied, 434 U.S. 1040 (1978).

[7] *See* Policy and Rules Concerning Rates for Competitive Common Carrier Services and Facilities Authorizations, 85 FCC 2d 1 (1980); Policy and Rules Concerning Rates for Competitive Common Carrier Services and Facilities Authorizations, 91 FCC 2d 51 (1982); Policy and Rules Concerning Rates for Competitive Common Carrier Services and Facilities Authorizations, 48 Fed. Reg. 46791 (Oct. 14, 1983); Policy and Rules Concerning Rates for Competitive Common Carrier Services and Facilities Authorizations, 95 FCC 2d 554 (1983); Policy and Rules Concerning Rates for Competitive Common Carrier Services and Facilities Authorizations, 98 FCC 2d 1191 (1984); Policy and Rules Concerning Rates for Competitive Common Carrier Services and Facilities Authorizations, 99 FCC 2d 1020 (1985).

[8] *MCI Telecommunications Corp. v. Federal Communications Commission,* 765 F.2d 1186, 1192 (D.C. Cir. 1985); *AT&T v. Federal Communications Commission,* 978 F.2d 727 (D.C. Cir. 1992); *MCI Telecommunications Corp. v. AT&T,* 114 S.Ct. 2223 (1994).

[9] 47 U.S.C. § 161.

[10] The following discussion relies heavily on the FCC's own description of the possibilities it considered. Policy and Rules Concerning Rates for Dominant Carriers, 2 FCC Rcd. 5208, 5215-21 (1987).

[11] A service may consist of several rate elements, or charges.

[12] MTS is message telephone service (i.e., the switched, metered, long-distance service used by most residential and small-business customers). Reach Out America was a so-called block of time offering: it provided for an hour of station-to-station, night/weekend calling per month for a fixed charge. The Reach Out America rates were "fixed without regard to the distance between the terminal points of calls, and night/weekend charges (were) fixed without regard to usage except insofar as they exceed(ed) 60 minutes within a given billing period." *MCI Telecommunications*

Corporation v. American Telephone and Telegraph Company, 104 FCC 2d 1283, 1384-85 (1986).

[13] 800 service, sometimes called Inward WATS, is used by businesses to encourage customer inquiries or placement of telephone orders. It permits callers from prescribed areas to reach the business without paying for the calls.

[14] WATS, or wide area telecommunications service, is a bulk-rate long-distance service used chiefly by medium or large-sized businesses. The WATS subscriber is connected to the service provider over a dedicated access line. 800 service sometimes is called Inward WATS; Outward WATS permits the subscriber to place outbound calls to a wide variety of distant locations at a bulk rate.

[15] Software defined networks are customized voice and data networks provided to major customers over AT&T facilities.

[16] Wideband private lines are dedicated facilities suitable for high-speed data transmissions.

[17] Policy and Rules Concerning Rates for Dominant Carriers, 4 FCC Rcd. 2873, 3054 (1989).

[18] *Id.*

[19] *Id.*

[20] Motion for Reclassification of American Telephone and Telegraph Company as a Non-Dominant Carrier, CC Docket No. 79-252, filed September 22, 1993.

[21] See Policy and Rules Concerning Rates for Competitive Common Carrier Services and Facilities Authorizations, 85 FCC 2d 1 (1980).

[22] Motion of AT&T Corp. to be Reclassified as a Non-Dominant Carrier, 11 FCC Rcd. 3271, 3307 (1995).

[23] The Commission declined to find AT&T nondominant in interexchange markets.

[24] *Id.* at 3281.

[25] 47 U.S.C. § 203(a). The judicial decisions rejecting the FCC's detariffing initiatives are cited at endnote 8, supra.

[26] 47 U.S.C. § 161.

[27] Policy and Rules Concerning the Interstate, Interexchange Marketplace, Implementation of Section 254(g) of the Communications Act of 1934, as amended, 11 FCC Rcd. 20730 (1996).

[28] *MCI WorldCom, Inc. v. Federal Communications Commission,* No 96-1459 (D.C. Cir. 2000).

[29] IXCs also must continue to comply with Section 254(g) of the Communications Act, which requires that "rates charged by providers of interexchange telecommunication services to subscribers in rural and high-cost areas shall be no higher than the rates

charged by each provider to its subscribers in urban areas"; and that "a provider of interstate interexchange telecommunications services shall provide such services to its subscribers in each State at rates no higher than the rates charged to its subscribers in any other State." 47 U.S.C. § 254(g).

[30] IXCs also must comply with regulatory obligations that apply to all telecommunications carriers, including contributions to universal service and other funds and accessibility for persons with disabilities.

7

Pay Telephones and Operator Services Providers

Pay telephones, as we all know, are used by transient customers to place local and long-distance calls. Calls placed through pay telephones are of three kinds. First, customers place many calls to local telephone numbers, usually by depositing coins in the telephone. Second, customers sometimes pay for long-distance calls with deposited coins. (These are called "1+" calls.) Third, customers often place calls on a collect, third-party-number or calling-card basis. (These are often called "0+" calls) [1].

However routine they may seem to the customer, these calls and the service arrangements that permit them are complex transactions involving several players. One participant owns or controls the premises at which the payphone is placed; another may own the payphone; someone else (the ILEC or CLEC) provides the payphone with access to the local telephone network; another company may provide interexchange long-distance and operator services; and still other parties provide the calling cards used by customers to place calls billed to those cards [2]. The terms under which these entities contribute their pieces of the puzzle, and the relationships among them, are subject to varying degrees of regulation. Perhaps the best way to untangle all of this is to look at each player in turn.

I. The Business or Premises Owner

The business or premises owner chooses the payphone vendor whose equipment will be placed at his or her establishment. This makes the business or premises owner a key economic player. Payphone providers solicit the premises owner's business, offering to pay a fee to locate a telephone at their premises. Where premises providers also are in a position to choose the presubscribed carriers for 1+ and 0+ calls, IXCs and operator service providers offer them commissions, based on a percentage of the revenue generated by calls placed through the presubscribed carrier, as an inducement to presubscribe the phones to them [3].

II. The Payphone Provider

Until divestiture, all pay telephone service was provided by incumbent local exchange carriers. Intrastate service offered through ILEC pay telephones was (and is) regulated by the states as a common carrier service. After divestiture, most state PUCs authorized non-ILEC entities—known as customer-owned coin-operated telephone providers, or COCOTs—to provide payphones for the completion of intrastate calls.

The state commissions exercise varying degrees of control over the terms and conditions of competitive pay telephone service. Some require COCOTs to apply for permission to offer intrastate service and to give notice of intention to discontinue service; they also regulate the local coin rate, exercise some control over intrastate toll rates, and often regulate access to 911, the ability of COCOTs to pass along ILEC charges for directory assistance calls, and other issues [4].

Payphone providers also have certain rights and obligations under the Communications Act and the FCC's regulations. Among those obligations, payphone providers are required to post information concerning the rates charged for calls placed from their telephones and the identity of the provider of long-distance and operator services. Among the rights of payphone providers is the right to be compensated for so-called "dial-around" calls. This right, which is of critical importance to the payphone industry and has been the subject of extensive litigation, requires a word of explanation.

Payphone providers earn much of their income from the long-distance and operator-service providers to which their payphones are presubscribed. When a customer makes a 1+ call through a payphone, the presubscribed carrier transports the call in exchange for coins deposited in the phone.

When a customer makes a 0+ call, the presubscribed operator-service provider (OSP) not only provides for transport of the call, but also performs the services associated with collect, third-party number or credit card billing. The OSP shares its revenues from those calls with the payphone provider in the form of commissions.

Many customers, however, avoid the payphone's presubscribed carrier by dialing access codes or toll-free numbers that connect the customer directly with the IXC of the customer's choice [5]. When customers place these dial-around calls, the presubscribed carrier earns no revenue and the payphone provider earns no commissions. In recognition of this fact, the FCC determined that interexchange carriers should compensate payphone providers for dial-around calls through a separate funding mechanism.

The first of the Commission's dial around compensation mechanisms was limited by the inability, at that time, of IXCs to track the number of dial-around calls they received from each pay telephone. Accordingly, the Commission decided that payphone owners would simply be paid $6.00 per month per pay telephone, regardless of the volume of dial-around calls placed through those telephones. These payments were to be funded by all IXCs that provided operator services and earned in excess of $100 million annually. The amount each IXC paid into this fund would be based upon the IXC's percentage of total revenues earned by all IXCs [6].

Later, the Telecommunications Act of 1996 required the Commission to "establish a per-call compensation plan to ensure that all payphone service providers are fairly compensated for each and every completed intrastate and interstate call using their payphone" [7]. Pursuant to this congressional mandate, the Commission tried repeatedly to arrive at a reasonable per-call compensation rate.

In its first effort under the 1996 Act, the Commission adopted a market-based rate of $.35 per completed call for per-call compensation in cases in which a different rate had not been negotiated between the IXC and the payphone provider [8]. This decision was appealed, and the U.S. Court of Appeals for the D.C. Circuit determined that the Commission had acted arbitrarily, and without regard to cost data presented by commenters, when it arrived at its "market-based rate" [9].

Following the judicial rejection of its $.35 rate, the Commission calculated a new rate by taking the local coin rate as a starting point and deducting an amount that supposedly represented the difference in cost between handling coin and noncoin calls [10]. The appellate court rejected this calculation, as well, on the ground that the Commission had not sufficiently justified its assumption that a cost-based rate could be derived by deducting certain costs from the market-based coin rate [11].

Finally, in 1999, the Commission entered an order that established a "bottom up" compensation rate of $.24 per call, based upon the incremental costs incurred by payphone providers to carry coinless calls [12]. This decision, too, was appealed, but the D.C. Circuit Court of Appeals found that the Commission had sufficiently justified its methodology and rejected the challenges to the decision as without merit [13].

III. Incumbent Local Exchange Carriers

Historically, ILEC payphones were provided under technical, business, and regulatory arrangements that differed sharply from those of COCOTs. Notably, ILECs historically subsidized their pay telephone services with revenues from other services. Also, ILEC payphones were provided as part of a tariffed, end-to-end service, rather than treated as customer-premises equipment, and those ILEC payphones obtained network access through a special "coin line" arrangement that located the intelligence of the service at the central office. At the same time, ILECs provided COCOT telephones with tariffed interconnections that were little more than ordinary business access lines, for which a recurring monthly charge was assessed. The ILECs also imposed usage charges for local calls from COCOT phones on a per-message or measured-rate basis; and imposed extra charges for optional services (including overseas call blocking) intended to reduce the incidence of toll fraud against the COCOTs [14].

COCOTs complained that the different economic and technical arrangements for ILEC and non-ILEC payphones were discriminatory. Specifically, the COCOTs argued that ILECs should not be permitted to subsidize their payphone services with revenues earned from local exchange and exchange access services; and that ILECs either should provide COCOTs with the same coin line that ILECs furnish to their own payphones, or classify the ILEC payphones as customer-premises equipment (as COCOT phones are classified) and pay the same charges for access that the COCOTs pay.

The Telecommunications Act of 1996 expressly addressed these concerns. Specifically, Section 276 of the 1996 Act provides that a "Bell operating company that provides payphone service shall not subsidize its payphone service directly or indirectly from its telephone exchange service operations or its exchange access operations" and shall not "prefer or discriminate in favor of its payphone service" [15]. The Act also provided, more generally, that all ILECs (not just BOCs) must "discontinue the intrastate and interstate carrier access charge payphone service elements and payments in effect (on the date of enactment of the 1996 Act), and all intrastate and interstate payphone subsidies from basic exchange and exchange access revenues" [16].

The 1996 Act also addressed the inability of Bell operating companies and other ILECs to negotiate with premises owners and OSPs on the same basis as COCOTs. Specifically, the ILECs, because their payphone service was tariffed and subsidized, historically were not allowed to offer contractual incentives to premises owners to place payphones at those owner's premises. Also, under the terms of the AT&T divestiture decree, BOCs were not permitted to choose the presubscribed OSPs for their payphones and earn commissions from those OSPs. Congress determined that after BOC payphones were deregulated and the subsidies for those payphone services were removed, BOCs should no longer be subject to these disabilities. Accordingly, the Act directed the Commission to prescribe regulations that permit all payphone service providers, including BOCs, to have "the same right to negotiate with the location provider on the location provider's selecting and contracting with, and, subject to the terms of any agreement with the location provider, to select and contract with, the carriers that carry interLATA calls from their payphones, unless the Commission determines in the rulemaking pursuant to this section that it is not in the public interest" [17]. The Commission subsequently found that extending these rights to BOCs would serve the public interest, and enacted rules accordingly [18].

Finally, the Commission's orders deregulating the payphone industry addressed another disparity between BOCs and COCOTs—specifically, the fact that BOCs historically were not permitted to receive commissions from presubscribed OSPs for 0+ calls placed through BOC payphones. Although the 1996 Telecommunications Act effectively remedied this restriction, by permitting BOCs to negotiate contracts with IXCs and OSPs, the Act also "grandfathered" existing contracts between premises owners and presubscribed OSPs at BOC payphone locations [19]. So long as those contracts were in effect, the BOCs were unable to negotiate commission arrangements with the presubscribed OSPs. In order to correct this disparity, the FCC decided that BOCs also were entitled to receive per-call compensation for calls placed from their payphones, payable by the OSP to which the premises owner had presubscribed the telephone [20]. This entitlement would not be effective, however, until the BOCs had reclassified their payphones as customer-premises equipment and eliminated subsidies for their payphone services, as required by the 1996 Act. This entitlement also would apply only "so long as [BOCs] do not otherwise receive compensation for use of their payphones in originating 0+ calls."

IV. Operator Service Providers

Before divestiture, there were no independent OSPs; all operator services were provided by the Bell system and other ILECs at tariffed rates. After divestiture, a number of long-distance companies (some transporting traffic over their own facilities, others reselling the interexchange services of other carriers) approached premises owners and offered to provide operator services to the telephones located on their premises. These OSPs, often called "alternate operator services" companies, or AOSs, proposed that the premises owners presubscribe their telephones to them. In exchange for the premises owner's agreement to route all calls to them, the AOSs agreed to pay the premises owner a percentage of their revenues from 0+ calls [21].

Before Congress and the FCC stepped in, the rates and practices of the AOS industry were unregulated and prompted a heavy volume of consumer complaints. The practices that caused the most concern were the high charges imposed by some AOSs for 0+ calling, and the blocking of customer access to other carriers. Users of payphones, hotel guests, and other customers at aggregator locations found that when they dialed 0+, their call was routed automatically to the AOS chosen by the premises owner. Often the AOS did not identify itself or advise the caller of the charges that would be imposed (even though these were likely to be much higher than the caller expected). If the customers tried to reach their preferred carriers by dialing 10XXX access codes, 950-XXXX numbers, or through any other method, the AOS simply blocked the call. If the customer tried to use a calling card issued by a particular carrier and the AOS honored that card, the call still would be carried by the AOS, rather than the carrier that issued the card.

Consumers felt trapped and abused by these practices, and their complaints were heard by Congress. In 1990 Congress responded with the Telephone Operator Consumer Services Improvement Act (TOCSIA), which required all operator service providers (AT&T as well as the AOSs) to identify themselves to consumers (a practice called "branding") and quote their rates upon request [22]. The Act also required aggregators to unblock access to other carriers and post certain disclosures on or near each telephone. Congress directed the FCC to write rules implementing the Act and to resolve specific issues that were left to the Commission's expertise [23]. The

Commission eventually adopted regulations implementing TOCSIA and resolving the questions committed to it [24].

Endnotes

[1] As we discuss below, 0+ calls require the participation of a provider of operator services (OSP). The caller may access the OSP simply by dialing 0 and the called number, in which case operator services will be provided by the OSP to which the payphone owner has presubscribed the payphone. The caller also may dial an access code or toll-free number to reach the OSP of the caller's choice, in which case the presubscribed carrier is bypassed.

[2] In the days before divestiture, the unified Bell system performed every function on this list except ownership of the business or premises at which the payphone was located.

[3] Until recently, Bell operating companies that provided payphones were not permitted to selected the presubscribed OSPs, which were selected by the premises owners. As we discuss later in this chapter, the Telecommunications Act of 1996 permits Bell operating companies to negotiate with ISPs for presubscription to Bell payphones, and permits the Bells to accept commissions from OSPs.

[4] *See*, e.g., Investigation by the Commission on its Own Motion into Customer Owned Coin Equipment in the Public Telephone Market, Case No. 7841 II, Proposed Order of Hearing Examiner (Public Service Commission of Maryland 1986).

[5] Customers also avoid the presubscribed carrier when they dial toll-free numbers to reach providers of products and services. Interexchange carriers that provide toll-free numbers must pay dial-around compensation for these calls, as well.

[6] *See* Policies and Rules Concerning Operator Service Access and Pay Telephone Compensation, 6 FCC Rcd. 4736, 4745 (1991); Order on Reconsideration, 11 FCC Rcd. 21233.

[7] 47 U.S.C. § 276(b)(1)(a).

[8] Implementation of the Pay Telephone Reclassification and Compensation Provisions of the Telecommunications Act of 1996, 11 FCC Rcd. 20541 (1996) (First Payphone Order).

[9] *Illinois Public Telecomm. Ass'n v. FCC,* 117 F.3d 555 (D.C. Cir. 1997).

[10] Implementation of the Pay Telephone Reclassification and Compensation Provisions of the Telecommunications Act of 1996, 13 FCC Rcd. 1778 (1997).

[11] *MCI Telecomm. Corp. v. FCC,* 143 F. 3d 606 (D.C. Cir. 1998).

[12] Implementation of the Pay Telephone Reclassification and Compensation Provisions of the Telecommunications Act of 1996, 14 FCC Rcd. 2545 (1999).

[13] *American Public Communications Council v. FCC,* 215 F. 3d 51 (D.C. Cir. 2000).

[14] In various proceedings brought before the FCC, COCOT interests contended that the antifraud and other features that ILECs provided to COCOT payphones were inferior to similar features that ILECs provided to their own payphones.

[15] 47 U.S.C. § 276(a).

[16] *Id.* § 276(b)(1)(B).

[17] *Id.* § 276(b)(1)(D).

[18] *See* First Payphone Order, *supra.*

[19] 47 U.S.C. § 276(b)(3).

[20] First Payphone Order, 11 FCC Rcd. at 20569.

[21] Not all of the premises owners had payphones at their locations. The AOSs also were interested in hotels, hospitals, and universities that provided telephones for the use of their room guests, patients and students. All such providers of telephone service to transient populations (known in the industry and to the FCC as "aggregators") were potential customers of the AOSs.

[22] OSPs also were required to refrain from billing for uncompleted calls in most locations, were required to withhold payment of commissions to aggregators that had not unblocked access for dial-around calls, and were required to file information tariffs with the FCC.

[23] The state commissions had treated the AOS industry with great skepticism and in some cases had refused to let AOSs handle intrastate calls. The states were unable to act against intrastate call handling by AOSs, however, and such calls were the bulk of the AOSs' business. *See* Report 101-439 of the 101st Congress, 2d Session, Committee on Commerce, Science and Transportation to accompany S. 1660 at 4.

[24] *See* Policies and Rules Concerning Operator Service Access and Pay Telephone Compensation, 6 FCC Rcd. 4736 (1991).

8

Mobile Telephone Companies

Some telephone services use radio, rather than wireline telephone facilities, to establish the "last mile" connection between the public switched telephone network and the customer. Because they are radio-based, these services are subject to the FCC's authority to allocate radio spectrum to specified uses, and to award individual licenses for use of that spectrum. To the extent they offer services to the public and connect those services to the public switched telephone network, they also are subject to regulation as common carriers.

In order to understand the regulatory problems of mobile telephone service, it is helpful to understand the technology and network architecture on which the service is based. After providing that background, we discuss the two principal areas of regulatory involvement in the mobile telephone industry (i.e., licensing of mobile telephone companies to use radio frequency spectrum, and state and federal regulation of mobile telephone companies that operate as common carriers).

I. Mobile Telephone Technology

The most important and popular mobile telephone services—whether marketed as cellular service, personal communications service (PCS), specialized mobile radio service (SMR), or mobile satellite service (MSS), and whether based on digital or analog technology—rely on the innovation commonly described as cellular radio.

When cellular radio was introduced, it was "not so much a new technology as a new idea for organizing existing technology" [1]. Its genius, both simple and elegant, is the way it uses radio spectrum.

Imagine for a moment that you are operating a car telephone system in the precellular era. You have a broadcasting tower that covers your entire city, its suburbs, and adjacent rural areas; only twenty frequencies are available for simultaneous use at any time. When one of your customers makes a call, he or she ties up one of those frequencies, throughout your system, for the duration of the call. Whenever twenty conversations are in progress, your other subscribers get a busy signal until someone hangs up. The only way to provide your customers with decent service, under these conditions, is to limit the number of subscribers severely—which is why there once were waiting lists of several years for car telephones in major metropolitan areas [2].

A few inventive souls began to see this problem, not as a lack of frequencies, but as a wasteful use of the existing frequencies. Suppose the single broadcasting tower, with its enormous range, was replaced by dozens of small towers (or "cells") with a far more limited range? When a customer dialed a call in a system of this kind, he or she still would tie up one of the twenty frequencies, but only within the small area covered by the nearest cell. Another customer, in another part of town served by another cell, could simultaneously use the same frequency for a different conversation; and as these customers passed out of range of the cells through which they began their conversations, they would free up a channel on each of those cells for new calls. (At the same time, the cellular company's switch would "hand off" their calls to a closer cell without interrupting their conversations.) The cellular idea, taking advantage of the mobility of radio-telephone customers, lets a few channels do the work of many.

While the concept of cellular service dates from the 1940s, cellular systems were not operational until the 1980s. Today, tens of millions of customers subscribe to mobile telephone service in the United States, and the service has expanded to include messaging, Internet access, and other value-added functions.

The enterprises that provide this service are in some ways similar to LECs. Like LECs, the mobile companies use a switch (housed in a mobile telephone switching office, or MTSO) to link their customers in a network. Instead of connecting to their customers over LEC-style fixed-access lines, however, the mobile telephone companies connect their MTSOs to the dispersed cell sites, and those sites establish radio links with the mobile customers' telephones. Other links connect the MTSOs to the local LECs, so that mobile customers can call anyone on the public switched telephone network, and any user of the public switched telephone network can call them.

II. Licensing Mobile Telephone Companies

Like other users of radio transmission equipment in the United States, mobile telephone companies must secure radio licenses from the FCC. The FCC's authority to award radio licenses, in turn, is based on two stubborn physical realities. The first of these realities is that the radio frequency spectrum useable for radio service is limited. The second is that if multiple transmitters emit energy on the same frequencies in the same geographic area, each of the transmitters' signals may interfere with and degrade the others' signals.

Because of these physical limitations, the United States and all other countries decided long ago that government must regulate the use of radio frequency spectrum. Such regulation is of two kinds. First, regulators *allocate* spectrum by deciding which frequency ranges will be available to particular types of radio-based service. Second, regulators *assign* particular frequencies, within the ranges of frequencies allocated to particular services, to individual providers of those services.

The first of these regulatory processes (i.e., allocation of frequencies) is a matter of global as well as national concern. In order to avoid crossborder electromagnetic interference and provide for the more efficient deployment of radio services and equipment, a treaty organization called the International Telecommunication Union (ITU) works out allocations of spectrum to particular services on an international (global or regional) basis [3]. Once those international allocations are made, national regulators will, for the most part, allocate spectrum domestically in accordance with—or in ways not flatly incompatible with—the ITU allocations.

The FCC is both an allocating and an assigning agency. The FCC allocates nongovernmental spectrum among radio-based services in the United States, and assigns spectrum to particular users through exercise of the licensing authority conferred by Title III of the Communications Act [4].

The FCC's licensing authority is described generally in Section 301 of the Communications Act, which states the intent of Congress to "provide for the use of (radio) channels, but not the ownership thereof, for limited periods of time" [5]. More specifically, other sections of Title III give the Commission express authority to grant, condition, and revoke licenses as it sees fit, subject only to a general requirement that the licensing authority be exercised in the public interest.

In carrying out its mandate to award radio licenses in the public interest, the Commission has taken a number of approaches. For example, in assigning spectrum to radio and television broadcast stations, and in the early rounds of cellular licensing, the Commission has relied on trial-type comparative hearings at which competing applicants presented evidence of their ability to operate in the public interest. In later mobile telephone licensing rounds, the Commission conducted lotteries among applicants. Finally, after Congress granted statutory authority in 1993, the Commission conducted auctions of mobile telephone spectrum.

As the following discussion shows, the history of mobile telephone licensing in the United States involves a variety of technologies and regulatory approaches, and presents some unusually stark examples of licensing processes gone right and gone wrong.

A. A Brief History of Mobile Telephone Spectrum Assignment

The Commission first allocated domestic spectrum for cellular radiotelephone service in 1974, but did not assign that spectrum to individual users until the early 1980s.

When the Commission entered its first cellular allocation decision, it declared that it would license only one carrier in each market and accepted applications only from the monopoly telephone companies already serving each market. The FCC based its decision on reasoning that was typical of the monopoly era (i.e., that the cellular market and available frequencies would not support competition) [6]. Later, the Commission decided to award two licenses per market—one to the existing, monopoly wireline carrier and the other to a wireless service provider [7]. This decision left the Commission with a task that proved one of the most vexing in the history of its spectrum management function (i.e., deciding which applicants would obtain access to the nonwireline licenses for mobile telephone services).

The initial cellular-licensing process was made enormously complex and inefficient by the Commission's reliance on its traditional comparative hearings approach. Although the Commission streamlined the process somewhat by conducting only "paper" proceedings, the first three rounds of cellular licensing attracted a total of over 1,100 applicants. The task of reviewing these applications, which often ran to over 1,000 pages of application documents and supporting exhibits, caused years of delay [8].

Despite the obvious inefficiency of the comparative-hearing process, the Commission was not free to abandon that approach without express congressional authority to do so [9]. Limited congressional relief came in 1981, when an amendment to the Communications Act empowered the Commission to assign certain radio licenses by lottery [10]. Unfortunately, the lottery process attracted applicants that had no intention of providing mobile telephone service to the public, and "application mills" that assisted speculators wishing to obtain spectrum and resell it to prospective service providers [11]. The winners of these lotteries created a secondary market in cellular, SMR, and other spectrum that eventually was purchased and aggregated by entities with an intention to provide service. On the whole, the Commission's experience with lotteries demonstrated that as a means of placing spectrum promptly in the hands of those most qualified to use it, with the minimum necessary expenditure of the Commission's resources, lotteries had little to recommend them over the comparative-hearings approach.

The Commission's experience with comparative hearings and lotteries gave new life to a view that had been circulating in academic and governmental circles for years (i.e., the notion that radio frequency spectrum should not be given away but should be auctioned to the highest bidders) [12]. Although the full economic case for auctions is somewhat complex, the principal public-interest argument is that the entity that puts the most money at risk for a license has the greatest incentive to put its licensed spectrum to profitable use with the least delay. Accordingly, the high bidder will have or acquire the expertise and experience that profitable use of its spectrum requires, will identify those services that the public is most willing to purchase and will move as efficiently as possible to deploy those services. In short, spectrum that is assigned through a well-designed auction not only should bring new revenue to the public treasury, but should produce the greatest public benefit in the shortest time.

Congress decided to put these views to the test in 1993, when it passed Section 309(j) of the Communications Act [13]. Under Section 309(j), the Commission is empowered to award spectrum by auction when competing applications for that spectrum are mutually exclusive. Pursuant to that authority, the Commission adopted a procedure known as the Simultaneous Multiple Round Auction. Under that procedure, a group of licenses is put up for bid over a series of rounds. At the end of each round, the high bid for each license helps to fix the minimum acceptable bid for the next round and becomes the winning bid if no higher bids are received in later rounds. Bidders that fail to bid in a round and have not submitted a minimum number of high bids in earlier rounds may be ineligible to continue. In order to prevent collusion among bidders, entities that have applied for the right to bid in the same markets are prohibited from collaborating, discussing their bidding strategies, or disclosing their bids. Exceptions to these rules are made for discussion with entities with which bidders have joint ventures, partnerships, or certain other agreements, so long as those arrangements are disclosed.

For the most part, the Commission's use of its auction authority has been a success. The only notable exception is the difficulty the Commission has encountered in carrying out Congress's mandate to promote the award of licenses to rural telephone companies, minority-owned and female-owned firms, and small-business entities [14]. Pursuant to this congressional requirement, the Commission created a special auction, known as the C block auction, in which only these "designated entities" would participate. C block participants were to be eligible for a number of benefits not available to bidders in the regular auctions, including bidding credits and installment payment plans.

Three days before the deadline for initial applications for the C block auction, however, the U.S. Supreme Court decided the case of *Adarand Constructors, Inc. v. Pena* [15]. In that decision, the Court found that when government confers benefits and burdens on the basis of a racial classification, that action must serve a compelling governmental interest and must be narrowly tailored to serve that interest. In light of that decision, the Commission decided to revisit its C block auction rules to ensure their compliance with the Court's strict scrutiny standard.

As a result of its review of the C block procedures, the Commission decided to eliminate eligibility requirements based expressly on race and gender and confer special incentives and benefits on all participants with revenues below certain defined thresholds. In the Commission's view, because the class of small businesses should significantly overlap with the categories of businesses owned by minorities and women, conferring C block benefits on small businesses would serve the congressional goal of favoring minority and female-owned businesses without attracting strict judicial scrutiny under the Adarand standard [16].

Unfortunately, the financial incentives granted to small-business participants in the C block auctions, coupled with the unrealistically high levels of the winning bids, led some successful bidders to report that they would default in their installment payment obligations. (Some winning bidders even declared bankruptcy.) These developments, and the efforts of the Commission to salvage the situation, did not undermine the rationale for competitive bidding as a means of assigning spectrum. The C block experience did, however, call into question the wisdom of creating incentives that favored less experienced and well-financed participants, however worthy the social goals that inspired those efforts.

III. Regulation of Mobile Telephone Licensees

Besides the method of assignment of radio frequency spectrum, the most significant question the Commission has faced in its management of mobile telephone services is the proper regulatory classification of those services. Specifically, the Commission must decide whether mobile telephone service providers, which in many ways resemble LECs, should be subject to some or all of the common carrier obligations to which LECs are subject.

Prior to 1993, the Commission recognized two categories of mobile, land-based radio services. Private land mobile services "were predominantly dispatch services such as those operated by police departments, fire departments, and taxicab companies, for their own purposes" [17]. Public mobile service was "radio-telephone service (including cellular) which interconnected with existing telephone systems" [18]. Of these two categories only the second was subject to common carrier obligations such as tariffing of rates, providing service upon reasonable request and avoidance of "unjust or unreasonable discrimination in charges, practices, regulations, facilities or services" [19].

This scheme of classification began to show some strain when the Commission authorized new categories of mobile service. Notably, in 1991 the Commission permitted a company called Fleet Call to use SMR frequencies, which historically had been used for private-fleet dispatch services, as a means of providing an interconnected service called "wide area SMR" that potentially would compete directly with cellular. The following year, the Commission initiated the allocation of frequencies to PCS, which also would compete with cellular service. The Commission had to make a choice of regulatory schemes for these new services: it could regulate them as private carriage (in which case they would be free of the common carrier obligations to which competing cellular providers were subject); or it could ensure regulatory parity among like services by classifying cellular, SMR, and PCS uniformly as private or common carriage.

The 1993 Budget Act gave the Commission the flexibility it needed to reach a practical resolution of this question. Specifically, the Budget Act created two new categories of mobile radio services: commercial mobile radio service (CMRS) and private mobile radio service (PMRS). The Budget Act defined CMRS as "any mobile service that is provided for profit and makes interconnected service available (A) to the public, or (B) to such classes of eligible users as to be effectively available to a substantial portion of the public" [20]. PMRS was defined as "any mobile service that is not a commercial mobile service or the functional equivalent of a commercial mobile service" [21]. Congress classified CMRS service providers as common carriers, but gave the Commission discretion to forbear from applying most of the common carrier requirements of the Communications Act to those services if the Commission found such forbearance to be in the public interest [22]. The Budget Act also preempted state regulation of rates and entry for both CMRS and PMRS providers (although states could petition the FCC for continuing authority to regulate mobile services in certain circumstances) [23].

Pursuant to this new authority, the Commission undertook an exhaustive review of mobile radio services. The Commission's review concentrated on two tasks: (1) the appropriate classification of services into the new CMRS and PMRS categories; and (2) the extent to which CMRS services should enjoy forbearance from common carrier regulation.

As to the first task, the Commission decided that the CMRS category must include "any mobile service that is provided with the intent of receiving compensation or monetary gain" [24], that makes service "broadly available through use of the public switched network" [25], and that is offered to the public "without restriction on who may receive it" [26]. Defined in this way, CMRS includes cellular, PCS, wide-area SMR, and MSS service provided to the public, as well as store-and-forward paging services that are not operated exclusively for the licensee's internal communications.

As to the second task, the Commission decided that the CMRS marketplace was sufficiently competitive to support forbearance from many elements of common carrier regulation. Notably, the Commission declared that CMRS providers need not file tariffs or request the Commission's permission to enter a market or add or remove a line. The Commission retained, however, the general requirements that CMRS carriers provide service to all who reasonably request it, at rates and other terms that are just, reasonable, and not unreasonably nondiscriminatory. The Commission also found that CMRS providers are subject to the bringing of formal complaints before the FCC by persons aggrieved by the CMRS providers' violations of the Communications Act. Subsequent to the CMRS decision, the Commission has granted forbearance from a number of other common carrier requirements [27].

Finally, although exempt from most common carrier regulations, CMRS providers must observe a number of other requirements that are peculiar to their industry or to particular participants within that industry. Notably, until January 1, 2002, nonrural ILECs may provide in-region CMRS services only through separate affiliates [28]. CMRS providers also must provide emergency E911 services with an Automatic Location Identification capability [29]; must permit intersystem roaming from customers that leave their home service areas [30]; must permit resale of their services until November 24, 2002 [31]; and may not hold more than 45 MHz of broadband PCS, cellular, and SMR spectrum in a common geographic area [32].

IV. Interconnection Between Mobile Telephone Licensees and Wireline Carriers

The Commission has long asserted its plenary jurisdiction over the physical terms of interconnection between LECs and radio-telephone companies. Since passage of the 1993 Budget Act, with its broad grant of preemption authority, the Commission has taken plenary jurisdiction over the rates charged for such interconnection as well.

A. Physical Terms of Interconnection

As early as 1986, the Commission determined that each LEC must offer independent cellular carriers a form of interconnection that is at least as good as "that used by the wireline cellular carrier" [33]. This rule relied on the FCC's practice of licensing one telephone company affiliate (or wireline carrier) as a cellular carrier in each service area; the rule made the wireline carrier's interconnections a benchmark for judging the adequacy of the connections furnished to the nonwireline carrier.

Second, the Commission required LECs to negotiate with cellular carriers, upon request, for interconnections different from those offered to the wireline carrier. This rule recognized that the local wireline's arrangements with the LEC would not always meet the nonwireline carrier's requirements, and that cellular carriers should not be "locked into the specific interconnection arrangements requested by a wireline carrier" [34].

Third, the FCC specifically required LECs to provide an arrangement known as Type 2 interconnection upon request [35]. This rule requires some explanation.

Type 2 interconnection is the more efficient alternative to Type 1 arrangements, which treat the mobile carrier's switch as though it was a PBX rather than an end office [36]. The LEC provides Type 1 access through a facility connecting the mobile switch (a MTSO) to a single LEC end office; all traffic between the public switched telephone network and the MTSO must pass through that single end office.

The FCC found that Type 1 interconnections were relatively costly and inefficient. When a call from a wireline telephone is placed to a cellular telephone by way of a Type 1 interconnection, the call goes first to the Class 4 office serving the end office to which the cellular company's MTSO is connected [37], then to the end office, and then to the MTSO for delivery to the called party. According to the Commission, this roundabout delivery method caused loss of signal quality and increased expense [38].

Type 2 interconnections, on the other hand, link the MTSO directly to the LEC Class 4 office. Under this arrangement, a call from the landline network to a cellular customer is routed to the Class 4 office, and directly from there to the MTSO, without passing through an intermediate Class 5 office. The MTSO has access to end offices served by the Class 4 office; and the cellular company also may arrange to connect with any IXCs served by that Class 4 office [39].

The Commission's view is that cellular companies, as cocarriers, are entitled to connections that achieve the best feasible levels of efficiency and customer-service quality [40]. Type 2 connections appear to satisfy this standard, and LECs may not refuse to provide them.

The FCC's fourth requirement was that LECs furnish the cellular carriers with telephone numbers for assignment to their customers, and that they do so on the same basis as they furnish numbers to independent telephone companies [41]. This requirement has been superseded by the transfer of number allocation and assignment functions from the LECs to an independent entity.

Finally, the FCC determined that LECs may not restrict the resale of cellular service by the carriers to which they furnish interconnections [42].

After the Budget Act of 1993 was enacted, the Commission adopted Section 20.11 of its regulations, which requires LECs to provide reasonable interconnection to CMRS providers on request [43]. Later, after the Telecommunication Act of 1996 defined interconnection rights and obligations among all telecommunications carriers, the Commission amended Section 20.11 to clarify that LECs and CMRS providers must comply with all interconnection obligations set out in Sections 251 and 252 of the Act. Under those provisions, CMRS providers have all of the interconnection obligations imposed upon telecommunications carriers under Section 251(a) of the Act;

CLECs owe CMRS providers all of the interconnection obligations imposed upon LECs in Section 251(b); and ILECs owe CMRS providers all of the interconnection obligations imposed by Section 251(c) of the Act [44].

B. Economic Terms of Interconnection

Before enactment of the Telecommunications Act of 1996, compensation of CMRS providers for transport of calls placed by LEC customers was governed by Section 20.11 of the Commission's rules, which required LEC and CMRS providers to pay reasonable compensation for transport and termination of the other carrier's calls [45]. This rule, however, was widely ignored by LECs, which did not compensate CMRS providers for transport and termination of LEC customers' calls, even when those same LECs were charging CMRS providers for transport and termination of calls placed by CMRS customers to ILEC customers [46]. In a proceeding that was overtaken by passage of the 1996 Act, the Commission proposed to resolve the LEC-CMRS reciprocal compensation impasse by imposing an interim bill-and-keep mechanism (i.e., an arrangement under which neither carrier paid the other for transport and termination services) [47].

After the new Act was passed, in its Local Competition Order, the Commission determined that "CMRS providers are telecommunications carriers and, thus, reciprocal compensation obligations under Section 251(b)(5) (of the 1996 Act) apply to all local traffic transmitted between LECs and CMRS providers" [48]. In deciding that LEC-CMRS reciprocal compensation was covered by Section 251(b)(5), the Commission effectively ceded to the states the responsibility for establishing the level of compensation between providers. In setting out a framework for state arbitration of LEC-CMRS reciprocal compensation disputes, the Commission declined to mandate bill-and-keep, primarily because the record did not establish that LEC-CMRS traffic flows are balanced and that the per-unit cost of LEC-CMRS interconnection is *de minimis* [49]. In fact, the record in the Local Competition proceeding "contain[ed] no estimates of the cost of CMRS termination" [50]. Accordingly, the Commission determined that "although states may rely on bill-and-keep for particular pairs of firms based on the circumstances prevailing between them, we conclude that we are correct in not adopting bill-and-keep as a single, nationwide policy that would govern all LECCMRS transport and termination of traffic" [51].

The compensation framework adopted by the Commission for reciprocal compensation between LECs and nonpaging CLECs is the same as that for ILECs and CLECs. The formula is based on the language of Section 252(d)(2), which provides that reciprocal compensation shall reflect "a reasonable approximation of the additional costs of terminating calls." The Commission found that the category of "additional costs" under Section 252(d)(2) does not include the nontraffic-sensitive costs of local loops and switch ports, but includes only "the forward-looking, economic cost of end-office switching that is recovered on a usage-sensitive basis" [52]. The Commission also established a presumption that states would set transport and termination rates for non-ILECs, including those for nonpaging CMRS providers, at the same level as the rates paid to ILECs [53]. This presumption of symmetry may be rebutted if a carrier submits a forward-looking cost study to its state commission that justifies a departure from symmetrical rates [54].

V. Mobile Satellite Service

It is fair to say that some remote areas of the earth's surface never will play host to conventional, cellular-type MTSOs and cell towers. For mobile telephone service to these areas, including Indian reservations and remote areas of the United States, the most practical option may be direct communication between a customer's mobile telephone terminal and an orbiting satellite.

Service of this kind, known as MSS, is made especially practical by the use of low-Earth orbit (LEO) satellites rather than geostationary satellites to establish the space link for an MSS system. Geostationary satellites, which orbit at 22,235 miles above the Earth's surface, have the advantage that they appear stationary to a ground-based observer and can be in permanent contact with any ground station within their coverage area. However, the launch costs of a geostationary satellite can be 20 times greater than that of a LEO satellite. Also, geostationary satellites operate at such a great distance from the earth that transmissions to and from those satellites must be made at higher power, or involve more sensitive and expensive ground stations, than LEO satellites demand. Accordingly, it can be more efficient to launch an array of MSS satellites than to provide the service from a single, geostationary satellite [55].

After an international allocation of new mobile-service frequencies was made at the 1992 World Administrative Radio Conference, the FCC moved aggressively to provide spectrum for MSS services in the United States. The principal, resulting allocations include spectrum for so-called little LEO services, which provide nonvoice services such as data messaging and position determination; and spectrum for low- and middle-Earth orbiting systems, such as those of Globalstar and ICO, which provide combinations of voice and data services.

Because MSS systems operate on a transnational basis and implicate orbital assignment and international coordination problems that terrestrial mobile telephone services need not address, the story of MSS regulation is especially complex. Those complexities are beyond the scope of this book, but have been ably addressed elsewhere [56].

Endnotes

[1] Calhoun, G., Digital Cellular Radio 39 (1988).

[2] The reality could be worse than our example. "(I)n New York City in the 1970s (a megalopolis of nearly 20 million people over a 1000 square miles or more) the Bell mobile system could support just 12 simultaneous mobile conversations. The 13th caller was blocked." *Id.* at 40.

[3] *See* Codding, Jr., G. A., *The International Telecommunication Union: An Experiment in International Cooperation*, New York: Arno Press, 1972; Codding, Jr., G. A., and A. M. Rutkowdki, *The International Telecommunications Union in a Changing World*, Norwood: Artech House, 1982.

[4] Spectrum allocated to the federal government is managed by the National Telecommunications and Information Administration (NTIA).

[5] 47 U.S.C. § 301.

[6] An Inquiry Into the Use of the Bands 825-845 and 870-890 MHz for Cellular Communications Systems and Amendment of Parts 2 and 22 of the Commission's Rules Relative to Cellular Communications Systems, 86 FCC 2d 469 (1981), *modified,* 89 FCC 2d 58 (1982), *further modified,* 90 FCC 2d 571 (1982), *appeal dismissed sub nom.* United States v. FCC, No. 82-1576 (D.C. Cir. 1987).

[7] This decision was challenged in the courts but was upheld as within the Commission's discretion. National Association of Regulatory Utilities Commissioners v. FCC, 525 F. 2d 630, 173 U.S. App. D.C. 413 (D.C. Cir. 1976), *cert. denied sub nom.* National Association of Radiotelephone Systems v. FCC, 425 U.S. 992 (1976).

[8] *See* Fritts, B. C., Private Property, Economic Efficiency, and Spectrum Policy in the Wake of the C Block Auction, 51 Fed. Comm. L. J. 849 (May, 1999).

[9] *See Ashbacker Radio Corp. v. FCC,* 326 U.S. 327 (1945), finding that under the public interest standard of the Communications Act, the Commission must hold comparative hearings to resolve mutually exclusive, *bona fide* radio license applications.

[10] Omnibus Budget Reconciliation Act of 1981, Pub. L. No. 97-35, 95 Stat. 736-737, amended, Communications Amendment Act of 1982, Pub. L. No. 97-259, § 115, 96 Stat. 1087.

[11] The speculative character of the lottery process was made possible by the FCC's abandonment of its effort to prescreen applicants and allow only qualified entities to participate. The prescreening process resulted in delays reminiscent of the comparative hearings process, and in 1987 the Commission abandoned prescreening and opened the lotteries to all comers.

[12] *See* Coase, R. H., The Federal Communications Commission, 2 J. L. & Econ. 1, 13 (1959).

[13] Omnibus Budget Reconciliation Act of 1993, Pub. L. No. 103-66, § 6002, 107 Stat. 312, 387-392, adding provisions to the Communications Act that are codified at 47 U.S.C. § 309(j) (Budget Act).

[14] In Section 309(j)(b) of the Communications Act, added as part of the 1993 Omnibus Budget Reconciliation Act, the Congress required the Commission to "disseminat(e) licenses among a wide variety of applicants, including small businesses, rural telephone companies, and businesses owned by members of minority groups and women." 47 U.S.C. § 309(j)(B).

[15] 515 U.S. 200 (1995).

[16] Implementation of Section 309(j) of the Communications Act—Competitive Bidding: Amendment of the Commission's Cellular-PCS Cross-Ownership Rule, 11 FCC Rec. 136, § 42 (1995).

[17] Implementation of Sections 3(n) and 332 of the Communications Act: Regulatory Treatment of Mobile Services, 9 FCC Rcd. 1411 (1994) n. 8 (CMRS Order).

[18] *Id.*

[19] *Id.* at § 3. In the case of paging services classified as private land mobile carriers, the Commission decided to treat those services as nondominant and therefore subject to forbearance from most common carrier obligations. Prior to its 1994 CMRS Order, however, the Commission classified cellular service providers as dominant and subject to all of the usual common carrier requirements.

[20] Budget Act, codified at 47 U.S.C. § 332(d)(1).

[21] *Id.*, codified at 47 U.S.C. § 332(d)(2).

[22] *Id.*, 47 U.S.C. § 332(c).

[23] *Id.*, 47 U.S.C. § 332(c)(3).

[24] CMRS Order at § 43.

[25] *Id.* at § 54.

[26] *Id.* at § 65.

[27] *See* Forbearance from Applying Provisions of the Communications Act to Wireless Telecommunications Carriers, WT Docket No. 98100, Release No. FCC00311 (First Report and Order rel. Aug. 21, 2000) at § 6.

[28] 47 C.F.R. § 20.20.

[29] *Id.* § 20.18.

[30] *Id.* § 20.12(c).

[31] *Id.* § 20.12(b).

[32] *Id.* § 20.6. In rural areas, the CMRS "spectrum cap" is 55 MHz.

[33] In the Matter of the Need to Promote Competition and Efficient Use of Spectrum for Radio Common Carrier Services, 59 Rad. Reg. 2d (P&F) 1275, para. 12 at 1278 (Memorandum Opinion and Order released March 5, 1986 at 1283) (Interconnection Order).

[34] *Id.* (Appendix B at para. 12), quoting In the Matter of an Inquiry Into the Use of the Bands 825845 MHz for Cellular Communications Systems; and Amendment of Parts 2 and 22 of the Commission's Rules Relative to Cellular Communications Systems, 89 FCC 2d 58, para. 51 at 82 (Memorandum Opinion and Order on Reconsideration released March 3, 1982) (Cellular Spectrum Order).

[35] Where immediate provision of Type 2 was technically infeasible, it was to be provided no more than six months after a request was received.

[36] Paging companies typically used Type 1 interconnections.

[37] An end office, also called a Class 5 office, connects directly to end users over subscriber access lines. A tandem, or Class 4, office is not directly connected to end users; instead, it switches traffic from one end office to another, and from end offices to IXCs that elect to interconnect at the tandem switch rather than at the end offices.

[38] FCC Order, *supra*, 2 FCC Rcd. para. 33 at 2914. Type 1 interconnections also do not support direct IXC billing of cellular customers, limit the design of the MTSO and require the cellular carrier to purchase operator services from the LEC. *Id.*

[39] In fact, the LECs offer two variants of Type 2: Type 2A and Type 2B. Type 2A is the arrangement already described, providing a facility linking the MTSO with a Class 4, or tandem, switch. Type 2B, usually purchased in combination with a Type 2A arrangement, connects the MTSO directly with an end office that has an especially heavy flow of traffic to and from the MTSO. (An MTSO in the Chicago area, for example, might want a 2B connection to an end office in the downtown business district.)

[40] "A cellular system operator is a common carrier, rather than a customer or end user, and as such is entitled to interconnection arrangements that 'minimize unnecessary duplication of switching facilities and the associated costs for the ultimate consumer.'" Interconnection Order, *supra*, 59 Rad. Reg. 2d (P&F) at 1284 (Appendix B at para. 2), quoting In the Matter of an Inquiry Into the Use of the Bands 825845 MHz for Cellular Communications Systems; and Amendment of Parts 2 and 22 of the Commission's

Rules Relative to Cellular Communications Systems, 86 FCC 2d 469, para. 56 at 496 (Report and Order released May 4, 1981)(825845 MHz Inquiry).

[41] Interconnection Order, *supra*, 59 Rad. Reg. 2d (P&F) at 1284 (Appendix B at para. 4).

[42] 825845 MHz Inquiry, 86 FCC 2d para. 105 at 511.

[43] 47 C.F.R. § 20.11.

[44] These obligations are described in detail in Chapters 3 and 5.

[45] 47 C.F.R. § 20.11.

[46] Implementation of the Local Competition Provisions in the Telecommunications Act of 1996, Interconnection Between Local Exchange Carriers and Commercial Mobile Radio Service Providers, 11 FCC Rcd. 15499 (1996) at § 1094 (Local Competition Order).

[47] Interconnection Between Local Exchange Carriers and Commercial Mobile Radio Service Providers, 11 FCC Rcd. 5020 (1996) at §§ 60-62.

[48] Local Competition Order at §1041.

[49] *Id.* at §§ 1098, 1117.

[50] *Id.* at § 1117.

[51] *Id.* at § 1118.

[52] *Id.* at §1057. The resulting regulation states that compensable termination includes "the switching of local telecommunications traffic at the terminating carrier's end office switch, or equivalent facility, and delivery of such traffic to the called party's premises." 47 C.F.R. § 51.701(d).

[53] *Id.* at §§ 1085-89. This presumption of symmetry also applies to CLEC transport and termination rates.

[54] *Id.*

[55] MSS service *is* possible from geostationary satellites, however, and the FCC has allocated spectrum for such services.

[56] *See* Zuckman, H. R., Corn-Revere, R. Frieden, and C. Kennedy, *Modern Communication Law*. Vol. 3 at Sec. 15.10, West Group 1999.

9

Internet Service Providers

Internet services are based upon telecommunications, in the specific sense that customers reach and interact with websites and other Internet services primarily over the local telephone network and long-haul facilities provided by telecommunications carriers. Internet services themselves, however, do not fit within the statutory definition of telecommunications (i.e., they are not merely "the transmission, between or among points specified by the user, of information of the user's choosing, without change in the form or content of the information as sent and received" [1]. Instead, Internet services generally are classified as information services, which the Communications Act defines as "the offering of a capability for generating, acquiring, storing, transforming, processing, retrieving, utilizing, or making available information via telecommunications" [2].

Despite this statutory distinction, the legal problems of Internet service providers and telecommunications service providers are difficult to disentangle. Often, a single company provides services of both kinds. Even where this is not the case, Internet service providers and telecommunications companies must work closely together in ways that are aided when each player understands the other's legal concerns. For these reasons, it is appropriate that a book on telecommunications law should also discuss the legal problems of the Internet.

In addressing Internet law, we enter a universe of problems very different from those of the telephone companies that have occupied us so far. Telephone companies, because they do not create, select, or modify the messages that their customers send and receive, are generally not liable for any harm (such as defamation, copyright infringement, or distribution of obscenity) that their customers' communications may cause. In fact, this freedom from responsibility for content has distinguished telephone companies, which are regulated historically under Title II of the Communications Act, from the radio and television broadcasters that are regulated under Title III of the Act. Because broadcasters create or select the information they make available to their listeners, they are subject to the usual kinds of liability imposed upon publishers and distributors of harmful speech [3].

Providers of Internet-based services, of course, are neither telephone companies nor radio and television broadcasters. In their relationship to the content they make available to their customers, they range from passive, "conduit" providers of Internet access to operators of Web sites on which all of the site's content is created by the website operator. In between these extremes lie portal services, hosts of bulletin boards and chat rooms, and other entities that vary widely in their knowledge of, and role in providing, the information available through their services. Not surprisingly, the law has had considerable difficulty in applying traditional notions of liability to the information transmitted and made available by means of these various services.

The legal difficulties of the Internet are further compounded by its use as a medium for the purchase and sale of goods and services. As merchants have moved on-line and e-commerce has become a genuine alternative to "brick and mortar" stores and mail-order catalog sales, the law has been forced to address new problems of jurisdiction, taxation, and contract formation.

Finally, in addition to the legal problems posed by harmful content and e-commerce, all on-line activities are affected by the specter of invasion of customer privacy. This complex set of issues, ranging from interception of e-mail to the use of personal information provided by on-line customers to service providers, is attracting increased legislative attention in the state capitals, Congress, and around the world.

This chapter addresses the principal legal problems of providers of on-line services, which we shall refer to generically as ISPs.

I. Liability for Harmful Content

Although freedom of speech is fundamental to the legal system of the United States, persons subject to U.S. law sometimes find themselves in legal trouble because of words, images, or other communications that they speak, write, or distribute. Notably, it is generally unlawful to publish falsehoods that harm the reputations of others. It also is unlawful to copy works, or use trademarks, that belong to others or to distribute obscene materials. Involvement in any of these activities may result in civil or criminal penalties.

Long before the advent of the Internet, the law faced the task of assigning liability for these kinds of harmful speech to persons who are not the primary or original authors of that speech. For example, the law distinguishes speakers and publishers of defamatory statements from mere distributors of those statements, such as bookstores and libraries. Similarly, the law has developed different rules for direct infringement of copyright, on the one hand, and mere contributory or vicarious copyright infringement on the other. Finally, the First Amendment to the U.S. Constitution protects unknowing distributors of obscene materials from the kind of criminal liability that is imposed upon knowing pornographers.

The Internet complicates this picture by introducing new kinds of information creators and distributors that do not always fit neatly within the traditional categories. It is not immediately clear, for example, whether Internet access providers, website hosts, and bulletin-board service (BBS) operators are publishers or distributors of defamatory statements placed on their services by others. Similarly, it is not clear how these entities should be classified for purposes of assigning liability for copyright infringement or obscene communications distributed through the use of their facilities.

The law has answered these novel questions with a combination of judicial, legislative, and regulatory responses. The following describes these responses as they affect the liability of ISPs for defamation, copyright infringement, trademark infringement, and obscene or indecent communications.

A. Liability of ISPs for Defamation on the Internet

Most of us are familiar with the common-law tort of defamation, which provides civil remedies for the publication of a false statement about another person that injures that person's reputation [4]. Specifically, a successful action for defamation requires proof of the following:

- A false, defamatory statement of or concerning the plaintiff;
- Publication of the statement;
- Resulting injury to the plaintiff;
- Fault, amounting at least to negligence, on the part of the defendant [5].

Persons who post derogatory material about others to Web sites, bulletin-board services, or other Internet locations where those statements will be available to others are subject to the law of defamation. If the statements are about public figures (i.e., persons generally well-known to the public or who have injected themselves into public controversies) then the speaker will not be liable unless he or she knew the statements to be false or was indifferent as to their truth or falsity [6]. Where a speaker's statements are not about public figures, however, mere carelessness concerning the truth or falsity of the statements may result in civil liability [7].

ISPs must be concerned about their possible classification as publishers or distributors of defamatory statements placed on their services and facilities by customers and other third parties. Put briefly, common-law publishers (i.e., newspapers, magazines, television news organizations, and other entities that both promulgate and take editorial responsibility for the works of authors) are no less liable than the authors themselves for statements that meet the elements of a cause of action for defamation [8]. Bookstores, news vendors, libraries, and other "passive conduits" for published material, however, are liable only for the distribution of defamatory material of which they know or have reason to know [9].

In order to apply these principles in cyberspace, it is important to be aware of a handful of published decisions and the provisions of Section 230 of the Communications Act [10]. Of these authorities, the most important is the Act and its declaration that "[no] provider or user of an interactive computer service shall be treated as the publisher or speaker of any information provided by another information content provider" [11]. This terse statutory language raises at least two issues: first, the scope of the class of persons and entities to which it applies; and second, the scope of the protection from liability that Section 230 affords.

The first question is answered by the Act itself, which defines an interactive computer service as "any information service, system, or access software provider that provides or enables computer access by multiple users to a computer server, including specifically a service or system that provides access to the Internet and such systems operated or services offered by libraries or educational institutions" [12]. This definition appears to embrace both "passive" ISPs that act only as conduits for Internet access, and operators of BBSs and similar services that facilitate postings of content by third parties.

The second issue (i.e., the legal significance of the Act's declaration that ISPs and their users are not publishers) is not resolved by the statutory language. Several courts, however, have held that the exemption of Section 230 must be read in sweeping terms.

The first reported decision to interpret Section 230 was *Zeran v. America Online, Inc.* (AOL) [13], in which the court dismissed a claim that AOL was liable for defamatory postings to its electronic bulletin board [14]. The plaintiff had alleged that AOL was liable as a distributor because it knew or should have known that the postings were defamatory. AOL did not dispute that the plaintiff had properly pleaded the elements of distributor liability, but argued that Section 230 of the Act exempted ISPs from such liability. The court agreed with AOL, finding that the Act's determination that ISPs are not publishers precluded, not just strict liability as a publisher, but distributor liability as well. According to the court, distributor liability is not a "common-law tort concept different from publisher liability." Rather, in the court's view, "distributor liability treats a distributor as a 'publisher' of third-party statements where that distributor knew or had reason to know that the statements were defamatory." Therefore, a finding of distributor liability against an ISP or ISP user would violate the Act by treating those entities as publishers. Accordingly, in the court's view, Section 230 of the Act must preempt any state law that mandates a contrary result. The Fourth Circuit Court of Appeals, in affirming the district court's decision, agreed entirely with this view of Section 230 of the Act [15].

The next decision to interpret Section 230 was that of the *U.S. District Court for the District of Columbia in Blumenthal v. Drudge and America Online, Inc.* [16]. In that case, an Internet publication carried on AOL's service had falsely accused Sidney Blumenthal, a White House aide, of physically abusing his wife. Blumenthal brought a defamation suit against Matt Drudge, the author of the article and publisher of the on-line report in which the article was contained. Blumenthal also sued AOL, alleging primarily that AOL's contractual relationship with Drudge gave AOL editorial control over Drudge's publication and made Drudge AOL's agent for purposes of liability. Although Blumenthal presented evidence to support this characterization of AOL's relationship with Drudge, the court nonetheless found that Section 230 posed a complete bar to Blumenthal's claims.

A similar result was reached by the U.S. Court of Appeals for the Tenth Circuit in *Ben Ezra, Weinstein, and Company, Inc. v. America Online Inc.* [17]. In that case, AOL was accused of publishing incorrect information concerning the plaintiff's stock price and share volume. The information had appeared in a "Quotes and Portfolios" area of the AOL service, but the information was collected and furnished to AOL by a third-party provider. The Court of Appeals upheld a district court ruling, dismissing the claims against AOL on summary judgment, on the ground that AOL did not create or develop the challenged information and therefore was protected from liability under Section 230 [18].

Although the *Zeran, Blumenthal,* and *Weinstein* decisions are good news for ISPs, those decisions' expansive interpretation of ISPs' rights still should be treated with some caution. It is at least equally plausible to interpret Section 230 as providing, not that ISPs and their users never can be indirectly liable for defamation, but rather that ISPs and their users cannot be strictly liable as publishers for defamation carried over their facilities. This interpretation, which is more readily supported in the literal language of the CDA, leaves plaintiffs free to prove, if they can, that ISP defendants knew or had reason to know of the defamatory nature of particular communications carried over their facilities.

Two cases, decided before enactment of the CDA, show how the traditional standards of publisher and distributor liability may be applied to ISPs accused of defamation. Taken together, these cases suggest that when the usual, common-law rules are applied, ISPs that exercise editorial control will be subject to liability for defamatory statements carried over their facilities, while ISPs that disclaim—and avoid exercising—editorial control may be liable only for defamatory statements of which they knew or had reason to know.

In *Cubby, Inc., et al. v. CompuServe, Inc., et al.,* ("Cubby") [19], a U.S. district court dismissed a libel claim brought against CompuServe by an electronic publisher of news and gossip concerning the broadcasting industry. The plaintiff alleged that a competing on-line gossip publication had defamed the plaintiff through statements carried over CompuServe's service. On a motion for summary judgment, the court found that CompuServe had no editorial control over defendants' publications. Specifically, the court found that the defamatory statements originated with one Don Fitzpatrick Associates (DFA) and were carried on an independently managed service called the Journalism Forum, managed by a company called Cameron Communications, Inc. (CCI). CompuServe had no contractual relationship with DFA, and had a contract with CCI that expressly gave CCI complete editorial control over the Journalism Forum. On these facts, the court found that CompuServe could be liable, at most, as a distributor of the alleged defamation; and that, in the absence of any evidence that CompuServe knew or had reason to know of the allegedly defamatory statements, the defamation claim against CompuServe must be dismissed.

Somewhat different facts were presented in the later case of *Stratton-Oakmont, Inc., et al., v. Prodigy Services Company, et al.,* ("Stratton-Oakmont")[20]. In Stratton-Oakmont, an investment firm alleged defamation by an unknown user of a computer bulletin board who posted statements accusing the plaintiff of criminal fraud. The bulletin board was operated by Prodigy, which employed a "board leader" who was instructed to screen material posted to the bulletin board according to editorial standards promulgated by Prodigy. The court found that Prodigy, unlike CompuServe in the Cubby case, "held itself out to the public as controlling the content of its computer bulletin boards" and "implemented this control through [an] automatic software screening program, and guidelines which board leaders are required to enforce." On these facts, the court found that Prodigy was a publisher and therefore liable for defamatory material that appeared on the bulletin board, regardless of its lack of actual knowledge of the nature of that material [21].

In summary, although some courts have found that Section 230 of the Communications Act provides a complete defense for ISPs to defamation suits based on third-party statements, ISPs still should be cautious about exercising editorial control over materials placed on their services by third parties.

B. ISP Liability for Copyright Infringement on the Internet

At any given time, a substantial quantity of material moving over the Internet is there in violation of someone's copyright. In order to understand how ISPs may avoid liability for copyright infringement, it is first necessary to offer a very brief summary of copyright law.

1. Some Copyright Basics

A copyright, as the name suggests, is the exclusive right to copy something. The "something" is any work of authorship—a category that includes literary works, paintings, songs, motion pictures, and other creative products—and the exclusive right to copy the work is created when the author first puts the work into tangible form (a process known in copyright law as "fixation"). Contrary to popular belief, copyright does not require registration of the work with the Copyright Office or placement of a notice on the work (although there are good reasons to do those things) [22]. Once the author has fixed his or her work in tangible form, the copyright exists. No one else may copy that work without the permission of the author or a subsequent holder of the copyright. More precisely, the Copyright Act provides that no one but the copyright holder may:

1. Reproduce the work;
2. Prepare derivative works based upon the work;
3. Distribute copies of the work to the public;
4. Display the work in public;
5. Perform the work in public [23].

Violations of copyright may be direct or indirect. Direct copyright infringement is a strict liability offense involving the unauthorized copying, distribution, display, or performance of a copyrighted work, or the unauthorized creation of a derivative work therefrom. Direct infringement is established when the plaintiff proves that he or she owned the copyright in a work, that the defendant had access to the work, and that the defendant either copied the work or (where direct evidence of copying is not available) that the allegedly infringing work is substantially similar to the original.

ISPs are unlikely to be successfully charged with direct infringement when they carry or distribute materials provided by others, but they may be charged with indirect violations under the doctrines of contributory and vicarious infringement. Contributory infringement is established when a defendant induces, causes, or materially contributes to an act of direct infringement committed by another, and does so with knowledge of the infringing activity. Vicarious infringement is an offense committed by a defendant who may lack knowledge of the infringing activity of another, but who has the "right and ability to supervise" that activity and has a direct financial interest in the act of infringement.

2. ISP Liability for Direct, Contributory, and Vicarious Infringement

ISPs provide facilities through which billions of bits of information move. Those bits may contain electronic mail, pictures, bulletin-board postings, poems, novels, or full-length movies. Imposition of direct liability on ISPs, where infringing material is posted to or transmitted through the ISPs' facilities by others, is somewhat controversial. Two U.S. district court opinions have suggested that such liability may be imposed, but the better view, adopted by at least two other courts, is that such liability is inappropriate.

In the 1994 decision in *Sega Enterprises Ltd. v. MAPHIA,* [24] the court found a substantial likelihood that the operator of a bulletin board had infringed the plaintiff's copyrights in video game software by soliciting users of the BBS to post Sega software to the service for others to download. Although the facts and the court's opinion supported a finding of contributory liability for the customers' infringing acts, there was dictum in the opinion to suggest that the court had found evidence of direct liability, as well. In *Playboy v. Frena,* [25] another U.S. district court stated that a BBS operator whose customers uploaded and downloaded copyrighted images scanned from magazines might be directly liable because the operator "supplied a product containing unauthorized copies of a copyrighted work" [26]. Taken together, the Sega and Playboy cases suggest, although without any analytical basis, that an ISP that merely provides a facility over which infringing information is transmitted, and that has no knowledge that the infringement is occurring, may nonetheless be liable for that infringement.

Two cases subsequent to Sega and Playboy, however, offer persuasive authority against this result. In *Religious Technology Center v. Netcom Online Communication Services, Inc.* ("Netcom") [27], a U.S. district court disagreed with the apparent rationale of the Sega and Playboy decisions. As the Netcom court quite sensibly pointed out, holding an ignorant ISP directly liable for infringement by a customer would erase the distinction between direct liability, on the one hand, and contributory and vicarious liability, on the other [28]. Similarly, in its 1996 decision disposing of the case on the merits, the Sega court determined that the BBS provider in that case was only contributorily (not directly) liable for the acts of infringement committed by others who uploaded and downloaded copyright game software through his service [29].

At present, therefore, liability of users and ISPs alike for copyright infringement is governed primarily by the rules of contributory and vicarious infringement. Under the rules concerning contributory liability for copyright infringement, ISPs must not knowingly permit their facilities to be used for distribution of copyrighted materials, and should insist that their customers discontinue infringing activities of which the ISPs become aware. Under the rules of vicarious liability for copyright infringement, ISPs should be aware that where their contracts with their customers give them the right to supervise the customers' use of ISP facilities, a copyright plaintiff may seek to prove that the ISP also had the practical ability to prevent transmission of infringing material and that the ISP in fact profited from such infringing transmissions. Evidence of the ISP's practical ability to supervise may involve testimony concerning the availability of blocking software that can screen for particular Internet users or key words contained in infringing information; evidence of financial benefit to the ISP might include proof that the ISP charges on a usage, rather than flat-rate, basis for Internet access, so that a large number of infringing transmissions will bring the ISP greater income.

3. The Digital Millennium Copyright Act

Certain provisions of the Digital Millennium Copyright Act (DMCA) give ISPs limited protection from liability for infringing materials that appear on their services and facilities [30]. To qualify for that protection, ISPs must adopt and implement policies providing for termination of users that repeatedly infringe copyrights, and must accommodate and not interfere with technical measures adopted by copyright holders to protect their works. So long as an ISP complies with these requirements, it will not be liable for passive transmission and routing of material originated by third parties. ISPs also will not be liable for infringing materials that they store, cache, or provide hypertext links to, so long as those ISPs comply with the DMCA's "notice-and-takedown" requirements.

The notice and takedown requirements are at the heart of the DMCA's ISP protection scheme. When an ISP receives specific written notification, made under penalty of perjury, that infringing material is stored on, cached on, or linked through the ISP's facilities, that ISP must disable or block access to the allegedly infringing material. The DMCA does not require the ISP to investigate or ascertain the merits of the notifying party's claim under copyright law; so long as the notification meets the formal standards of the statute, the ISP must block or disable access to the material and may not be sued by its customer for doing so.

The DMCA also gives the targets of notice-and-takedown complaints a limited opportunity to have access to their materials restored. Specifically, an ISP that receives notice of infringement under the DMCA must take prompt and reasonable steps to notify its customer that the ISP has blocked or disabled access to the customer's material. Upon receipt of that notice from its ISP, the customer may send the ISP a "counter notification" stating that the removal or blocking was the result of error or misidentification of the allegedly infringing materials. If the counter notification meets the formal requirements of the DMCA, the ISP is required to forward the counter notification to the copyright holder that sent the original takedown notice. Unless the copyright holder then notifies the ISP that the holder has filed a court action seeking to restrain the alleged infringement, the ISP must replace or unblock the material within 10 to 14 days of receiving the customer's counter notification.

For ISPs, the great value of the DMCA lies in the statute's clarification of an ISP's obligations when confronted with a claim of copyright infringement. Before enactment of the DMCA, ISPs often felt some pressure to investigate the merits of infringement claims concerning their customers, or simply to terminate the privileges of customers about whom complaints were received. The DMCA reduces the likelihood of frivolous copyright complaints, gives ISPs clear guidance as to the claims that require a response, and eliminates any need for ISPs to act as the arbiters of the merits of those claims under the Copyright Act.

At the same time, the DMCA does not immunize ISPs from liability for infringing activity in which these ISPs are involved in ways that meet the definitions of contributory or vicarious infringement. Accordingly, ISPs should avoid activities that actively contribute to infringement and should be prepared to terminate the privileges of customers that ISPs know to be involved in infringing activities.

C. ISP Liability for Trademark Infringement

Trademark law in the United States is primarily federal and based on a statute popularly known as the Lanham Act [31]. Trademark protection extends to any names, symbols, and other devices used by makers and vendors of goods to distinguish their products from those made and sold by others [32]. If a trademark is distinctive and does not present a likelihood of confusion with an existing trademark, then it may be registered with the Patent and Trademark Office. Once registered, and so long as the trademark is not abandoned, the trademark owner may bring an action against anyone who uses, in commerce, a copy or imitation of the trademark that is likely to cause confusion, mistake, or deception [33].

When a plaintiff brings an action for trademark infringement and establishes that its trademark is valid and enforceable, the principal issue is likely to be the "likelihood of confusion" caused by the defendant's use of a device identical or similar to the plaintiff's trademark. In order to determine whether a defendant's use of a trademark is likely to cause confusion, the trier of fact (i.e., the jury or, in the case of a trial not involving a jury, the judge) must consider:

- The similarity of the plaintiff's and defendant's marks;
- The similarity of the goods or services in connection with which the marks are used;
- The nature of the markets in which the plaintiff's and defendant's goods or services are sold;

- The strength of the plaintiff's mark;
- The defendant's intent [34].

Trademark problems have arisen on the Internet in two principal ways. First, some ISPs and ISP customers have posted trademarked materials to the Internet, raising questions of potential liability for trademark infringement and unfair competition. Second, persons have registered and used World Wide Web domain names that arguably violated trademarks belonging to others.

As with copyrights, ISPs have been accused of both direct and indirect infringement of trademark rights. The principal case involving a claim of direct infringement is *Playboy Enterprises, Inc. v. Frena* [35], in which subscribers to a bulletin-board service posted images from Playboy magazine on the BBS for others to download. The BBS operator stripped Playboy's trademark from some of the uploaded materials and replaced it with his own name and address, and left the Playboy trademark on other materials. According to the court, both actions violated the Lanham Act. When the BBS operator replaced the Playboy trademark with his name and address, he created confusion as to the origin of the photographs. It was held that when the BBS operator kept the Playboy trademark on other photographs, he wrongly suggested that Playboy had authorized posting of the photographs to the BBS [36].

ISPs also have been accused of indirect trademark infringement, based upon claims that an ISP knew or had reason to know of the infringing character of material transmitted through or posted to its facilities by its customers or other third parties. In order to sustain such a charge, however, a complainant must prove more than that the ISP received a claim of infringement. In *Watts v. Network Solutions, Inc.,* for example [37], a plaintiff charged Network Solutions, Inc. (NSI) with contributory trademark infringement when NSI refused to deactivate an allegedly infringing domain name. As the court in that case pointed out:

> Watts (the plaintiff) asserts that NSI knew or had reason to know by virtue of this repeated correspondence with NSI that Schwab (the registrant of the domain name) is engaged in trademark infringement and unfair competition. The undisputed facts show nothing of the sort. The evidence shows at most only that NSI knew that Watts had a dispute with Schwab, but that Watts had not taken any serious or effective steps to resolve that dispute directly with Schwab. Watts' bald assertion of

infringement imposed no obligation on NSI to referee their dispute or to grant what would be comparable to a private preliminary injunction against Schwab: "The mere assertion by a trademark owner that a domain name infringes its mark is not sufficient to impute knowledge of infringement to NSI" [38].

As the Watts decision makes clear, ISPs are not required to terminate their customers or take other corrective action on the basis of "bald assertion[s] of [trademark] infringement" unaccompanied by any reference to "serious or effective steps" taken against the alleged infringer. ISPs should avoid involvement with infringing activities of which they have specific knowledge and from which they obtain a financial benefit; but they are not required to "referee dispute[s]" between their customers or other third parties, on the one hand, and alleged trademark holders, on the other.

The *Watts v. Network Solutions* case brings us to the other, principal source of Internet-based trademark disputes (i.e., the registration of Web domain names that are alleged to infringe trademarks). These disputes have their origin in the practice of "cyberpiracy" or "cybersquatting" that is, the registration of valuable domain names, including the trademarks of well-known corporations, in order to secure payments from those corporations in exchange for surrendering the cybersquatters' registrations of these domain names.

When the practice of cybersquatting first appeared, some companies paid the cybersquatters rather than incur the cost and delay of litigating their trademark rights. Others demanded that NSI, at that time the sole registry of World Wide Web domain names, cancel the offending registrations or transfer those registrations to the alleged trademark holders. In response to these demands, NSI adopted a practice of holding domain names in abeyance until the contending parties could obtain a judicial resolution of their competing claims. This approach, which did not spare trademark holders the cost of litigating their rights and imposed costs of litigation on domain name holders that might not be infringers, satisfied no one.

As a result of these disputes, both the U.S. Congress and the newly created Internet Corporation for Assigned Names and Numbers (ICANN) adopted procedures for the summary resolution of domain name disputes. Under the ICANN procedure, trademark holders may invoke an arbitration process and obtain the limited remedies of cancellation or transfer of domain names that were registered in bad faith. Under the anticybersquatting statute enacted by Congress, trademark holders may obtain an award of damages and, where appropriate, an injunction against bad faith registration of an infringing domain name.

D. ISP Liability for Obscenity and Indecency on the Internet

Many people use the Internet, as they have used other media throughout history, as a means of transmitting and acquiring pornographic and otherwise offensive text and images. Both state and federal authorities have sought to impose liability for these activities on Internet users and content providers [39]. In order to make sense of the somewhat tangled condition of the law concerning pornography and indecency on the Net, it is important to know something about the law of obscenity and indecency as that law has developed in connection with more traditional media.

Most fundamentally, obscenity and indecency differ in that the latter enjoys First Amendment protection and the former does not. Obscenity, which is roughly coextensive with hard-core pornography, is not constitutionally protected [40]. Accordingly, state and federal authorities may criminalize and punish the distribution of obscene material. Indecent material, on the other hand, is merely offensive. Dissemination of indecent material may not be forbidden altogether; it may only be restricted to the extent necessary to protect minors or serve some other compelling state interest [41].

Both states and the federal government have effective laws against the posting, transmission, and retrieval of obscenity and child pornography through use of the Internet. However, as a result of judicial action to overturn two federal statutes, there is no effective federal law prohibiting the posting or transmission of merely indecent material on or over the Internet. In order to understand the state of the law concerning these kinds of offensive material, it is useful to review both the two failed federal statutes concerning Internet indecency, and the state and federal antiobscenity laws.

1. Indecent Material on the Internet: The Communications Decency Act and the Child On-Line Protection Act

Ever since the emergence of the Internet as a medium of widespread public communication, legislators have tried to create statutory barriers between children and the many kinds of indecent material that can be obtained on websites, through BBSs and chatrooms, from usenet newsgroups and from other on-line sources. At the federal level, these efforts resulted in two statutes: the Communications Decency Act (CDA) [42] and the Child On-line Protection Act (COPA) [43].

The first of these statutes, CDA, was passed into law in February, 1996 as part of the Telecommunications Act of 1996. The CDA provided that:

"(A) Whoever—

"(1) in interstate or foreign communications—

"(B) by means of a telecommunications device knowingly—

"(i) makes, creates, or solicits, and

"(ii) initiates the transmission of, any comment, request, suggestion, proposal, image, or other communication which is obscene or indecent, knowing that the recipient of the communication is under 18 years of age, regardless of whether the maker of such communication placed the call or initiated the communication;

"(2) knowingly permits any telecommunications facility under his control to be used for any activity prohibited by paragraph (1) with the intent that it be used for such activity, shall be fined or imprisoned not more than two years, or both" [44].

The framers of the CDA were aware that under the First Amendment, Congress could not lawfully prevent adults from having access to merely indecent (rather than obscene) materials, and the CDA accordingly permits ISPs to offer indecent material to adults if they take, in good faith, "reasonable, effective, and appropriate actions" to prevent access by minors [45] or take measures to confirm that persons accessing indecent materials are adults. For example, by requiring credit card access, adult access codes, or personal identification numbers [46].

The CDA was challenged in the federal courts almost immediately upon its enactment, but only as to those provisions that relate to indecent, rather than obscene, material. Two three-judge panels agreed that the CDA's indecency provisions violate the First Amendment [47], and the Supreme Court overturned the CDA to the extent it applied to indecent communication [48].

The second federal statute aimed at on-line indecency, COPA, was designed to cure some of the constitutional defects of the CDA. Notably, COPA included a detailed definition of the kind of material it proscribed, purported to apply only to commercial content providers and included affirmative defenses for sites that used "technologically and economically feasible method[s] to restrict the access of minors to harmful materials."[49] These arguments notwithstanding, a U.S. district court found that COPA did not represent the least restrictive means of regulating the access of minors to indecent on-line material [50]. The district therefore entered an injunction against enforcement of COPA, and that decision was upheld by the U.S. Court of Appeals for the Third Circuit [51].

Because of the judicial rejections of the CDA and COPA, there is no effective federal statute that restricts ISPs and content providers from providing merely indecent materials over the Internet.

2. Obscenity and Child Pornography on the Internet

The Supreme Court's rejection of the Communications Decency Act should not obscure the continuing liability of ISPs, under both state and federal law, for the distribution of obscene materials and child pornography.

The federal obscenity statute, as amended by provisions of the CDA that have not been overturned or challenged as unconstitutional, makes it a crime to do any of the following with respect to "obscene, lewd, lascivious, or filthy matter":

- Import such materials or knowingly use a common carrier or interactive computer service to carry such materials in interstate or foreign commerce [52];
- Take or receive such materials from a common carrier or interactive computer service [53];
- Use an interactive computer service in or affecting interstate or foreign commerce for the purpose of sale or distribution of such materials [54].

Under this federal statute, ISPs may be liable for any obscene materials they make available on BBSs, on Web sites, through usenet newsgroups, or otherwise. Also, as the courts have confirmed, ISPs that make obscene materials available are subject to prosecution according to the local community standards of any locality in which those materials are received [55].

Although the potential liability of content providers for obscene material and child pornography transmitted over the Internet is reasonably clear, the exposure to prosecution of ISPs that merely provide access for those content providers is less certain. Historically, the biggest obstacle to obscenity prosecution of ISPs has been the scienter requirement—that is, the constitutional requirement that persons accused of distributing obscene materials must have knowledge of the material they are accused of distributing. The Supreme Court has expressed the requirement in these words:

> The Court accepts the settled principle that traffic in obscene literature may be outlawed as a crime. But it holds that one cannot be made amenable to such criminal outlawry unless he is chargeable with knowledge

of the obscenity. Obviously the Court is not holding that a bookseller must familiarize himself with the contents of every book in his shop. No less obviously the Court does not hold that a bookseller who insulates himself against knowledge about an offending book is thereby free to maintain an emporium for smut. How much or little awareness that a book may be found to be obscene suffices to establish scienter, or what kind of evidence may satisfy the how much or the how little, the Court leaves for another day. [56]

Scienter is difficult to prove against an ISP that offers only a conduit to the Internet and neither provides content nor monitors the billions of bytes moving through its system to ascertain their content. Accordingly, Internet-related obscenity prosecutions have been aimed at content providers, who cannot plausibly claim ignorance of the nature of the materials they post and make and available for downloading.

Some antipornography advocates, however, have urged prosecutors to target ISPs for enforcement actions under the obscenity laws. These groups see prosecution of ISPs as an efficient strategy that will shut down many content providers at once—particularly because a few well-publicized cases might persuade ISPs to censor those services to which they provide access. In order to deal with the scienter requirement, these groups are urging prosecutors to bring cases alleging that ISPs have knowledge of the nature of obscene materials to which they provide access where:

1. The ISP accepts postings from hierarchies such as alt.binaries.pictures.alt.sex and alt.binaries.pictures.erotica, the names of which may be said to suggest the nature of the material posted to those newsgroups; or
2. The ISP has received a letter notifying it that obscene material is carried on its facilities.

Although criminal prosecutions based on scienter evidence of this kind will stretch the limits of the First Amendment and ultimately could fail for that reason [57], ISPs should be alert to any indication that they have been targeted for prosecution by antipornography activists or local prosecutors. ISPs also should consider whether they will reduce their risk of prosecution by refusing to host Web sites or accept postings from newsgroups that are likely to contain obscene materials or child pornography.

Purveyors of child pornography, like purveyors of pornography generally, have found the Internet an attractive medium. Distribution of child pornography is prosecuted primarily under the Protection of Children Against Sexual Exploitation Act of 1977, which makes it a felony knowingly to do any of the following:

- Transport or ship in interstate or foreign commerce by any means, including a computer, any visual depiction of conduct that involves the use of a minor engaging in sexually explicit conduct;
- Receive or distribute any visual depiction of conduct involving the use of a minor engaged in sexually explicit conduct where that depiction has been mailed, shipped, or transported in interstate or foreign commerce by computer or other means;
- Sell or possess with intent to sell any visual depiction involving the use of a minor engaged in sexually explicit conduct where that depiction has been mailed, shipped, or transported in interstate or foreign commerce;
- Possess books, magazines, periodicals, videotapes, or other matter containing any visual depiction of conduct involving the use of a minor engaging in sexually explicit conduct where those materials have been mailed, shipped, or transported in interstate or foreign commerce, or produced using materials mailed, shipped, or transported in interstate or foreign commerce, by computer or other means [58].

Under these statutes, ISPs face potential criminal liability if they make available on Web sites, or through usenet newsgroups or otherwise, any photographs of minors engaged in sexual activity. ISPs should particularly be aware that materials may violate the child pornography statute even where those materials do not satisfy the legal definition of obscenity. ISPs also should be aware that although proof of an offense under the statute requires evidence that the ISP knew the persons depicted to be minors [59], ISPs cannot protect themselves merely by failing to confirm the ages of the persons depicted. The prosecutor in a child pornography case may be required to prove no more than that the persons depicted in the offending materials appear to be under eighteen years of age.

Many states also have enacted statutes that regulate the distribution and receipt of obscene materials and child pornography within their borders. ISPs should be aware that when persons download materials from Web sites and BBSs, they give the states in which they are located jurisdiction over the ISPs that originated those materials for purposes of prosecution under state obscenity and child pornography laws.

Although liability for obscene Internet communications and child pornography are primarily concerns of ISPs and content providers, users should be aware of their potential liability when they access and download such materials to their own computers. First, as to accessing obscene communications posted to the Internet, the Supreme Court has held that persons may not be prosecuted for mere possession of obscene material. However, it is conceivable that a prosecutor might seek to show that an Internet user's act of downloading pornography constitutes an act of distribution and therefore can give rise to liability. And child pornography laws, which forbid the knowing receipt of pornographic images involving children, clearly may be applied against persons who download child pornography from the Internet.

E. Liability of ISPs for Spamming

Some entrepreneurs have discovered that mass, unsolicited electronic mail can be a cost-effective method of marketing products and services. Unfortunately, this marketing device, condemned as "spam" by most Internet users, is a source of irritation to its unwilling recipients and a burden on ISP facilities. Those who send spam, therefore, can expect resistance and should be aware of their rights and liabilities.

Most fundamentally, it is not unlawful per se to send spam, but spammers may be sued under various legal theories if their activities cause harm to ISPs or others. For example, if spam clogs an ISP's servers or prevents the ISP's subscribers from gaining access to other messages or services, the ISP may have a cause of action under state and federal antihacking statutes. Notably, under amendments to the federal Computer Fraud and Abuse Act (CFAA) adopted in 1994, certain claims may be brought without proof that a spammer gained access to an ISP's facilities without authorization [60]. So long as the spammer's actions cause at least $5,000 in economic damages during any one-year period—a threshold that should not be hard for a large ISP confronted with a major spamming campaign to meet—a private action may be brought against the spammer [61]. The relief available under the CFAA includes injunctions and other equitable remedies, as well as damages [62].

In addition to remedies available under antihacking statutes, spammers may be met with tort claims based upon alleged harm to ISP facilities, software, and customer goodwill. In one case, a federal district court entered an injunction preventing CyberPromotions, Inc. from sending any unsolicited e-mail to any e-mail address maintained by CompuServe [63]. In support of its decision, the court found that CompuServe could establish a claim against CyberPromotions for trespass to chattels, which is defined as the intentional use of, or intermeddling with, a chattel in the possession of another.

CyberPromotions responded to CompuServe's claim by arguing that trespass to chattels cannot be established unless a defendant "actually takes physical custody of the property or damages it" [64]. The court, however, found that a trespass occurs whenever a spammer's conduct diminishes the value of an ISP's property [65]. In the case before it, the court found that CyberPromotions' electronic transmissions to CompuServe's facilities were sufficiently tangible to constitute a trespass. The court further found that "the enormous volume of mass mailings that CompuServe receives places a tremendous burden on its equipment," with the result that data processing and storage resources are diverted from serving CompuServe's subscribers [66]. Similarly, the court found that receipt of unwanted e-mail caused harm to "plaintiff's business reputation and goodwill with its customers" [67]. These forms of injury, while concededly involving no physical damage to CompuServe's equipment, diminished the value of the equipment to CompuServe and satisfied the injury element of the tort.

Spammers should be aware that the courts have generally been sympathetic to recipients of spam and the ISPs that seek to prevent spammers from reaching their subscribers. Spammers also should be aware that claims of an ISP obligation to deliver spam, based on First Amendment grounds or an assertion that ISPs have common carrier obligations, have so far been rejected by the courts.

II. Jurisdiction over ISP Activities

Simply put, jurisdiction is the "power of the court to decide a matter in controversy" [68]. Because the Internet permits an entity to engage in numerous activities almost anywhere from almost anywhere, courts have faced difficult questions of jurisdiction over individuals or companies engaged in disputes concerning on-line activity. A number of cases, however, have shown that traditional notions of jurisdiction are sufficient to resolve these questions.

Jurisdiction is of two kinds: jurisdiction over the subject matter and jurisdiction over the person. For most cases involving the Internet, the difficult question is the ability of a court to take jurisdiction over a person that does not reside within the state where the court sits and, in many instances, has had only limited contacts with that state. A court may exercise jurisdiction in such cases only if: (1) the law of the state in which the court sits confers such jurisdiction, and (2) the exercise of jurisdiction comports with the Due Process clause of the Fourteenth Amendment [69]. The first requirement is typically satisfied, if at all, by a state's "long-arm statute," which describes the circumstances under which the state court may exercise jurisdiction over a nonresident [70]. Most states' long-arm statutes permit jurisdiction to be exercised to the full extent consistent with constitutional due process or are broadly written to confer jurisdiction in the majority of cases [71]. Accordingly, Due Process considerations rather than statutory language frequently become the decisional factors.

The Due Process test itself has two basic requirements: (1) the defendant must have certain "minimum contacts with [the forum state]," and (2) exercise of jurisdiction over the defendant must be consistent with "traditional notions of fair play and substantial justice"[72]. Under this "minimum contacts" approach, the defendant's contacts with the forum state "must justify a conclusion that the defendant should reasonably anticipate being hailed into court there" [73]. In addition, the contacts in question must have a basis in "some act by which the defendant purposefully avails itself of the privilege of conducting activities within the forum state, thus invoking the benefits and protections of its laws" [74].

Minimum contacts analysis takes into account both general and specific jurisdiction. Specific jurisdiction exists where the action arises from the defendant's purposeful forum-related activities (i.e., the activities establishing contact with the forum are the same as those over which suit was brought). For specific jurisdiction, the relationships between the forum, the defendant, and the litigation are examined to determine whether maintaining the suit is fair and just. General jurisdiction exists where the action at issue does not arise out of the defendant's contacts with the forum state, but where those contacts are otherwise continuous and systematic, supporting a reasonable exercise of jurisdiction by the court [75].

Applying these principles, a number of federal courts have held that commercial Internet activities available to citizens of a state were sufficient to create jurisdiction in the forum state. In *CompuServe, Inc. v. Patterson* [76], the defendant subscribed to CompuServe, an Ohio on-line service provider, and entered into an agreement with CompuServe to sell shareware products on the CompuServe system. Noting that although subscription to CompuServe by itself would be insufficient to establish jurisdiction, the court held that Patterson had availed himself of the privilege of doing business in Ohio by repeatedly sending software to Ohio for sale and by actually selling to Ohio residents.

Similarly, in *Inset Systems, Inc. v. Instruction Set, Inc.* [77], the court held that a toll-free phone number and advertisements on an Internet Web site were sufficient contacts to warrant jurisdiction in Connecticut. And in *Maritz, Inc. v. Cybergold* [78], the U.S. District Court for the Eastern District of Missouri determined that a Web site advertising future services that had been accessed by Missouri residents provided sufficient contacts to assert jurisdiction. In these last two cases, both trademark infringement actions, the courts each determined that "through their websites, the defendants consciously chose to transmit advertisements to all Internet users, including those in the forum state and purposefully availed themselves of the benefits and privileges of doing business within the forum state" [79].

In other instances, courts have found Web-based activities to be insufficient to establish jurisdiction. In *Bensusan Restaurant Corp. v. King* [80], the U.S. District Court for the Southern District of New York granted a motion to dismiss a trademark infringement case because activities on the Web did not establish sufficient contacts with the State of New York to support jurisdiction. The owner of the registered trademark Blue Note filed suit after learning that the owner of a Missouri nightclub had created a Web site promoting the club. The court found that "[c]reating a site, like placing a product into the stream of commerce, may be felt nationwide—or even worldwide—but, without more, it is not an act purposefully directed toward the forum state" [81]. The court also found that exercising jurisdiction based solely on the Web advertisements would not satisfy the minimum contacts requirement of the Due Process clause [82].

The cases decided to date "reveal that the likelihood that personal jurisdiction can be constitutionally exercised is directly proportionate to the nature and quality of commercial activity that an entity conducts over the Internet." [83] Accordingly, jurisdiction ordinarily will be appropriate over a defendant that transacts business over the Internet or enters into contracts with residents of a foreign jurisdiction that involve or require continual transmission of computer files over the Internet [84]. At the same time, jurisdiction should not be found where a defendant has simply posted information on an Internet website that is accessible by users in a foreign jurisdiction [85]. Interactive websites through which a user can exchange information with the host computer occupy the middle of the scale. The level of interactivity and the commercial or noncommercial nature of the site will determine the appropriateness of asserting jurisdiction in these cases [86].

III. ISPs and the FCC: Regulation of the Internet

Today, the Internet is not subject to economic regulation (i.e., no government agency now tells ISPs and Internet-based businesses how much they may charge for their services or requires them to take on the traditional obligations of telephone companies and other common carriers). But Internet users and service providers, and their attorneys, should know that there is persistent pressure to change this approach.

The FCC now classifies ISPs as enhanced service providers—a category that includes any telecommunications service that stores, retrieves, alters, or otherwise deals with information in a way that goes beyond mere switched or dedicated transmission [87]. The FCC distinguishes enhanced services from basic services, which simply transmit information chosen by the sender without changing the information's form or content. So, for example, local and long-distance telephone service are basic; voice messaging, on-line data processing, and Internet access are enhanced.

Because they are ESPs rather than providers of basic telecommunications services, ISPs are relieved of certain regulatory obligations that otherwise would apply. Notably, local telephone companies must treat ISPs as end-user customers rather than carriers. For this reason, ISPs do not pay the costly access charges that long-distance telephone companies pay the local telephone companies to originate and terminate calls. Also, because ISPs are regarded as customers rather than carriers, they have not been required to file tariffs setting out their rates and conditions of service, and are not scrutinized by the FCC to ensure that their rates are not "unreasonable" or "unreasonably discriminatory."

Of these various advantages of ESP status, by far the most important is the ESPs' freedom from the necessity to pay access charges. The long-distance companies have stated, without contradiction, that access charges paid to local telephone companies account for up to 40% of their cost of providing long-distance service. Imposition of these same charges on ISPs would drastically change the cost structure of the Internet access business.

Not surprisingly, the local telephone companies have argued strenuously that ISPs should pay access charges. As the Internet has grown in importance, local telephone facilities increasingly are used to connect customers with their ISPs. The telephone companies claim that the growth in time spent on-line is burdening their switches and imposing costs that the telephone companies cannot recover unless they are paid access charges by the ISPs.

The FCC so far has declined to extend access charge obligations to the ISPs, and for now ISPs will continue to be classified as enhanced service providers. The FCC has opened an inquiry into the future regulatory treatment of Internet access services, however, and at some point—probably after the FCC has completed its present effort to reform the access charge system and reduce the level of those charges—the FCC may consider extending a modified access charge payment system to the ISPs [88].

IV. Privacy and Data Security on the Internet

Privacy and data security are important values for most Internet users. Electronic mail may contain sensitive or even intimate communications intended only for the addressee. On-line transactions may involve transmission of credit card numbers and other commercially valuable information. And personal information stored in databases may be subject to on-line access, destruction, and alteration. For these and other reasons, users of the Internet and on-line services are justifiably concerned about their rights to protect stored and transmitted information.

This section considers privacy from two broad perspectives. First, we consider the circumstances under which Internet transmissions may be intercepted—that is, monitored or recorded as they are taking place. Second, we consider the circumstances under which governmental and private parties may use the Internet, or other avenues of on-line access, to acquire stored personal information. In the section of this chapter that deals with problems of electronic commerce, we discuss a third privacy-related question (i.e., the ability of on-line services to collect, use and disclose personal information concerning their customers.

A. Interception of Internet Communications

The circumstances under which persons may intercept Internet communications—that is, acquire those communications as they are being transmitted—are defined primarily by Title I of the Electronic Communications Privacy Act (ECPA) [89]. Title I of the ECPA prohibits anyone from intentionally intercepting an electronic communication, and defines "electronic communication" in a way that includes electronic mail, postings to BBSs, on-line commercial transactions, and all other electronic transmissions of information [90].

The ECPA's prohibition on interceptions of electronic communications also contains numerous exceptions, however, and some of those exceptions are quite important. Notably, it is not unlawful under the ECPA to intercept user information that identifies the parties to an on-line communication or transaction [91]. Also, ISPs are permitted to intercept, disclose, or use the communications of their customers (or any other communications carried over their facilities) as necessary for "the rendition of his service or to the protection of the rights and property of the provider of that service" [92]. The ECPA also permits persons to intercept communications that are readily accessible to the general public, and permits third parties to intercept a communication so long as one party to the communication consents to the interception [93].

The ECPA also permits state and federal law enforcement agencies to intercept electronic communications, so long as certain due process requirements are observed. Unfortunately for persons using electronic mail and other on-line services, the limitations on law enforcement interceptions of electronic communications are less stringent than those for voice (wire and oral) communications [94]. Specifically, interception orders aimed at electronic communications may be obtained by lower level officials, in connection with a wider range of suspected offenses, than interception orders for wiretaps of telephones and electronic listening devices [95]. Similarly, the ECPA does not require evidence seized through an unlawful interception of an electronic communication—as opposed to a wire or oral communication—to be suppressed at trial [96].

In summary, users may bring civil actions against those who intercept, use, or disclose the users' e-mail or other on-line communications, so long as those communications are not publicly available transmissions such as postings to unrestricted BBSs. Relief may not be available, however, where the ISP or other system operator claims that the interception was required to provide service or protect the operator's property, or where the interception was made in compliance with a properly obtained interception order. Relief also may not be available where one party to an e-mail, or other two-party communication, consented to the interception. Finally, users' rights under the ECPA may be waived or limited by the user agreement with the ISP or system operator.

B. Access to Computers and Stored Information

Communications over the Internet and on-line services permit persons to feed information to remote databases and retrieve information from those databases. They also permit persons to gain access to computers' operating systems, and in some cases to alter, damage, or destroy those systems. When users send personal information to remote databases, they probably hope that that information will be used only for its intended purpose, and that it will not be disclosed to others. Owners of computers, operating systems, and stored information also may hope that those valuable resources will be safe from on-line vandals and thieves. Unfortunately, these expectations often are frustrated, and legal protections are spotty at best.

1. The ECPA

We already have discussed those sections of the ECPA that protect against unauthorized interception of electronic communications. Other sections, contained in Title II of that Act, offer some protection against unauthorized access to, or disclosure and use of, communications after they have come to be stored in servers or computer databases [97]. As with the anti-interception provisions of Title I of the ECPA, however, the unauthorized access provisions of Title II are subject to significant exceptions.

Notably, Title II of the ECPA protects only stored communications—that is, stored information that came to be placed in a computer as a result of electronic transmission of that information [98]. The protected communication may be stored in the facilities of an electronic communication service [99], such as an electronic mail provider, or may be stored in a computer maintained by a remote computing service [100], such as a payroll or tax preparation service, by a customer of that remote computing service. So long as the stored information qualifies as a communication, Title II makes it a criminal offense to do any of the following:

1. For anyone, intentionally to access, without authorization [101], a facility through which an electronic communication service is provided and through such access obtain, alter, or prevent authorized access to a wire or electronic communication while it is in electronic storage [102].

2. For any provider of an electronic communication service or remote computing service to the public, to knowingly divulge the contents of a communication stored in, or carried or maintained on, that service [103].

Although these protections of Title II are stated comprehensively, the ECPA does not offer airtight protection against access to and disclosure of information stored in electronic communication services and remote computing services. Notably, the operator of such a service may divulge the contents of a communication where such disclosure is "necessarily incident to the rendition of the service or to the protection of the rights or property of the provider of that service" [104]. This exception generally is taken as giving employers, in particular, the right to examine and divulge the contents of e-mail and other communications transmitted by employees over employer-provided communications systems. Similarly, the ECPA does not affect the ability of system operators to limit the privacy rights of their customers by contract.

Title II also defines the circumstances under which law enforcement agencies may gain access to stored communications. These provisions, which are quite complex, reserve their strongest protections for e-mail and similar communications that have been stored in a system operator's server for fewer than 181 days. Somewhat weaker protections against disclosure to government are provided for communications stored in an electronic communication service for more than 180 days, and communications stored in a remote computing service.

2. The CFAA and State Antihacking Statutes

The federal government and a number of states have enacted statutes that protect computer systems from disruption and protect information stored on computer systems from unauthorized access or destruction.

The most important of these antihacking measures is the federal CFAA [105], enacted in 1984 and substantially amended in 1986, 1994, and 1996 [106]. The passage of the CFAA reflected Congress's realization that existing statutes, such as the federal prohibition on wire fraud, did not extend to computer hacking [107]. As amended, the CFAA criminalizes and creates private causes of action for a wide variety of conduct involving abuse of both public and private computer systems.

Subsection (a)(1) of Section 1030 of the CFAA makes it a felony offense for anyone "knowingly" to access the computer of another without authorization (or to exceed an authorization) where the person committing the offense has "reason to believe that such information so obtained could be used to the injury of the United States, or to the advantage of any foreign nation" [108]. The scienter requirement of this subsection is met by proof "that the actor was aware of a high probability of the existence of the circumstance" [109].

The CFAA also makes it an offense to "intentionally" access a computer without authorization (or by exceeding an authorization) and thereby obtain information contained in the financial records of a financial institution [110], card issuer, or consumer reporting agency; or to obtain information from any department or agency of the United States or "from any protected computer if the conduct involved an interstate or foreign communication" [111]. The scienter standard in this subsection requires a "clear intent to enter, without proper authorization, computer files and data belonging to another" [112], and does not cover mere "mistaken, inadvertent, or careless" actions resulting in unauthorized access [113].

The CFAA also criminalizes intentional, unauthorized access to "any nonpublic computer of a department or agency of the United States" or any "computer of that department or agency that is exclusively for the use of the government of the United States" [114]. The prohibition of this subsection also applies to nonpublic government computers that are used for nongovernmental purposes, so long as the computer is used by or for the government and the act of unauthorized access affects the U.S. government's use of the computer [115].

The CFAA also reaches the use of computers to perpetrate fraud. Specifically, it is an offense for anyone "knowingly and with intent to defraud" to access a "protected computer" in a manner that "furthers the intended fraud" and thereby to "obtain anything of value" where the object of the fraud and the value of the use exceeds $5,000 in any one-year period. This subsection does not apply to fraudulent schemes in which a computer is used only to maintain records or in some other fashion that is "wholly extraneous" to the intended fraud; it applies only where the fraud is directly linked to the use of a computer [116].

Other provisions of the CFAA criminalize a broad range of computer misuse. Notably, it is unlawful under the CFAA to transmit "a program, information, code, or command, and as a result of such conduct, intentionally cause damage without authorization, to a protected computer" [117]. It also is a CFAA offense knowingly, and with intent to defraud, to traffic in computer passwords that may be used to access a computer without authorization [118].

The CFAA provides for both criminal penalties [119] and a civil cause of action. In order to obtain damages or injunctive relief, a person harmed by actions that violate the CFAA must have suffered at least $5,000 in damages for any one-year period unless the injury is to medical care, involves physical injury to any person or involves a threat to public health or safety [120]. The CFAA's civil remedy provisions also include a two-year statute of limitations.

In addition to the CFAA, a number of states have enacted antihacking statutes. Some of these statutes differ markedly from the CFAA in critical elements, such as their intent requirements and the kinds of computer access and abuse that will support findings of liability. Accordingly, ISPs and on-line service providers who believe they have been harmed by hacking activities should familiarize themselves with their rights under both state and federal laws [121].

3. Privacy Under the United States Constitution

As most people are aware, the Supreme Court has interpreted the Constitution as affording some recognition of a right to privacy, and the Fourth Amendment to the Constitution protects persons specifically against unreasonable searches and seizures. Where agents of government, rather than private persons, have gained access to an aggrieved person's on-line communications, and where more specific statutory protections are not available, these constitutional safeguards may in some cases provide a remedy. A review of the principal relevant judicial decisions, however, will demonstrate the limitations of constitutional privacy rights.

The constitutional right of privacy is a judge-made construct, not found in literal form in the Constitution, that has been interpreted to protect two kinds of interest: "the individual interest in avoiding disclosure of personal matters," and "the interest in independence in making certain kinds of important decisions" [122]. The latter interest is invoked when government restricts the ability of persons to make decisions concerning such matters as contraception and abortion [123]. The former interest, which is directly relevant to questions of informational privacy, applies when government engages in "unwarranted disclosure of accumulated private data" or administers a system under which such disclosures are not subject to constitutionally adequate "security provisions" [124].

The United States Supreme Court has not yet found a program of governmental accumulation of personal information to violate the right to informational privacy. Although the question was squarely presented in one case, the Court found that the state program at issue there, which required physicians to report all prescriptions they wrote for drugs that had both lawful and unlawful uses, served a legitimate purpose and contained sufficient protections against unauthorized disclosure [125]. Accordingly, we have little guidance as to the type of governmental data-gathering effort that will violate the constitutional right to privacy.

The Fourth Amendment also may apply to cases of governmental intrusion upon on-line communications—whether intercepted in the course of transmission or accessed while in computer storage. The principal Fourth Amendment case concerning interception of communications is *Katz v. United States* [126], in which the Supreme Court found that governmental interception of a communication is a "search and seizure," subject to the warrant requirement of the Fourth Amendment, if the government's surveillance involved an area in which the maker of the intercepted statements had a reasonable expectation of privacy.

Supreme Court decisions subsequent to *Katz,* however, underscore the limited utility of the Fourth Amendment as a source of informational privacy rights. For example, the doctrine of Katz was applied to the security of stored personal information in *United States v. Miller* [127]—a case involving a warrantless governmental seizure of a suspect's bank records. The Court in Miller found that a depositor has no reasonable expectation of privacy in records maintained by the depositor's bank because those records are not the depositor's personal papers but instead are business records of the bank. Similarly, in *Smith v. Maryland* [128] the Supreme Court found that a telephone company customer does not have a reasonable expectation of privacy in his or her telephone number.

At least one court, however, has found that persons have a reasonable expectation of privacy in their e-mail messages. In *United States v. Maxwell* [129], an Air Force colonel had engaged in sexually explicit on-line communications with another officer and had used his personal computer to download child pornography. Federal authorities served a warrant on the colonel's Internet access provider, which produced materials—including e-mail correspondence—not identified in the warrant. The appellate court found that the production of these communications violated the Fourth Amendment, and expressly stated that persons have a reasonable expectation of privacy in their e-mail because their "individually assigned passwords" create an objectively low risk that e-mail, either transmitted to other subscribers or stored in the service provider's computers, will be accessed by others.

Even where the Fourth Amendment applies to a particular interception or seizure of an on-line communication, legal challenges based on Fourth Amendment suffer from a number of inherent limitations. Notably, such challenges can be brought only where government carries out an interception or acquisition. Also, evidence seized in violation of the Fourth Amendment is not automatically excluded from evidence at trial, but only applies to criminal cases in which the government seeks to use illegally seized evidence to incriminate the victim of the unlawful search. Finally, the Fourth Amendment, unlike statutory remedies such as the ECPA, does not create a right to civil damages.

4. Tort Remedies

In addition to the statutory and constitutional causes of action we have described, certain common-law privacy rights may provide a remedy for persons whose Internet communications have been disclosed, intercepted, or acquired by their ISPs or third-party intruders. These common-law theories are subject to important limitations, however, that make their practical application problematic at best.

From its earliest beginnings, the common law has recognized a legitimate interest in the protection of persons and property from wrongful injury. Accordingly, a rich body of civil and criminal law has grown around such notions as assault, battery, wrongful homicide, trespass, conversion, theft, and fraud. The law took more slowly, however, to the notion that mental tranquillity, no less than property or the integrity of the body, might in some cases be a protectable interest. In fact, much of the common law of privacy is a product of the last one hundred years.

The common law of privacy often is dated from an article published in 1890 by Samuel Warren and Louis Brandeis [130]. Warren and Brandeis argued that individuals should have civil remedies for certain intrusions into their private lives, and their suggestion led in time to the development of a new set of torts. As defined by the state courts over the years, these torts are: (1) the intrusion upon another's seclusion, (2) the commercial appropriation of another's name or likeness, (3) the public disclosure of facts about another's personal life, and (4) publicity that places another in a false light [131].

The "privacy torts" continue to be controversial because of their perceived conflict with the First Amendment. As we explain briefly in the following sections, each of these theories also has limited utility in addressing the problem of informational privacy.

a. Intrusion upon Another's Seclusion

The tort of intrusion occurs when someone intentionally violates "the solitude or seclusion of another or his private affairs" [132]. The offending conduct must be highly objectionable to a reasonable person and any information gathered must not already be openly available to the public [133].

To the extent this tort can be used in the defense of informational privacy, it applies only to the methods used to gather information—not the maintenance, use, and disclosure of such information. A plaintiff also is unlikely to prevail merely by showing that information was gathered furtively or without the plaintiff's consent—that is, the conduct must be offensive and not merely surreptitious. The tort of intrusion, does not address most of the concerns people have about informational privacy in the computer age.

b. Misappropriation of Name or Likeness

This tort has its origins in the unauthorized use of the names and images of individuals to endorse products or otherwise reap a commercial benefit.

Most courts limit the reach of this cause of action to cases in which the exploitation of a name or image exploits the plaintiff's "personality" for financial gain, and defendants may defeat claims by asserting legitimate needs for the information or invoking their rights under the First Amendment. There is no need, however, to prove an expectation of privacy in order to establish the tort: even public figures may assert misappropriation claims, and in fact are the persons most likely to do so.

The tort of misappropriation, like the tort of intrusion, has limited application to informational privacy. The theory does not cover the gathering of information—only its dissemination and use. The tort also will not reach the dissemination of information about a person that amounts to less than a depiction of that person's "personality" [134]. So, for example, a compilation of purchases organized into a profile of a person's consumption habits might amount to a representation of that person's personality, while a credit card report would not.

c. Public Disclosure of Private Facts

Of the four recognized privacy torts, this is the one least likely to prove useful in protecting informational privacy [135]. Notably, this tort applies to information not generally available [136] and requires dissemination of the information to the general public [137]. The tort does not create a cause of action for disclosure of data—however personal—culled from public sources, or dissemination of information to a targeted group of users such as subscribers to credit reports.

d. Placing in a False Light

The "false light" theory, which protects persons from "publicity that places [them] in a false light before the public" [138], requires that the false information receive wide dissemination. Like the tort of public disclosure of private facts, therefore, this cause of action is not helpful as a remedy for interceptions or acquisitions of on-line information that is not subsequently disclosed to a large number of people.

V. Problems of Electronic Commerce

As the Internet's use as a means of selling goods and services has grown, on-line businesses have confronted legal limitations and uncertainties that differ somewhat from those of more traditional enterprises. Although an exhaustive discussion of the legal problems of e-commerce is beyond the scope of this chapter, two topics are of particular importance: (1) the problems posed by the information of enforceable on-line contracts; and (2) legal limitations on the use of personal information collected from customers of, and visitors to, Internet-based commercial services. The following discusses each of these issues in turn.

A. Making and Enforcing On-Line Contracts

When an on-line merchant takes an order from a customer for a product or service, the merchant ordinarily assumes that an enforceable contract with the customer has been formed. Specifically, the merchant reasonably will expect that it now is obligated to ship the product or provide the service, and that the customer is obligated to pay the agreed price for that product or service. The merchant further assumes that if the customer fails to pay or denies having placed the order, the courts will enforce the bargain the parties have made.

In the Internet world, these expectations may be undermined in several ways. First, the customer may deny that an on-line contract, which is in electronic rather than paper form, is legitimate and enforceable. Second, the customer may argue that the alleged contract requires a signature, and that a purported "signature" in electronic rather than physical form does not satisfy this requirement. Third, the customer may argue that he or she did not understand and agree to all terms of the supposed contract when it placed the on-line order. The efforts of the law to address these problems are very much a work in progress.

1. Electronic Documents and Signatures

When customers place orders or engage in other transactions on-line, they create electronic documents and authenticate those documents by typing in personal information such as their names, e-mail addresses, personal identification numbers, encrypted digital signatures, or other forms of electronic authentication.

When parties to these transactions attempt to enforce those bargains, the electronic character of the documents and signatures may present two types of problems. First, the party resisting an action to enforce the contract may claim that the document is inaccurate or inauthentic or that the electronic signature is not that of the party. Second, the party resisting enforcement may argue, not just that the document or signature is inauthentic or inaccurate as a factual matter, but that electronic documents or signatures are legally insufficient and may not be enforced even when genuine and accurate.

Claims of the first kind (i.e., that particular documents and signatures are false or corrupted) are best avoided by use of key encryption and other technologies that make tampering and forgery difficult to achieve and easy to detect. Although somewhat costly for use in mass-market consumer transactions, these technologies are a sound investment for enterprises that form high-value, business-to-business contracts. Although the use of encryption-based digital signatures and document-creation technologies cannot prevent claims of forgery and error altogether, they can make such claims much harder to sustain.

Claims of the second kind (i.e., that electronic documents and signatures are unenforceable as a matter of law) are much more easily foreclosed than claims of the first kind. Specifically, Congress and some state legislatures have enacted laws that, with certain exceptions, forbid courts from denying legal effect to documents and signatures simply because those documents and signatures are in electronic form.

The most important of these laws is the Electronic Signatures in Global and National Commerce Act (the ESign Act), which creates a federal rule that "a signature, contract, or other record may not be denied legal effect, validity, or enforceability solely because it is in electronic form." The ESign Act also provides that a contract relating to a transaction may not be denied legal effect simply because "an electronic record or signature was used in its formation."

Although the ESign Act gives on-line merchants broad protection from claims that Internet-based contracts and signatures are unenforceable, the statute contains a number of significant exceptions. Most importantly, the ESign Act recognizes that many consumer-protection laws require certain documents to be made available to consumers in writing. Where those laws apply, the ESign Act states that the requirement of a written document will be satisfied by an electronic record only where the consumer has "affirmatively consented" to the use of the electronic record. The statute also requires the on-line merchant to make a number of disclosures to the consumer before the consumer's consent is obtained.

In addition to the ESign Act, the enforceability of electronic records and signatures is recognized by the Uniform Electronic Transactions Act (UETA) and the Uniform Computer Information Transactions Act (UCITA). These are so-called uniform acts, adopted by the national Conference of Commissioners on Uniform State Laws, that become law only when enacted by the legislatures of individual states.

2. Enforcing Clickwrap and Other Mass Market On-Line Contracts

In addition to the question of legal equivalency between electronic and paper documents and signatures, the Internet has presented problems of enforcement of contracts that are presented to customers on-line and may not be carefully read by those customers before the purchase is made. For example, a company that sells software on-line may present the purchaser with a lengthy, detailed license agreement that appears on the customer's computer screen before the purchase is completed. The merchant may place a message at the bottom of the screen on which the contract appears, saying: "By clicking below you agree to be bound by the terms of this license." On the icon below the text may appear the words "I agree."

The problem presented by a contract of this kind is not new or unique to the Internet. For many years now, the law has struggled with the enforceability of mass-market contracts that are presented to the public on a "take it or leave it" basis. For example, consumers purchase insurance policies by signing "binder" documents that omit many, and often most, of the policies' most important terms. Similarly, customers purchase cars and other durable goods with only the vaguest knowledge of the terms of the warranties that accompany those purchases. In the case of the insurance binder, the detailed terms of the policy will be provided to the customer by mail, several days or even weeks after the binder document has been executed. In the case of the durable goods purchase, the warranty may be provided to the customer at the time of purchase, but no one seriously suggests that the customer has a realistic opportunity to review the dense boilerplate of those warranties. In fact, "under the conditions of modern contracting buyers are usually not reasonably expected even to read the forms before they become bound" [139].

The courts, recognizing the efficiencies realized by standard-form contracting in consumer markets, have enforced the terms of standard contracts, except where the results are "unconscionable" or fail to fulfill "reasonable expectations." The unconscionability analysis involves scrutiny of both the substantive fairness of the contested contract term and the procedural fairness of the process by which the term was—or was not—made available for the customer's review before the customer became bound [140]. The reasonable expectations analysis, which the courts first developed in insurance cases, measures the disputed terms against those that a consumer reasonably would expect to encounter in a contract of the type under review [141]. Where the terms of a standard form contract are found to be unconscionable or contrary to reasonable expectations, a court may decline to enforce those terms.

These are principles apply equally to formation of mass-market on-line contracts. On-line merchants should ensure that the terms of licenses and other contracts are within reasonable industry practice for the goods and services involved, and that customers have a reasonable opportunity to review the terms of those contracts before becoming bound by their terms [142].

B. Protecting Customers' Personal Information

On-line merchants collect substantial amounts of personal information from visitors to their sites and customers that purchase goods and services through those sites. This information can be highly valuable. From data collected in this way, on-line enterprises can assemble mailing lists for their own advertising campaigns or for sale to others.

Consumer advocates within and outside government have become increasingly concerned about the possible collection, distribution, and use of personal information without the knowledge and consent of the persons to whom that information relates. Although no comprehensive legislation directed at these concerns has been enacted in the United States, on-line merchants should be aware of three areas of potential legal exposure: (1) enforcement efforts of the Federal Trade Commission (FTC); (2) the COPPA; and (3) the impact of the European Union Data Protection Directive.

The FTC's involvement in on-line privacy arises out of its general authority to protect consumers from unfair, deceptive, and abusive practices. Consistent with that mandate, the FTC has taken a strong interest in Internet-based companies that violate privacy commitments they have made to their customers. Although the FTC does not yet have the legislative authority to require on-line services to establish privacy policies, it will act vigorously against companies that adopt, and then violate, such policies.

COPPA became law in October, 1998 and is enforced by the FTC under regulations that became effective in April, 2000 [143]. Under COPPA, operators at websites directed at children under 13, or that have actual knowledge that they are collecting personal information from children under 13, must obtain "verifiable parental consent" before collecting such personal information. Services subject to COPPA also must post prominent notices of their data collection, use, and disclosure practices, permit parents to review and obtain correction of information collected from their children, and comply with a number of other stringent data protection and security requirements.

Finally, the European Union Data Directive imposes stringent obligations on all persons and entities that engage in the automated collection of personally identifiable information on EU countries. Although not directly applicable to ISPs that collect personal data in the United States, the Directive affects U.S. business by forbidding the transfer of personal data from an EU country to another country that does not provide "adequate" data protection as defined by the EU. Because the EU does not regard the United States as having adequate data protection laws in most sectors of its economy, personal data may be transferred from EU countries only to U.S. entities that comply with "Safe Harbor" regulations negotiated between the EU and the U.S. Department of Commerce.

Conclusion

Although Congress, the FCC, and state legislatures routinely disclaim any intention to regulate the Internet, regulation of ISP activities will be among the most active areas of legislation for years to come. For this reason, ISPs must remain aware of their evolving rights and obligations under both foreign and domestic law.

Endnotes

[1] 47 U.S.C. § 153(43).

[2] *Id.* § 153(20).

[3] In fact, because they are licensees of a scarce public resource (i.e., the electromagnetic spectrum), broadcasters are subject to content restrictions that would be unlawful under the First Amendment to the U.S. Constitution if applied to other media. *See,* e.g., *Red Lion Broadcasting v. Federal Communications Commission,* 395 U.S. 367 (1969).

[4] *See,* e.g., RESTATEMENT (SECOND) OF TORTS § 558 (1977).

[5] *Id.*

[6] *See New York Times Co. v. Sullivan,* 376 U.S. 254, 84 S.Ct. 710, 11 L.Ed.2d 686 (1964).

[7] For cases involving nonpublic figures, the state courts define the level of fault that will establish a claim of defamation. Most states have found that a defendant's negligence as to the truth or falsity of a statement will suffice. *See,* e.g., *Miami Herald Corp. v. Ane,* 458 So.2d 239 (Fla. 1984).

[8] *See, e.g., Miami Herald Publishing Co. v. Tornillo,* 418 U.S. 241, 94 S.Ct. 2831, 41 L.Ed.2d 730 (1974); RESTATEMENT, SECOND TORTS §578 (1977).

[9] There is a constitutional basis for the requirement of fault in cases of distributor liability. Specifically, the Supreme Court has found that imposition of strict liability for distributor defamation would violate the speech and press freedoms guaranteed by the First Amendment and made applicable to the states by the Fourteenth Amendment. *Smith v. California,* 361 U.S. 147, 152-53, 80 S.Ct. 215, 4 L.Ed.2d 205 (1959).

[10] 47 U.S.C. § 230.

[11] 47 U.S.C. § 230(c)(1).

[12] *Id.* §230(e)(2).

[13] 958 F. Supp. 1124 (E.D. Va. 1997).

[14] The postings at issue were particularly scurrilous. They consisted of phony advertisements, purporting to be those of the plaintiff, for T-shirts with slogans praising the 1995 Oklahoma City bombing and ridiculing the victims of that attack. The postings resulted in hate mail and threats against the plaintiff.

[15] *Zeran v. America Online, Inc.,* 1997 U.S. App. LEXIS 31791 (4th Cir. 1997).

[16] 9992 F. Supp. 44 (D.D.C. 1998).

[17] 206 F. 3d 980 (10th Cir 2000).

[18] As we discuss further under the heading of obscenity, another court has found that Section 230 immunizes an ISP for its involvement in child pornography as well as defamation. *Doe v. America Online,* 718 So. 2d 385 (Ct. App. Fla. 1998).

[19] 776 F. Supp. 135 (S.D.N.Y. 1991).

[20] 1995 N.Y. Misc. LEXIS 229 (1995).

[21] The legislative history of § 230 of the CDA includes an express declaration of congressional intent to overrule *Stratton-Oakmont.* H.R. Rep. No. 104-458 at 194 (1996). ("One of the specific purposes of this section is to overrule *Stratton-Oakmont v. Prodigy* and any other similar decision which have treated providers and users as publishers or speakers of content that is not their own because they have restricted access to objectionable material. The conferees believe that such decisions create serious obstacles to the important federal policy of empowering parents to determine the content of communications their children receive through interactive computer services.")

[22] Specifically, registration of a copyright is required before the copyright owner may bring an infringement action, and also is necessary in order for the plaintiff in an infringement action to receive an award of statutory damages and attorneys' fees.

[23] 15 U.S.C. § 106.

[24] Supra n. 91.

[25] Supra n. 12.

[26] 839 F. Supp. at 1556.

[27] 907 F. Supp. 1361 (N.D. Cal. 1995).

[28] "Where the infringing subscriber is directly liable for the same act, it does not make sense to adopt a rule that could lead to the liability of countless parties whose role in the infringement is nothing more than setting up and operating a system that is necessary for the functioning of the Internet." *Id.* at 1372.

[29] *Sega Enterprises Ltd. v. MAPHIA,* 948 F. Supp. 923 (N.D. Cal. 1996).

[30] Pub. L. No. 105304, 112 Stat. 2877 (1998), codified at 17 U.S.C. § 512.

[31] 15 U.S.C. §§ 1051-1127.

[32] The law also protects service marks, which are used "to identify and distinguish the services of one person from the services of others and to indicate the source of the services." 15 U.S.C. § 1127.

[33] 15 U.S.C. § 1114(1). Trademarks that have not been registered with the Patent and Trademark Office also may be protected, but with much greater difficulty. Notably, registration of a trademark makes the trademark incontestable after five years' continuous use, permits the registrant to bring an enforcement action in federal court and protects against infringement anywhere in the country. 15 U.S.C. § 1065.

[34] *See,* e.g., *White v. Samsung Electronics of America, Inc.,* 971 F.2d 1395 (9th Cir. 1992); Electronic Design and Sales, Inc. v. Electronic Data Systems Corp., 954 F.2d 713 (Fed. Cir. 1992). The intent element of the cause of action does not logically affect the possibility of confusion, but may tip the balance of a close case and may be dispositive where a plaintiff seeks equitable relief.

[35] 839 F. Supp. 1552 (M.D. Fla. 1993) *(Playboy v. Frena).*

[36] ISPs should be aware that the proscriptions of the Lanham Act go beyond strict trademark infringement to reach other unfair trade practices that may or may not be trademark-related. Notably, as *Playboy v. Frena* shows, Section 43(a) of the Lanham Act reaches the common-law tort of "palming off," or representing one's own goods as those of another, or any false statements about a product or service that are likely to deceive consumers and injure a plaintiff. *See,* e.g., *Potato Chip Institute v. General Mills, Inc.,* 333 F. Supp. 173 (D. Neb. 1971), aff'd, 461 F.2d 1088 (8th Cir. 1972).

[37] 1999 U.S. Dist. LEXIS 20605 (S.D. Ind. 1999).

[38] *Id.* (citation omitted).

[39] *See United States v. Thomas,* 74 F.3d 701 (6th Cir. 1995); *Florida v. Cohen,* 696 So.2d 435 (Ct. App. Fla. 1997); *New York v. Barrows,* 1997 N.Y. Misc. LEXIS 473 (Sup. Ct. N.Y. 1997).

[40] The U.S. Supreme Court has found that a work is obscene if "the average person, applying contemporary community standards" would find that the work, taken as a whole: (1) appeals to the prurient interest; (2) depicts or describes, in a patently offensive way, sexual conduct specifically defined by applicable state law; and (3) taken as a

whole, lacks serious literary, artistic, political, or scientific value. *Miller v. California,* 413 U.S. 15, 93 S.Ct. 2607 (1973).

[41] Indecent speech has been defined by the Federal Communications Commission as "language that describes, in terms patently offensive as measured by contemporary community standards for the broadcast medium, sexual or excretory activities and organs." See *Federal Communications Commission v. Pacifica Foundation,* 438 U.S. 726, 98 S.Ct. 3026 (1978). The Supreme Court has upheld limitations on indecent speech only where those restrictions "promote a compelling (governmental) interest (and the government chooses) the least restrictive means to further the articulated interest." *Sable Communications of California, Inc. v. FCC,* 492 U.S. 115, 126, 109 S.Ct. 2829, 2836 (1989).

[42] Codified at 47 U.S.C. § 223.

[43] To be codified at 47 U.S.C. § 231.

[44] 47 U.S.C. § 223(a).

[45] *Id.* § 223(e)(5)(A).

[46] *Id.* § 223(e)(5)(B).

[47] *American Civil Liberties Union v. Reno,* 1996 U.S. Dist. LEXIS 7919 at *63, 67-68; Shea v. Reno, 1996 U.S. Dist. LEXIS 10720.

[48] *Reno v. American Civil Liberties Union,* 117 S.Ct. 2329, 138 L.Ed.2d 874 (1997).

[49] *American Civil Liberties Union v. Reno,* 31 F. Supp. 2d 473 (E.D.Pa. 1999).

[50] *Id.*

[51] *American Civil Liberties Union v. Reno,* 217 F. 3d 162 (3d Cir. 2000).

[52] 18 U.S.C. § 1462 (1996).

[53] *Id.*

[54] 18 U.S.C. § 1465.

[55] In *United States v. Thomas,* supra, the court confirmed that where providers of content on the Internet make that information available for downloading, those content providers are subject to prosecution for obscenity according to the standards of any community in which someone might access and download that information. See *United States v. Bagnell,* 679 F.2d 826, 830 (11th Cir. 1982), cert. denied, 460 U.S. 1047, 103 S.Ct. 1449, 75 L.Ed.2d 803 (1983); *United States v. Peraino,* 645 F.2d 548, 551 (6th Cir. 1981).

[56] *Smith v. California,* 361 U.S. 147, 161 (1959).

[57] A court probably would classify a scienter claim based only on the name of a newsgroup or an ISP's receipt of a warning letter as a claim of constructive, rather than actual, knowledge of the nature of the offending material. The Supreme Court has not yet decided whether a conviction based on constructive knowledge would satisfy the

requirements of the Constitution. *Cargal v. Georgia,* 438 U.S. 906 (1978); *Ballew v. Georgia,* 436 U.S. 962 (1978).

[58] 18 U.S.C. § 2252.

[59] *United States v. X-Citement Video, Inc.,* 115 S.Ct. 464 (1994).

[60] 18 U.S.C. § 1030(a)(5)(A). If access was not unauthorized, however, the damage caused by the intruder must be the result of intent to cause harm rather than mere recklessness. 18 U.S.C. § 1030(a)(5)(B).

[61] 18 U.S.C.§ 1030(e)(8)(A).

[62] *Id.* Although no court has ruled on the application of antihacking statutes to spamming, at least one reported case included CFAA claims against both a spammer and an ISP who returned the spammer's mailings to the originating facilities. *CyberPromotions, Inc. v. America Online, Inc.,* 948 F. Supp. 436, 438-39 (E.D. Pa. 1996).

[63] *CompuServe, Inc. v. CyberPromotions, Inc.,* 962 F. Supp. 1015 (S.D. Ohio 1997).

[64] 962 F. Supp. at 1022.

[65] *Id.*

[66] *Id.*

[67] *Id.* at 1023.

[68] Black's Law Dictionary 853 (6th ed. 1990).

[69] *See* e.g., *Ham v. LaCiennega Music Co.,* 4 F.3d 413 (5th Cir. 1993).

[70] *See* e.g., NY CPLR § 302 (1997).

[71] *See Id.,* see also, *Dotmar, Inc. v. Niagara Fire Ins. Co.,* 533 N.W. 2d 25, 29 (Minn.), cert. denied, 116 S.Ct. 583, 133 L. Ed. 2d 504 (1995). (The Minnesota long-arm statute, Minn. Stat. § 543.19, "permits courts to assert jurisdiction over defendants to the extent that federal constitutional requirements of due process will allow.")

[72] *International Shoe v. Washington,* 326 U.S. 310, 316 (1945).

[73] *World-Wide Volkswagen v. Woodson,* 444 U.S. 286, (1980).

[74] *Asahi Metal Industry Co. v. Superior Court of California,* 480 U.S. 102, (1987) (citing *Burger King Corp v. Rudzewicz,* 471 U.S. 462, (1985)).

[75] *Helicopterous Nacionales de Columbia v. Hall,* 466 U.S. 408 (1984).

[76] 89 F.3d 1257 (6th Cir. 1996).

[77] 937 F. Supp. 161 (D. Conn. 1996).

[78] 947 F. Supp. 1328 (E.D.Mo. 1996).

[79] *Haelen Products, Inc. v. Beso Biological, et. al,* 1997 U.S. Dist. LEXIS 10565 at *11 (E.D.La. 1997) (finding jurisdiction). See also Maritz, 947 F. Supp. at 1329-34; Inset, 937 F. Supp. at 164-65.

[80] 937 F. Supp. 295 (S.D.N.Y. 1996).

[81] *Id.*

[82] *Id. See also, Hearst Corporation v. Goldberger,* 1997 U.S. Dist LEXIS 2065 (S.D.N.Y. 1997).

[83] *Zippo Manufacturing Co. v. Zippo Dot Com, Inc.,* 952 F. Supp. 1119, 1124 (W.D.Pa 1997).

[84] E.g. CompuServe, 89 F.3d 1257.

[85] E.g. Bensusan, 937 F. Supp. 295.

[86] E.g. Maritz, 940 F. Supp. 96.

[87] The FCC first distinguished basic from enhanced services in its so-called *Computer II* and *Computer III* decisions, in which the Commission set the ground rules under which the established telephone companies would be permitted to offer enhanced services. See Amendment of §64.702 of the Commission's Rules and Regulations, CC Docket No. 85-229, Phase I, 104 F.C.C.2d 958 (1986) (Phase I Order), modified on recon., 2 FCC Rcd. 3035 (1987) (Phase I Reconsideration Order), further modified on recon., 3 FCC Rcd. 1135 (1988) (Phase I Further Reconsideration Order), second further recon., 4 FCC Rcd. 5927 (1989) (Phase I Second Further Reconsideration), Phase I Order and Phase I Reconsideration Order vacated, *California v. FCC,* 905 F.2d 1217 (9th Cir. 1990) (California I); Phase II, 2 FCC Rcd. 3072 (1987) (Phase II Order), recon., 3 FCC Rcd. 1150 (1988) (Phase II Reconsideration Order), further recon., 4 FCC Rcd 5927 (1989) (Phase II Further Reconsideration Order), Phase II Order vacated, California I, 905 F.2d 1217 (9th Cir. 1990); *Computer III* Remand Proceedings, 5 FCC Rcd. 7719(1990) (ONA Remand Order), recon., 7 FCC Rcd. 909, petitions for review denied, *California v. FCC,* 4 F.3d 1505 (9th Cir. 1993) (California II); *Computer III* Remand Proceeding: Bell Operating Company Safeguards and Tier I Local Exchange Company Safeguards, 6 FCC Rcd. 7571(1991) (BOC Safeguards Order), recon. dismissed in part, Order, CC Docket Nos. 90-623 and 92-256, FCC 96-222 (rel. May 17, 1996); BOC Safeguards Order vacated in part and remanded, *California v. FCC,* 39 F.3d 919 (9th Cir. 1994) (California III), cert. denied, 115 S.Ct. 1427 (1995).

[88] A Staff report of the FCC, The Digital Tornado, however, strongly suggests that the Commission will not choose to require ISPs to pay access charges to the local telephone companies.

[89] 18 U.S.C. §§ 2510-2522.

[90] The ECPA defines electronic communication as "any transfer of signs, signals, writing, images, sounds, data or intelligence of any nature transmitted in whole or in part by a wire, radio, electromagnetic, photoelectronic or photo-optical system." 18 U.S.C. § 2510(1). The ECPA also prohibits providers of electronic communication service to the

public from intentionally disclosing or divulging the contents of any communication while in transmission on that service to anyone other than an addressee or intended recipient of that communication. 18 U.S.C. §§ 2511(1)(c) and 2511(3)(a). System operators, however, may divulge the contents of communications where necessary to render service, protect the operator's rights or properties or where certain other conditions apply. 18 U.S.C. §§ 2511(2)(a)(i), 2511(3)(b)(i), 2511(1)(2)(a)(ii), 2511(3)(b)(ii), 2511(3)(b)(iii), 2511(3)(b)(iv) and 2517.

[91] Only interceptions of the "contents" of communications are forbidden. The ECPA defines "contents" to include "information concerning the substance, purport, or meaning of (a) communication." 18 U.S.C. § 2510(8). This represents a change from the earlier Wiretap Act, which had defined "contents" to include information concerning the identity of the parties to a communication.

[92] 18 U.S.C. §§ 2511(2)(a)(i) and 2511(3)(b). This exception, which refers to the rights of a broad category of "system operators," is relied upon by employers as giving them legal authority to intercept the electronic mail of their employees when that electronic mail is carried on an employer-provided network.

[93] 18 U.S.C. § 2511(2)(c)-(d).

[94] The ECPA reserves its strongest protections for wire communications, which essentially are wireline or wireless voice telephone conversations, and oral communications, which are oral statements made in circumstances suggesting that the speaker had a "reasonable expectation of privacy" when the statements were made. E-mail and other electronic communications enjoy the lowest level of privacy protection under the ECPA.

[95] 18 U.S.C. §2516(3).

[96] 18 U.S.C. § 2515. Users also should be aware that when e-mail is seized in storage in the service provider's server, that seizure is not an interception but at most a case of unauthorized access to stored information. *Steven Jackson Games, Inc. v. United States Secret Service*, 36 F.3d 457 (5th Cir. 1994); see also *Wesley College v. Pitts*, No. CIV. A. 95-536 MMS (D. Del. Aug. 11, 1997).

[97] 18 U.S.C. § 2701 et seq.

[98] For example, Title II does not appear to protect information placed in a database by an on-site operator, unless that information came to be in the operator's possession as a result of a protected electronic communication.

[99] The ECPA defines an electronic communication service as "any service which provides to users thereof the ability to send or receive wire or electronic communications." 18 U.S.C. § 2510(15).

[100] The ECPA defines a remote computing service as "the provision to the public of computer storage or processing services by means of an electronic communications system." 18 U.S.C. § 2710(2).

[101] It also is an offense for a person who is authorized to access an electronic communication service intentionally to exceed that authorization, where that person then uses that

access to obtain, alter or prevent authorized access to a stored wire or electronic communication.

[102] 18 U.S.C. § 2701(a).

[103] 18 U.S.C. § 2702(a).

[104] *See* 18 U.S.C. § 2702(b)(6).

[105] Counterfeit Access Device and Computer Fraud and Abuse Act of 1984, Pub. L. 98-473, §§ 2101-03, 98 Stat. 2190 (1984), codified at 18 U.S.C. § 1030.

[106] *See* Pub. L. 99-474, § 2, 100 Stat. 1213 (1986); Pub. L. 100-690, § 7065, 102 Stat. 4404 (1988); Pub. L. 101-73, § 962(a)(5), 103 Stat. 502 (1989); Pub. L. 101-647, §§ 1205(e), 2597(j), 3533, 104 Stat. 4831, 4910, 4925 (1990); Pub. L. 103-322, § 290001(b)-(f), 108 Stat. 2097-99 (1994); Pub. L. 104-294, § 201, 110 Stat. 3488 (1996).

[107] In *United States v. Seidlitz*, 589 F.2d 152 (4th Cir. 1978), cert. denied, 441 U.S. 922, 99 S.Ct. 2030, 60 L.Ed.2d 396 (1979), the Government successfully used the wire fraud statute to prosecute a former government contractor for "hacking" a computer of the Federal Energy Administration. The prosecutors were forced to base the wire fraud claim, not on the hacking activities themselves, but on the defendant's making of interstate telephone calls in connection with his activities.

[108] 18 U.S.C. § 1030(a)(1).

[109] *See* 1984 U.S. Code Cong. & Admin. News at 3702.

[110] "Financial institution" is defined at 18 U.S.C. § 1030(e)(4).

[111] 18 U.S.C. § 1030(a)(2)(B)-(C). A "protected computer" is a computer "exclusively for the use of a financial institution or the United States Government, or, in the case of computer not exclusively for such use, use by or for a financial institution or the United States Government and the conduct constituting the offense affects that use by or for the financial institution or the Government." The definition also includes any computer "used in interstate or foreign commerce or communication."

[112] 1986 U.S. Code Cong. & Admin. News at 2484.

[113] *Id.* at 2483.

[114] 18 U.S.C. § 1030(a)(3).

[115] *Id.*

[116] 1986 U.S. Code Cong. & Admin. News at 2487.

[117] 18 U.S.C. § 1030(a)(5)(A).

[118] 18 U.S.C. § 1030(a)(6). The password trafficking must affect interstate or foreign commerce, or the computer to which the password gives access must be "used by or for the Government of the United States." 18 U.S.C. § 1030(a)(6).

[119] 18 U.S.C. § 1030(c).

[120] 18 U.S.C. § 1030(e)(8).

[121] *See,* e.g., West's Ann. Cal. Penal Code § 502 (West 1988); Minn. Stat. Ann. §§ 270B.01 et seq. (West 1989); 21 Okl. Stat. Ann. § 1953 (West 1983); Wis. Stat. Ann. § 943.70 (West 1996).

[122] *Whalen v. Rowe,* 429 U.S. 589, 599-600, 97 S.Ct. 869, 51 L.Ed.2d 64 (1977).

[123] *See,* e.g., *Roe v. Wade,* 410 U.S. 113, 93 S.Ct. 705, 35 L.Ed.2d 147 (1973); *Griswold v. Connecticut,* 381 U.S. 479, 85 S.Ct. 1678, 14 L.Ed.2d 510 (1965).

[124] *Whalen v. Rowe,* supra, 429 U.S. at 596, 97 S.Ct. at 874-75, 51 L.Ed.2d at 71.

[125] *Id.* For a lower court case finding that a governmental information-gathering program violates the right to privacy, *see Merriken v. Cressman,* 364 F. Supp. 913 (E.D. Pa. 1973). This case involved a public school program in which students were asked intrusive questions about their family relationships in order to identify students who might be at risk for drug use.

[126] 389 U.S. 347 (1967).

[127] 425 U.S. 435, 96 S.Ct. 1619, 48 L.Ed.2d 71 (1976).

[128] 442 U.S. 735, 99 S.Ct. 2577, 61 L.Ed.2d 220 (1979). In Smith, police had not obtained a warrant or interception order before placing a pen register—a device that records all numbers dialed from a telephone—on a suspect's telephone line. The Supreme Court found that even if the suspect had a subjective belief in the privacy of his telephone number, that belief was not reasonable in light of the fact that telephone numbers are captured and used by the telephone company for billing purposes.

[129] 42 MJ 568 (Ct. Military App. 1995), rev'd in part on other grounds, 45 MJ 46 (1996).

[130] Warren, S., and L. Brandeis, "The Right to Privacy," 4 Harv. L. Rev. 193 (1890).

[131] See RESTATEMENT (SECOND) OF TORTS § 652 (1977).

[132] *Id.* § 652(B).

[133] *Id.*

[134] *See,* e.g., *Freihofer v. Hearst Corp.,* 480 N.E.2d 349, 353 (1985); *Bartholomew v. Workman,* 169 P.2d 1012, 1014 (Okl. 1946); *Goodyear Tire & Rubber Co. v. Vandergriff,* 184 S.E. 452, 454 (Ga. 1936); RESTATEMENT (SECOND) OF TORTS § 652(c) CMT c (1977).

[135] For a discussion of this tort, see RESTATEMENT (SECOND) OF TORTS § 652(D) (1977).

[136] *See,* e.g., *Gill v. Hearst Publishing Co.,* 253 P.2d 441 (Cal. 1953).

[137] *See,* e.g., *Polin v. Dun & Bradstreet, Inc.,* 768 F.2d 1204 (10th Cir. 1985); *Senogles v. Security Benefit Life Ins. Co.,* 536 P.2d 1358 (Kan. 1975).

[138] RESTATEMENT (SECOND) OF TORTS § 652(E)(1977).

[139] Slawson, W. D., "The New Meaning of Contract: The Transformation of Contracts Law by Standard Forms," 46 U. Pitt. L. Rev. 21 (1984).

[140] *See Weaver v. American Oil Co.,* 276 N. E. 2d 144 (Ind. 1971); see also *Henningsen v. Bloomfield Motors, Inc.,* 161 A. 2d 69 (N.J. 1960).

[141] *See* Gray Zurich Ins. Co., 419 P. 2d 168, 172 (Cal. 1966).

[142] The Uniform Computer Information Transaction Act (UCITA), mentioned earlier, expressly recognizes clickwrap contracts and provides for their enforcement if not unconscionable and customers have had a reasonable opportunity to review their terms. At this writing, however, only two states have enacted UCITA.

[143] Pub. L. No. 105277, 112 Stat. 2681728, codified as amended at 15 U.S.C. §§ 650106.

10

Universal Service

Ever since state and federal regulation of the U.S. telecommunications industry began, regulators have pursued a policy of "universal service." The goal of this policy is to make basic telephone service available throughout the United States at rates that are both affordable and relatively uniform.

The difficulty with this policy is that the cost of serving different customers varies drastically. Notably, customers in rural and thinly populated areas are more costly to serve than customers in urban areas. The reasons for the disparity involve both differences in the costs of facilities and differences in the numbers of customers from whom those costs may be recovered. The first factor may be illustrated by comparing the cost of a 10-mile-long access line, connecting a farm family to a distant central office, with the cost of an urban access line that runs from a central office to a customer living two blocks away. The second factor may be illustrated by comparing a central office switch that serves 200 rural customers with a switch of similar size and cost that serves 10,000 urban customers. The 10-mile-long access line is far costlier that the 2-block-long access line, and the cost of the rural central office must be recovered from one-fifth as many customers as are served by the urban central office.

In an unregulated environment, in which rates of service are set at a level that recovers the associated costs, the rural customers in our examples would pay dramatically higher rates for telephone service than the urban customers. In the real world as it has existed for many decades, however, elaborate subsidy schemes ensure that customers' rates are not based on the cost of providing their service.

Until recently, most of these subsidies were hidden, or "implicit," rather than open or "explicit." Under the rate-of-return regime that characterized most regulation before the late twentieth century, rates were set at levels that produced a permitted return on the telephone company's overall investment, with no attempt to determine—much less base rates upon—the costs of providing particular services or serving particular customers. Accordingly, although some services and customers unquestionably subsidized other services and customers, the extent of those subsidies was impossible to determine.

Although these implicit subsidies could not be reliably quantified, no one doubted the directions in which they flowed. Historically, urban ratepayers subsidized rural ratepayers, business services subsidized residential services, long-distance service subsidized local service, and optional features, such as call waiting and call forwarding, subsidized basic telephone service. So long as competition was prohibited in the markets that provided the subsidies, this complex system of noncost-based pricing could be sustained.

In addition to implicit subsidies, regulators also endorsed a number of support programs that were open and explicit. The most important of these were: (1) the support mechanisms for high-cost and small telephone companies; and (2) the Lifeline and Linkup programs for low-income consumers. A complete understanding of the current universal service system requires some familiarity with these programs [1].

I. Explicit High-Cost Support Programs Before the 1996 Act

Historically, telephone companies that served rural and high-cost areas were subsidized by three federal support programs: the Universal Service Fund (USF), the Dial Equipment Minutes (DEM) Weighting program, and the Long-Term Support (LTS) program.

The USF fund (sometimes called the high-cost assistance fund) provided for payments to telephone companies with local loop costs that exceeded 115% of the national average of local loop costs. The calculation of these payments was tied closely to the mechanism by which telephone companies recovered the portion of their local loop costs attributed to interstate services. Specifically, the Commission's rules assigned a percentage of telephone companies' local loop costs to the interstate jurisdiction, and permitted the companies to recover those costs through their interstate access charges [2]. (At the time of enactment of the 1996 Act, the standard allocation of local loop costs to the interstate jurisdiction was 25%.) If a company's loop costs exceeded 115% of the national average, the Commission assigned a larger percentage of the company's local loop costs to the interstate jurisdiction and allowed the company to recover some or all of the additional cost by drawing payments from the USF fund [3]. The cost of this fund, prior to the 1996 Act, was borne entirely by the interexchange carriers.

DEM weighting was a method of defraying some of the switching costs of smaller telephone companies. Under DEM weighting, companies with fewer than 50,000 access lines allocated a larger percentage of their switching costs to the interstate jurisdiction than larger companies allocated. The DEM weighting subsidy did not require a demonstration by the recipient telephone company that its switching costs were, in fact, higher than those of larger companies: the program simply assumed that switching technology was not sufficiently "scalable" to offer smaller companies the same economies of scale that larger companies enjoyed.

Finally, the LTS program provided a mechanism under which all LECs could charge the same common carrier line access rates to IXCs, regardless of the actual cost of individual LECs' access lines. The LTS system was managed by the National Exchange Carrier Association (NECA), which projected the annual revenue requirement (i.e., the revenues required to yield an allowed return on investment) for the access lines of LECs that participated in the program. NECA then subtracted from this revenue requirement the amounts that LECs would earn from subscriber line charges and common carrier line access charges, and assessed the difference in the form of LTS contributions required from LECs with below-average access line costs. LECs with above-average access line costs drew from this pool the amounts needed to subsidize their carrier common line access charges and maintain those charges at a nationally averaged level.

II. Explicit Support for Low-Income Subscribers

In addition to the system of explicit subsidies for high-cost telephone companies, the Commission and participating states administered two programs that directly supported the telephone service of low-income subscribers. The more comprehensive of these was the Lifeline program, under which eligible subscribers' monthly telephone bills were reduced by an amount equal to some or all of the residential subscriber-line charge. The other program was Link Up, which paid one-half the installation charges when low-income subscribers initiated telephone service. Both programs required subscribers to establish their eligibility under means tests established and administered by their states.

III. Universal Service Requirements of the 1996 Act

Even before passage of the Telecommunications Act of 1996, the Commission and much of the telecommunications industry expressed dissatisfaction with the patchwork of mechanisms that supported the goal of universal service. Although few questioned the purpose of these programs, it was increasingly clear that the system was inconsistent with deregulation and competition.

The so-called implicit subsidies were especially vulnerable to criticism. At a time when regulators were permitting competition in many of the service markets that subsidized basic service, LECs and IXCs complained that above-cost rate structures threatened their ability to compete. Similarly, regulators became concerned that as incumbent carriers dropped their basic service rates in response to competition, rates for basic residential service inevitably would rise.

The explicit subsidy programs also came in for criticism. Notably, IXCs complained that they should not be singled out as the sole source of support for the USF fund, and the larger LECs complained that they should not be the exclusive source of LTS subsidies for the access charges of high-cost carriers. Also, few commentators doubted that the USF and DEM weighting programs, which based support on reported and assumed costs, respectively, subsidized inefficiency on the part of smaller telephone companies. Finally, the status of LECs as the sole eligible recipients of USF funds would make rural and high-cost areas much less attractive to potential competitors, in the event that Congress should open the local exchange services market to competition.

Against this background, the Telecommunications Act of 1996 set out a framework for comprehensive reform of the system. The Act did not call into question the goals of the system; in fact, it reconfirmed those goals when it required the Commission, with the assistance of a Federal-State Joint Board, to adopt policies and rules that would promote "[q]uality services at just, reasonable, and affordable rates," including access to advanced services for all regions of the United States and "access to telecommunications services and information services for customers in rural, insular and high-cost areas that are reasonably comparable to those services provided in urban areas and that are available at rates that are reasonably comparable to rates charged for similar services in urban areas" [4]. At the same time, Congress called upon the Commission and the Joint Board to replace the old system, which combined implicit subsidies with explicit supports funded by only a subset of telecommunications carriers, with "specific, predictable, and sufficient" mechanisms that would draw support from "[e]very telecommunications carrier that provides interstate telecommunications services" [5]. The Act also provided that universal service support should be available, not just to incumbent telephone companies, but to any entity that a state commission designated as an "eligible telecommunications carrier" according to requirements set out in the Act [6].

The 1996 Act also authorized a new set of subsidies for schools, health care providers, and libraries. Specifically, Congress determined that telecommunications carriers should be required to provide discounted services, for educational purposes, to elementary schools, secondary schools, and libraries, with the amount of the discount to be reimbursed through offsets against the carriers' universal service contribution obligations or through explicit support payments through the universal service support mechanism [7]. Congress also determined that telecommunications carriers should be required to provide telecommunications services that are necessary to healthcare "to any public or nonprofit health care provider that serves patients who reside in rural areas at rates that are reasonably comparable to rates charged for similar services in urban areas in that state" [8].

These provisions of the Act gave the Commission and the states a broad mandate to fashion a new universal service system. In order to carry out that mandate, however, the Joint Board and the FCC were required to answer a number of hard questions that Congress had left unresolved. For example, it was up to the regulators to decide which services must be provided, and subsidized, under the heading of universal service. It also was up to the regulators to decide how the funds that supported the system would be calculated and how—and from whom—those funds would be collected. Finally, the Commission and the Joint Board were required to decide how the amounts payable to recipients of universal service support would be determined. These questions, and others, were the subject of intense contention that has continued long after passage of the 1996 Act.

IV. The Commission's Post-1996 Universal Service Rules

In their decisions and recommended decisions implementing the universal service provisions of the 1996 Act, the Joint Board and the Commission took an incremental approach [9]. Notably, the Commission found that it had neither a statutory mandate, nor the practical ability, to eliminate all implicit subsidies—many of which were embedded in intrastate rate structures over which the Commission had no jurisdiction. Accordingly, the Commission's reforms have consisted primarily of efforts to improve the system of explicit subsidies, based as much as possible on the mechanisms already in place before the 1996 Act was passed.

A. Defining Universal Service

Congress required the Commission to define those telecommunications services that would be supported by the universal service system [10]. In doing so, the Commission was to consider the extent to which the included services: (1) are essential to education, public health, or public safety; (2) have, through the operation of market choices by consumers, been subscribed to by a substantial majority of residential customers; (3) are being deployed in public telecommunications networks by telecommunications carriers; and (4) are consistent with the public interest, convenience, and necessity [11].

Based upon the recommendation of the Joint Board, the Commission adopted a definition that includes: "voice-grade access to the public switched network, with the ability to place and receive calls; dual tone multifrequency (DTMF) (i.e., touch-tone) signaling or its functional equivalent; single-party service; access to emergency services, including in some instances, access to 911 and enhanced 911 (E911) services; access to operator services; access to interexchange services; access to directory assistance; and toll-limitation services for qualifying low-income consumers" [12]. In order to ensure that this definition was not static and would not slight different technologies by which telecommunications services may be provided, a Joint Board was to revisit this definition not later than January 1, 2001.

B. Eligibility to Receive Universal Service Support

Before the 1996 Act became law, the class of carriers eligible for universal service support was readily identified. Local telephone service was provided by established telephone companies with monopoly franchises to serve a defined area. Because there were no other sources of local telephone services, all subsidies for basic, local service flowed to the established carriers.

In the 1996 Act, Congress recognized that if competition was to reach high-cost areas, the base of eligible carriers must be expanded to include competitors of the established telephone companies. Accordingly, Section

214 of the Act requires the states to designate any common carrier that meets certain statutory requirements as an "eligible telecommunications carrier for a service area designated by the State commission" [13]. Also, except in areas served by rural telephone companies as defined in the Act, the states could not refuse to designate more than one eligible telecommunications carrier for a service area if additional carriers met the statutory requirements [14]. With these provisions, Congress tried to ensure that competitors could enter high-cost markets on the same terms as incumbents.

Based on the Joint Board's recommendation, the Commission found that it did not have discretion, under Section 214(e)(1) of the Act, to add to the requirements set out in the statute for designation as an eligible telecommunications carrier. Accordingly, the Commission merely confirmed the provisions of Section 214(e), which requires eligible telecommunications carriers to "offer and advertise all the services supported by federal universal service support mechanisms throughout their service areas using their own facilities or a combination of their own facilities and resale of another carrier's services" [15].

These statutory terms did, however, present some questions of interpretation. Notably, commenters were divided concerning the extent to which eligible telecommunications carriers were required to use their own facilities to provide subsidized services. The statute made it clear, and the Joint Board confirmed, that eligible telecommunications carriers could not provide supported services through resold ILEC services alone. Less clear, however, was whether carriers that purchased unbundled network elements from ILECs were using their own facilities within the meaning of Section 214. Some commenters, hoping to slow the entry of subsidized competition, argued that Congress meant to limit eligibility to carriers that had made at least some investment in facilities not obtained from other carriers. The Commission, however, decided that Congress intended to give competitors discretion to choose the most efficient mode of facilities-based entry. Accordingly, the Commission determined that "a carrier that offers the federally supported services through the use of unbundled network elements in whole or in part, satisfies the facilities requirement of Section 214(e)(1)" [16].

The Commission also gave considerable attention to the service areas within which the states should designate eligible telecommunications carriers. Notably, the Joint Board and some commenters were concerned that states might discourage competition if they required carriers to serve unreasonably large areas as a condition of designation as eligible telecommunications carriers, or if they retained or designated service areas for which wireless carriers might have difficulty configuring their services. Accordingly, the

Commission recommended that the states take these concerns into account, and particularly recommended that "state commissions not [simply] designate service areas that are based on ILECs' study areas" [17]. The Commission also established expedited procedures to change study-area definitions in areas served by rural telephone companies, in which the statute required the service areas to coincide with the rural carriers' study areas.

C. Who Contributes to Universal Service Support?

Defining the class of universal service contributors became one of the Commission's most contentious tasks. Before enactment of the 1996 legislation, IXCs were the exclusive supporters of the high-cost fund, with each IXC contributing "through a tariffed interstate charge that was based on the number of subscriber lines presubscribed to the IXC" [18]. The 1996 Act expanded the class of mandatory contributors to include "every telecommunications carrier that provides interstate telecommunications services" [19]. The Act also gave the FCC discretion to require contributions from "[a]ny other provider of interstate telecommunications if the public interest so requires," and permitted the Commission to exempt carriers or classes of carriers where those carriers' contributions to the system would be *de minimis* [20].

On the recommendation of the Joint Board, the Commission construed the categories of mandatory contributors broadly so as to ensure the broadest possible base of support for the system. Accordingly, the Commission found that contributions were mandatory from all providers of the following categories of interstate services: (1) cellular telephone and paging services; (2) mobile radio services; (3) operator services; (4) PCS; (5) access to interexchange service; (6) special access; (7) wide-area telephone service (WATS); (8) toll-free services; (9) 900 services; (10) message telephone service (MTS); (11) private-line service; (12) telex; (13) telegraph; (14) video services; (15) satellite services; (16) resale services; (17) common carrier videoconferencing; (18) common carrier channel service; (19) common carrier video-distribution services to cable head-ends [21]. For each of these categories, the Commission determined that service is interstate, and therefore subject to universal service contributions, when the "communication or transmission originates in any state, territory, possession of the United States, or the District of Columbia and terminates in another state, territory, possession or the District of Columbia" [22].

The categories of services from which contributions are required are as important for what they leave out as for what they include. Notably, because the definition of "telecommunications service" in the Act includes the requirements that those services be provided "for a fee" and "directly to the public," the Commission concluded that only services provided on a common carrier basis give rise to mandatory universal service contribution obligations [23]. Accordingly, private networks and services provided only to a limited, defined group of users are not mandatory contributors (although the Commission exercised its discretion to require certain noncommon carrier providers of telecommunications to contribute) [24]. More controversially, the Commission concluded that information services, which are defined in the Act separately from telecommunications services, are neither mandatory nor permissive universal service contributors under the Act.

The exemption of information service providers from the universal service system was revisited by the Commission in 1998, after Congress directed the Commission to examine that and other issues raised by its reform of the universal service system. Notably, Senator Stevens of Alaska and others were concerned that as information services became a larger factor in the communications carried over telecommunications services and facilities, the information services exemption might erode the base of universal service support.

In April of 1998, the Commission issued a Report to Congress that confirmed the exemption of information services from universal service support obligations. After carefully parsing the Act's definitions of "telecommunications," "telecommunications service," "telecommunications carrier," and "information service," the Commission concluded that information services and telecommunication services are mutually exclusive categories under the Act. Accordingly, providers of information services do not fall within the statutory categories of mandatory or permissive contributors to the universal service system. The Commission did confirm, however, that providers of telecommunications services over which information services are provided must contribute to the system, and opined that some voice services provided by means of the Internet protocol are potentially subject to contribution obligations.

D. How Support Payments Are Calculated

Even before enactment of the Telecommunications Act of 1996, the Commission was considering reform of the method of calculation of universal service support payments. The existing program, which permitted recipients of USF support to recover on the basis of their reported local loop costs, did nothing to encourage efficiency among program participants.

In accordance with the recommendation of the Joint Board, the Commission decided to move toward a system in which high-cost carriers would be compensated, not on the basis of their embedded costs actually incurred to provide local loops, but on the basis of forward-looking economic cost. The Commission also determined that instead of compensating each carrier on the basis of its own, reported forward-looking economic costs, the system would base compensation on cost models that calculate average local loop costs in all areas presenting similar terrain and other cost-causing factors. The Commission first implemented such a model for nonrural LECs, with a view to eventual implementation of such a model for rural high-cost carriers, as well [25]. The Commission also has pledged to continue studying the possibility of establishing an "auction" program under which eligible carriers would bid for high-cost support by pledging to provide service at lower cost to the program than other bidders.

E. The Lifeline and Link-Up Programs

Upon the recommendation of the Joint Board, the Commission did not eliminate the Lifeline and Link-Up programs, which the Joint Board concluded were valuable sources of explicit support to low-income subscribers. Accordingly, the Commission adopted rules that would expand Lifeline to all states and make Lifeline support available directly to carriers on a competitively neutral basis. The Commission also determined that Link-Up funding should be removed from the jurisdictional separations process and made available to any eligible telecommunications carrier [26].

F. Support for Schools and Libraries

The Schools and Libraries program proved to be one of the Commission's most controversial initiatives. Pursuant to Congress's broad direction, the Commission adopted a program under which nonprofit public and private elementary and secondary schools, with endowments of not more than $50 million, would be eligible to receive discounts on certain services. Similarly, discounts would be available to independent, nonprofit libraries. Under the program, those institutions may purchase discounted telecommunications services, Internet access, and internal connections. Discounts range from 20% to 90%, depending upon the recipient institution's location (urban or rural) and degree of economic disadvantage. Eligible institutions would pay only the percentage of charges for discounted services remaining after the discount is applied, and the service provider would be compensated from a portion of the federal universal service fund.

The controversy concerning the Schools and Libraries program arises principally from its magnitude and the degree to which the public was unaware of, and unprepared for, the appearance of these charges on their telephone bills. The FCC set an initial cap on the program of $2.25 billion annually (a number that later was slightly reduced). The funds to collect this revenue, like universal service contributions, generally were collected from providers of interstate communications services as an assessment against their revenues from those services. When some carriers translated those charges into a line item on their customers' bills, however, the resulting controversy was so serious that the FCC tried to discourage carriers from reflecting the charge on their invoices. Whatever its merits as a matter of public policy, the Schools and Libraries program is a classic example of a legislative and regulatory exaction imposed upon an ill-informed and ill-prepared public.

Conclusion

The universal service program, with its elaborate system of subsidies for politically favored services and ratepayers, remains the great exception to the FCC's ongoing program of deregulation of common carrier telecommunications services. Although the post-1996 reforms have created a system that is at last open and quantifiable, the controversy surrounding the Schools and Libraries program shows that openness may erode, rather than increase, public support for programs of this kind.

Endnotes

[1] For additional information on universal service and related programs, see the section on "Contributions to Universal Service and Other Funds" in Chapter 5, which describes state funds, the telecommunications relay service (TRS) fund, support for numbering administration and other mechanisms to which telecommunications carriers must contribute.

[2] The process of allocating costs to the interstate and intrastate jurisdictions is called the jurisdictional separations process: 47 C.F.R, Part 36. As part of the universal service policy of supporting local service through interstate charges, the separations rules have deliberately overallocated costs to the interstate jurisdiction.

[3] Eligible carriers recovered the full, embedded cost of their local loops.

[4] 47 U.S.C. § 254(b)(1)-(3).

[5] *Id.* § 254(d).

[6] *Id.* §§ 254(e), 214(e). In order to be designated an eligible telecommunications carrier, an entity must, throughout the service area for which it is designated, "offer the services that are supported by Federal universal service support mechanisms, either using its own facilities or a combination of its own facilities and resale of another carrier's services; and advertise the availability of such services and the charges therefor using media of general distribution. *Id.* § 214(e)(1).

[7] *Id.* § 254(h)(B).

[8] *Id.* § 354(h)(A).

[9] Implementation of the universal service provisions of the 1996 Act has required a series of rulemaking proceedings. The usual procedure has been for the Federal-State Joint Board, which includes federal and state regulators and a consumer advocate, to conduct its own proceedings, take comments from interested parties and furnish the FCC with recommended decisions. The Commission then conducts its own rulemaking proceedings and adopts regulations. For the most part, the Commission's decisions have adopted the recommendations of the Joint Board.

[10] 47 U.S.C. § 254(c)(1).

[11] *Id.* § 254(c)(1)(A)-(D).

[12] Federal-State Joint Board on Universal Service, 12 FCC Rcd. 8776 (1997) § 22.

[13] 47 U.S.C. § 214(e)(2).

[14] *Id.*

[15] *Id.* § 24.

[16] *Id.*

[17] Federal-State Joint Board on Universal Service, 12 FCC Rcd. 8776 at § 25. A study area "is generally an incumbent LEC's pre-existing service area in a given state." Federal-State Joint Board on Universal Service, 12 FCC Rcd .87 at n. 541.

[18] Federal-State Joint Board on Universal Service, CC Docket No. 96-45, Report to Congress (April 10, 1998) § 108 (Report to Congress).

[19] 47 U.S.C. § 254(d).

[20] *Id.*

[21] Federal-State Joint Board on Universal Service, 12 FCC Rcd. 8776 at §§ 780-781.

[22] *Id.*, 12 FCC Rcd. at § 778. For any private line or WATS line, the service is classified as interstate if the line carries more than 10% of its traffic as interstate transmissions. *Id.*

[23] *Id.*, 12 FCC Rcd. at § 786. Services provided to other carriers, such as exchange access service, are included in this category and give rise to universal service contribution obligations. *Id.*

[24] These include "private network operators that lease excess capacity on a noncommon carrier basis for interstate transmissions" and "payphone aggregators." (Report to Congress at § 117.)

[25] The Commission eliminated the DEM weighting program, and included in local switching costs assigned to the interstate jurisdiction an amount based on a modified DEM weighting factor. The Commission also made substantial changes to the LTS system. Federal-State Joint Board on Universal Service, 12 FCC Rcd. at §§ 304-306.

[26] *Id.* §§ 326409.

11

International Services

The Communications Act gives the FCC jurisdiction over both "interstate and foreign communication by wire or radio" [1]. Accordingly, all providers of telecommunications service between the United States and foreign destinations, regardless of the medium or technology by which that service is provided, must comply with the FCC's regulations concerning licensing, ownership, and terms and conditions of service. The Commission's jurisdiction over international services is exercised primarily under Section 214 of the Act, which governs licensing of carriers and the initiation and discontinuation of service; and Section 310, which limits foreign ownership of radio licenses in the United States.

I. The Section 214 Process

The Communications Act requires all carriers to seek permission from the FCC before extending lines to provide telecommunications service or, once service has commenced, to discontinue that service. Specifically, Section 214 of the Act provides that "[n]o carrier shall undertake the construction of a new line or of an extension of any line, or shall acquire or operate any line, or shall engage in transmission over or by means of such additional or extended line, unless and until there shall first have been obtained from the Commission a certificate that the present or future public convenience and necessity require or will require the construction" [2].

For domestic communications, the requirement to obtain previous FCC authority has been eliminated as part of the FCC's efforts to reduce the regulatory burden on carriers. Today, a carrier's Section 214 authority to provide domestic service is subsumed in its facilities license (satellite, microwave, or cable), while no specific authority or license is needed to resell domestic service.

The FCC has retained the licensing requirement in the international arena, however, because of the Commission's perception that international services present ongoing competitive issues. Because it is Section 214 of the Communications Act of 1934 that requires carriers to seek FCC permission, that license is known as "Section 214 authority," and the application that precedes it is commonly known as a "Section 214 application".

The requirement to obtain authority under Section 214 applies only to providers of basic telecommunications service on a common carrier basis. Accordingly, service providers that will offer only information services or will provide service on a private basis need not comply with Section 214.

When an application does propose to offer international telecommunications service on a common carrier basis, the Commission will grant the application if it is found to be in the public interest. The primary, although not exclusive, focus of this inquiry is the potential effect, if the application is granted, on competition in the U.S. market for international telecommunications services. Historically, the Commission's competitive analysis is triggered when the applicant is a foreign carrier, or a U.S. carrier affiliated with a foreign carrier that operates in a destination market. In these cases, the Commission has scrutinized the market power, if any, of the foreign or affiliated carrier to determine whether that carrier can discriminate in favor of the applicant in pricing of, or access to, gateway, transport, or local network facilities in a destination country.

In recent years, the Commission's competitive analysis has been dramatically simplified by the adoption, among member nations of the World Trade Organization (WTO), of a Basic Telecommunications Agreement under which the United States and 68 other countries have pledged to open their domestic telecommunications markets to competition [3]. After the WTO Agreement was concluded, the Commission relaxed the standards for U.S. market entry for foreign carriers from WTO countries and U.S. carriers affiliated with carriers from WTO-member countries.

A. Section 214 Applications of U.S. Carriers

Under the Commission's post-WTO Agreement rules, most U.S. carriers that apply for Section 214 authority will obtain approval on a "streamlined" basis. Specifically, an applicant may obtain a streamlined authorization for all routes on which the applicant is not affiliated with a carrier that operates in the destination market, or where any of the following criteria are met:

- The applicant is affiliated with a carrier that operates in the destination market, but the Commission previously has determined that the foreign carrier lacks market power;
- The applicant is affiliated with a carrier that operates in the destination market, but the foreign carrier "has less than a 50% market share in the international transport and local access markets in the destination country";
- The applicant is affiliated with a carrier that operates in the destination market, but the foreign carrier is from a country that is a member nation of the WTO and "the applicant seeks to serve that country solely by reselling the switched services of unaffiliated U.S. international carriers"; or
- The applicant is affiliated with a carrier that operates in the destination market, but the foreign carrier is from a WTO-member country and the applicant states that it will comply with certain FCC requirements applicable to "dominant" carriers [4].

An application that meets one of the criteria for streamlined treatment, if deemed acceptable for filing, will be placed on public notice and will be granted automatically, 14 days after the date of the public notice, unless the Commission first notifies the applicant in writing that the application has been removed from streamlined treatment. During the 14-day period, the Commission and Executive Branch agencies will review the application to determine whether it presents national security, law enforcement, foreign policy, or trade policy concerns.

If the application is not eligible for streamlining, the Commission will take comments and review the application under its public interest standard, with particular attention to the ability of the applicant's foreign affiliate to harm competition in the U.S. Market [5].

B. Section 214 Applications of Foreign Carriers

The FCC's regulations concerning Section 214 applications from foreign carriers also reflect the liberalizing influence of the WTO Telecommunications Agreement. Specifically, the Commission applies a strong presumption in favor of granting all applications from WTO-member nation carriers [6]. The presumption may be overcome only by "serious concerns raised by the Executive Branch regarding national security, law enforcement, foreign policy, or trade issues or, in the exceptional case where a carrier's entry presents a very high risk to competition" [7]. Where a WTO applicant's entry would present a very high risk to competition, the Commission may deny the application or impose "additional reporting requirements, prior approval for circuit additions, or other measures designed to ensure that a carrier with the ability to exercise market power in a relevant foreign market does not use that power to harm consumers in the U.S. Market" [8].

Where a Section 214 application is received from a foreign carrier from a non-WTO country, however, that application will receive greater scrutiny. Specifically, the Commission reviews a non-WTO application under the effective competitive opportunities (ECO) test, "which requires, as a condition of foreign carrier entry into the U.S. market, that there be no legal or practical restrictions on U.S. carriers' entry into the foreign carrier's market" [9]. This standard permits a wide-ranging inquiry into the competitive opportunities afforded to U.S. carriers in a foreign applicant's home market. Obviously, foreign administrations that continue to protect monopoly telephone companies and restrict new entry into their markets are unlikely to pass the ECO test.

C. Dominant Carrier Safeguards

The FCC also distinguishes between dominant and nondominant carriers on international routes, and imposes regulatory burdens on the former that it does not impose on the latter. Specifically, the FCC will classify a U.S.-licensed carrier as dominant on a particular route if that carrier is affiliated with a foreign carrier that possesses market power in a relevant market on the foreign end of the route [10]. In determining whether a carrier meets this definition of dominance, the Commission applies a "rebuttable presumption" that carries with less than a 50% market share on the foreign end of a route do not present a substantial risk of harm in the U.S. market. That presumption may be overcome only by specific evidence of the carrier's ability to harm competition.

If a carrier is classified as dominant under the Commission's rules, its application to provide international service will be granted subject to certain reporting requirements, including the requirement that it file quarterly traffic and revenue reports for the routes on which the carrier is dominant.

II. Foreign-Ownership Restrictions

Applicants for licenses to provide domestic and international services in the United States also are subject, to the extent they will use common carrier radio licenses issued by the FCC for any part of their service, to the foreign-ownership restrictions applied to broadcast and common carrier radio licensees under Section 310(b) of the Communications Act. Specifically, Section 310(b) provides that:

> No broadcast or common carrier or aeronautical en route or aeronautical fixed radio station license shall be granted to or held by—
>
> 1. Any alien or the representative of any alien;
>
> 2. Any corporation organized under the laws of any foreign government;
>
> 3. Any corporation of which more than one-fifth of the capital stock is owned of record or voted by aliens or their representatives or by a foreign government or representative thereof or by any corporation organized under the laws of a foreign country;
>
> 4. Any corporation directly or indirectly controlled by another corporation of which more than one-fourth of the capital stock is owned of record or voted by aliens, their representatives, or by a foreign government or representative thereof, or by any corporation organized under the laws of a foreign country, if the Commission finds that the public interest will be served by the refusal or revocation of such license.

The first three subsections of Section 310(b) permit the Commission no discretion in their application. Specifically, the Commission may not grant radio licenses to non-U.S. citizens or to their representatives, or to entities controlled by such persons. Similarly, under Section 310(b)(2), the Commission may not grant a radio license to a corporation organized under the laws of a foreign country. Finally, Section 310(b)(3) bars the FCC from granting a license to any corporation of which more than 20% of the capital stock is directly owned or voted by aliens, their representatives, a foreign government or a foreign corporation.

Of the four subsections of Section 310(b) only subsection (b)(4) allows the Commission some discretion in its application. Under that subsection, the Commission may—but is not required to—deny a broadcast or common carrier radio license to any corporation directly or indirectly controlled by another domestic corporation if that other corporation has more than one-fourth of its stock owned or voted by aliens, their representatives, a foreign government, or a foreign corporation. The Commission applies a public interest test to applications that exceed the Section 310(b)(4) benchmark, and sometimes finds that foreign stock ownership substantially in excess of the benchmark will not harm the public interest.

Although subsections 310(b)(2)–(4) refer to corporations, the Commission has applied those restrictions to a variety of noncorporate entities. For example, the Commission has applied the restrictions to limited partnerships and irrevocable trusts. The Commission also applies a "multiplier" to diminish the interest of a foreign entity that holds an interest in, but does not control, a domestic entity that holds an interest in a licensee. For example, if the multiplier is used, a 40% interest in a company that holds 25% of a licensee may be deemed only a 10% (40% × 25%) interest in the licensee itself. The use of a multiplier permits the Commission to gauge more accurately the level of actual ownership interest a foreign entity has over a licensee for purposes of determining whether the statutory benchmarks have been exceeded.

The foreign-ownership restrictions of Section 310(b), which were enacted in the belief that the national security might be compromised by foreign control of U.S. radio stations, seem anachronistic in an age of increasing globalization. Nonetheless, those restrictions remain binding on the Commission and must be taken seriously by anyone seeking a broadcast radio, television, common carrier microwave, CMRS, or other common carrier or broadcast license in the United States.

III. The International Settlements Policy

One of the more arcane artifacts of the monopoly era is the system of international settlements and accounting rates. Accounting rates are the means by which carriers at each end of an international switched telephone call compensate each other for transporting and terminating the call. So, for example, if two carriers agree that the accounting rate for calls between their home countries will be $4.00 per minute, to be divided on a 50/50 basis, then each carrier will receive $2.00 from the other when it terminates a call. Under this arrangement, if minutes transmitted in one direction equal minutes transmitted in the other direction, each company will be paid the same amount by the other. If one company terminates more minutes than the other, however, the company that terminates more traffic will make more money from the arrangement.

The FCC has perceived two problems with the system of settlements and accounting rates. First, when U.S. carriers seek traffic termination arrangements with monopoly carriers in foreign countries, they may be subject to "whipsawing" (i.e., demands that the U.S. carrier pay a higher accounting rate or agree to something other than a 50/50 split between the two companies). Second, the FCC long ago concluded that monopoly foreign carriers and their governments (often, these are the same entity) use above-cost accounting rates as a means of extracting monopoly rents from U.S. telephone ratepayers.

The Commission dealt with the first problem by adopting its Uniform Settlements Policy (USP), which required U.S. carriers to serve identical foreign points under identical accounting rates, settlement arrangements, and divisions of tolls. The USP also required U.S. carriers to accept return traffic from a foreign correspondent only in the same proportion as it sent traffic to that correspondent.

To address the second problem, the Commission recently adopted a set of rules intended to enforce reductions in foreign carriers' accounting rates. Specifically, the FCC has prescribed "benchmark" settlement rates for calls to foreign countries, which vary with the destination country's level of economic development. The Commission also factors accounting rate reform into its evaluation of Section 214 applications from foreign and foreign-affiliated carriers, by requiring such carriers seeking authorization to provide facilities-based switched or private-line service to demonstrate that at least 50% of all settled traffic on the routes for which the carriers seek authorization are settled at or below the applicable benchmark rate [11].

Finally, in 1999 the Commission substantially modified its USP [12]. Pursuant to the new policy, U.S. carriers may negotiate varying agreements for the exchange of traffic with foreign carriers, so long as the foreign carriers are nondominant in their home markets or operate in liberalized markets. In order for this policy to apply, the foreign correspondent carrier must lack market or U.S. carriers must be able to terminate at least 50% of their traffic in the destination market at 75% or less of the applicable benchmark rate [13].

Conclusion

Although the FCC has substantially liberalized its regulation of foreign services, especially since the signing of the WTO Basic Telecommunications

Agreement, the Commission remains concerned about possible abuse of U.S. consumers by dominant foreign carriers. For this reason, at this writing, the Commission still retains dominant carrier reporting requirements and other regulations intended to monitor and restrain carriers with market power, and has not detariffed international services (although it may do so shortly).

Finally, anyone seeking authorization to provide international services should be aware that the survey of issues offered in this chapter is far from exhaustive. Notably, this discussion does not include such subjects as international simple resale (ISR), callback, and the requirements for licensing of international submarine cables. As with every subject discussed in this introductory book, persons facing real-world decisions in this area should obtain legal advice from a specialist in the field.

Endnotes

[1] 47 U.S.C. § 152(a).

[2] *Id.* § 214(a).

[3] *See* Rules and Policies on Foreign Participation in the U.S. Telecommunications Market. 12 FCC Rcd. 23891 (1997) § 2.

[4] 1998 Biennial Regulatory Review of International Common Carrier Regulations, 14 FCC Rcd. 4909 (1999) § 22.

[5] All carriers with FCC licenses to provide international services also must observe the "no special concessions" rule, which prohibits U.S. international carriers from "entering into exclusive arrangements with any foreign carrier affecting traffic or revenue flows to or from the United States." The FCC has modified this rule, however, to apply only to "dealings between U.S. carriers and foreign carriers that possess market power in a relevant market on the foreign end of an international route." Rules and Policies on Foreign Participation in the U.S. Telecommunications Market, Market Entry and Regulation of Foreign-Affiliated Entities, 12 FCC Rcd. 23891, 23957 (1997) (Foreign Participation Order); 47 C.F.R. § 63.14.

[6] Foreign Participation Order.

[7] *Id.* at 23922.

[8] *Id.* at 23914.

[9] *Id.* at 23895.

[10] *Id.* at 22391. Affiliation is defined as an ownership interest of one carrier in another of more than 25%. *Id.* at 22392.

[11] If a foreign carrier reseller is from a WTO-member country, that carrier may operate in the U.S. without meeting the benchmark conditions.

[12] 1998 Biennial Regulatory Review—Reform of the International Settlements Policy and Associated Filing Requirements, 14 FCC Rcd. 7963 (1999).

[13] *Id.* at § 6.

Appendix A
The Economic Background of Telecommunications Law

Many of the legal and regulatory problems discussed in this book are also economic problems. Some of those issues (i.e., the predatory pricing claims brought against the Bell system and the FCC's rationale for its cost allocation rules) have to do with finding the economically optimal relationship between the cost of a service and the rates charged for that service. Other issues (i.e., the FCC's distinction between dominant and nondominant carriers, and the efforts of state commissions to identify markets that safely can be deregulated) require an analysis of market structures and the economic power of various players within those markets.

So far we have discussed each of these topics with only cursory attention to the underlying economic theories. While this approach is adequate as far as it goes, any student or practitioner who wants to participate fully in the debates that will decide the future of regulation must know some basic economics.

Fortunately, the necessary vocabulary of economic principles is simple, powerful, and well worth the time needed to master it. This appendix, which assumes no prior knowledge of the subject, builds the rudiments of that vocabulary one step at a time.

I. Competition and Monopoly

We can focus our discussion of economics by thinking of telecommunications law as a set of rules shaped by the tension between *competition* and *monopoly*. While this view suffers from some oversimplification, it is a useful and largely accurate generalization.

In the age of traditional regulation, for example, the commissions acknowledged (and in fact protected) the monopoly status of common carriers, but adopted elaborate rules to prevent those carriers from extracting monopoly profits or providing inferior service. Put another way, one objective of traditional regulation has been to force monopolists to behave as though they faced competition.

When competition first was permitted in selected telecommunications markets, a new kind of tension between competition and monopoly arose. Now, both the commissions and the antitrust courts find themselves restraining the established carriers from behaving like monopolists in their dealings with the new entrants, while at the same time leaving those carriers some freedom to compete.

More recently, many commissions have adopted incentive plans that combine monopoly with some of the efficiency incentives usually associated with competition. The commissions gradually are deregulating some markets, but only after satisfying themselves that the established carriers lack monopoly power in those markets (i.e., that competition in those markets is effective).

In keeping with the pervasive importance of this theme, the first part of this appendix will examine at some length how competitive and noncompetitive markets differ, and how the behavior of firms in both types of markets affects the welfare of consumers. After we have laid that groundwork, we apply these principles to particular controversies in telecommunications law and regulation.

Consumer Welfare: The Economic Case for Competition

Before we explain how competitive and monopolistic markets work, we should understand the sense in which economists regard competition as superior to other forms of market organization.

When economists say that competition is to our benefit, they are not talking about us in all of our human richness: they are talking about us as consumers [1]. They are saying that competition best permits us to satisfy our various desires for goods and services—a condition they refer to as *consumer welfare*. To understand this concept, we must know what it includes and what it leaves out.

First, the consumer-welfare model does not describe any ideal distribution of incomes. It takes consumers as they are, with their sharply differing abilities to acquire the goods and services they desire.

Second, the consumer-welfare model does not describe a method for increasing the capacity of a society to produce goods and services. Here, too, the economist takes the world as it is, with each economy's productive capacity limited by the resources and technology available to it.

Finally, the consumer-welfare model does not prescribe the things that consumers ought to desire. The economist may think that consumers would benefit from spending their money on postgraduate education rather than beer, but it is not part of the economist's task to say so. Consumers' desires, and the things they are willing to give up to satisfy those desires, are taken as given.

What consumer welfare does describe is a condition in which scarce resources are allocated in a way that best satisfies the desires of consumers, given the limitations of their resources and those of their society. While this ideal never is perfectly achieved, economists generally agree that markets approach this ideal to the extent they are competitive.

Under competition, firms reduce their costs as much as possible and charge consumers no more than is necessary to cover those costs [2]. This enhances consumer welfare in two ways: no more of society's resources than necessary are used to make the product (economists call this *productive efficiency*), and consumers pay no more for the product than the cost of making it (economists call this *allocative,* or *pricing, efficiency*). When both kinds of efficiency are maximized, consumers can satisfy more of their material wants than they could satisfy with any alternative allocation of resources.

Monopoly, on the other hand, encourages producers to set their prices above the associated costs. Monopolists earn economic profits (i.e., a return in excess of the cost of capital). In so doing they divert excessive societal resources into the making of their products (lowering productive efficiency), and leave consumers of their products with fewer resources with which to satisfy other needs (lowering allocative efficiency).

As we have said, many of the rules of telecommunications law and regulation try to encourage competition and discourage monopoly, or make monopolistic markets behave like competitive ones. Polices of this kind cannot reliably be advanced, and rules based on those policies cannot be evaluated, without some knowledge of *how* competition and monopoly achieve their effects on consumers.

Unfortunately, the real world affords few laboratory specimens of markets in pure competition or pure monopoly; and even if we could find them, it would be impossible to isolate them from the influences of the larger economy. When we study competition and monopoly, therefore, we build theoretical models that exclude all factors except those that are hypothesized, *all other things being equal*, to make markets more or less competitive. Once we understand these models of pure competition and pure monopoly, we are better able to recognize those real-world conditions that can be expected to enhance or undermine consumer welfare.

In the following pages, therefore, we describe a market in pure competition and show how it achieves pricing efficiency and productive efficiency. Then we show how monopoly undermines those goals.

A. What Is a Competitive Market?

A perfectly competitive market is one in which individual producers have no power to affect the market price or the total supply of a product. (In the real world, markets tend to be imperfectly competitive, with suppliers enjoying at least some latitude in pricing.)

Let us imagine, therefore, a perfectly competitive market for the purchase and sale of hang gliders. Such a market (if we could find it) would have the following characteristics:

- Each supplier's glider is exactly like every other supplier's glider: there are no differences in quality, and customers are unwilling to pay a premium to obtain a particular brand [3].

- All buyers are fully informed, at all times, about the prices at which hang gliders are available from different suppliers [4].

- There are many buyers and many sellers of hang gliders, so that no single buyer's purchasing decisions, and no single supplier's production choices, can affect the price or total supply of the product.

- There are no restrictions on entry to, or exit from, the hang glider market. New suppliers can enter the market readily.

In the discussion that follows, we explain the workings of this theoretical market from two perspectives: first, we show how competition forces producers to earn no more than their costs of production (i.e., to achieve pricing efficiency); and second, we show how competition forces producers to minimize those costs (i.e., to achieve productive efficiency).

1. How Firms in Competition Achieve Pricing Efficiency

Pricing efficiency, as we have said, assures that consumers will pay no more for goods and services than it costs to produce them. It is important to know that when we speak of the costs of production, we include in those costs a normal return on capital (i.e., a return to investors that is comparable to the return offered on investments presenting comparable risk). Firms that recover their costs of production in this sense are making a normal profit; earnings above that level are called *economic profit*, and prices that include economic profit are not optimally efficient.

So how does competition assure that customers will pay only the cost of production and that suppliers will not enjoy economic profit? *The key is that in pure competition the market—not the supplier—sets the price for the product, and suppliers adjust their output to the level that earns them the most profit.* Those profits, in turn, always are maximized when producers choose that level of output at which their costs equal the revenue they earn by selling at the market price. This process, when in a state we call *long-run equilibrium* (a term we define shortly), squeezes out economic profit relentlessly.

To understand this process better we must explore two phenomena. The first is the mechanism by which supply and demand (rather than the actions of individual producers) set the price in a competitive market. The second is the process by which producers find their profit-maximizing level of output.

a. How Supply and Demand Set the Market Price

A market is no more than the coming together of producers of goods and the consumers who want to obtain those goods [5]. The ongoing commerce between buyers and sellers in a competitive market yields, at any moment, a market price for the product.

To show how this works we begin with demand, which is nothing more than the amount of a product any consumer (or group of consumers) is willing to purchase at any particular price or range of prices. For most consumers and most products, the willingness to buy varies inversely with price: consumers will buy more when the price is lower, and less when the price is higher [6].

Consumer demand for a product can be expressed in a table or (more usefully) in a graph. Take a look at Table A.1 and Figure A.1: both show the weekly market demand for hang gliders at prices ranging from $1,300 to $1,800. We call Table A.1 the demand *schedule*, and Figure A.1 the demand *curve*.

Table A.1
Market Demand Schedule for Hang Gliders (Per Week)

Price	Quantity Demanded
$1,300	2,500
1,400	2,400
1,500	2,300
1,600	2,200
1,700	2,100
1,800	2,000

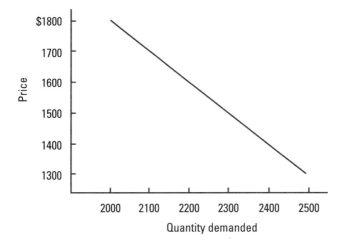

Figure A.1 Market demand curve for hang gliders (per week).

If we turn to the supply side of the hang glider market, we find the mirror image of the demand side. The willingness of suppliers to produce, like the willingness of consumers to buy, varies with price; but the willingness of suppliers to produce declines, rather than rises, as prices fall.

Supply can be stated for the whole market or for a single producer. The Sudden Downdraft Hang Glider Works (SD Gliders), for example, assembles gliders in a small shop and sells them directly to consumers. As Table A.2 shows, SD will make 21 gliders a week if it can sell them for $1,300, 22 gliders if it can sell them for $1,400, and so on to a maximum output (so far as this table tells us) of 26 gliders a week at $1,800 [7]. Figure A.2 plots the same data as a curve [8].

If we now make the simplifying assumption that the hang glider industry consists of 100 small shops just like SD Gliders, and that those shops all have the same supply curve, the supply schedule and supply curve for the entire market are easy to plot. At each price between $1,300 and $1,800, the industry will supply 10 times SD's output. The results are shown at Table A.3 and Figure A.3.

If we now place the demand and supply curves for hang gliders on a single graph, as we do at Figure A.4, we find that they intersect at $1,500. This means that $1,500 is the *market*, or *equilibrium*, price (i.e., the price at which the quantity producers are willing to supply just equals the quantity consumers are willing to buy). As the graph shows, at any other price there will be more, or less, of the product available than consumers are willing to buy. This means that at any price other than $1,500, suppliers will forego sales that they are willing to make, or will be stuck with unsold inventory.

This graph suggests why producers in a competitive market cannot set the price for their products. Imagine, for example, that one hang glider maker sets its price at $1,600: no one will have any reason to pay that amount, and the company will be left with a stock of unsold gliders until it

Table A.2
Weekly Supply Schedule for SD Gliders

Price	Quantity Supplied
$1,300	21
1,400	22
1,500	23
1,600	24
1,700	25
1,800	26

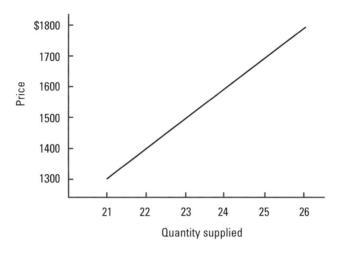

Figure A.2 Weekly supply curve for SD gliders.

relents and lowers its price to $1,500 [9]. At the same time, each producer can sell as many gliders as it wants at the market price.

What happens when demand changes? Suppose hang-gliding enjoys a sudden surge in popularity and overall demand rises?

What happens is shown in Figure A.5. More hang gliders are wanted at $1,500, and at every other price; the demand curve shifts to the right. If we

Table A.3
Market Supply Schedule for Hang Gliders (Per Week)

Price	Quantity Supplied
$1,300	2,100
1,400	2,200
1,500	2,300
1,600	2,400
1,700	2,500
1,800	2,600

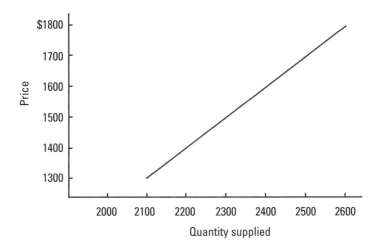

Figure A.3 Market supply curve for hang gliders (per week).

compare the new curve with the old, we find that at each price the quantity demanded is greater than before the demand curve shifted.

What happens if demand declines? In that case the demand curve shifts to the left: consumers will buy less at every price and the market, or equilibrium, price will fall.

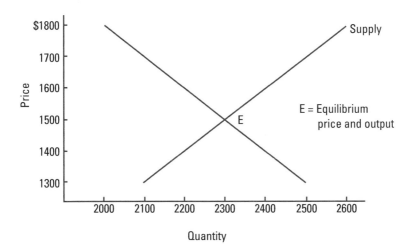

Figure A.4 Equilibrium market price for hang gliders.

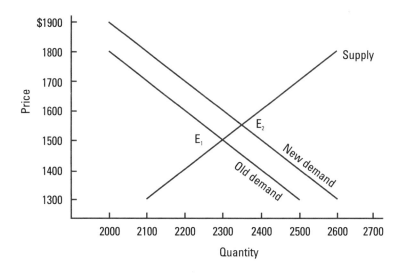

Figure A.5 Effect of an increase demand.

b. How Competitive Firms Maximize Profits

We have shown how the market sets the equilibrium price and why (when many suppliers compete for business) individual producers cannot alter that price. Now we explore how, under these circumstances, producers maximize their profits by selling the quantity of their product at which their costs of production equal the market price. In the course of this discussion we learn exactly how the hang glider market's supply curve is determined.

Using Marginal Cost to Find the Most Profitable Output

Profit is a relationship between revenue and cost. Since cost (for an economist) includes a normal return on capital, a firm makes a normal profit when its revenues equal its costs, and earns an above-normal profit when its revenues exceed its costs.

Costs, in turn, vary with output. Each time a supplier produces an additional unit of its product, it incurs some additional cost that it would have avoided if it had not made that additional unit. When the supplier sells that additional unit of output, it earns revenue. In pure competition, that additional revenue is always the market price; but the additional cost may be less than, equal to, or greater than the market price, depending on how efficient the firm is at that level of output [10].

The difference between the total cost incurred at one level of production and the total cost at a higher level of production is called the *marginal cost* of the decision to produce at the higher level. Marginal cost is a vitally important concept: as we demonstrate below, firms in competition achieve pricing efficiency when they find the level of output at which their marginal cost equals the market price.

Table A.4 takes us back to SD Gliders for a demonstration of marginal cost analysis. Note, for example, the information shown at the 21-glider output point. At that output your total cost of production is $28,686. The revenue you will earn at that level of output is $31,500 (21 gliders times the market price of $1,500). If we make and sell 21 gliders, therefore, we earn total profits of $2,814.

Will our profits be higher if we make 22 gliders? To find out we determine how much cost will be added by making 22 gliders instead of 21, and subtract that additional cost from the revenue that will be realized when you sell the additional glider [11]. In this case, making the 22nd glider will add $1,384.25 to your total cost, which is $115.75 less than the $1,500 you will earn when you sell it. Making glider number 22, therefore, is a profitable decision.

We can continue to increase production until we reach the first level of output at which marginal cost is greater than the market price. When we reach that point, our profit-maximizing decision is to reduce production to the previous level of output [12]. Applying this principle to Table A.4, we find that SD Gliders will lose profits if it increases production to 24 gliders, but will maximize its profits if it holds production to 23 gliders [13].

Now we know how individual producers in purely competitive markets set their level of output: they increase output until marginal revenue equals marginal cost, or up to the level at which the next unit of output would exceed marginal cost.

Having said all of this, we should caution that business people do not typically calculate marginal cost for each level of output before deciding how much of their product to make. Because real-world markets are not purely competitive and change constantly, businesses lack the information necessary to make such an analysis. Nonetheless, and all other things being equal, firms facing competition will succeed to the extent their output decisions approximate this ideal.

Table A.4
Cost and Revenue Schedule for SD Gliders

Q	A/C	TC	MC	MR	TR	Profit
21	$1,366	$28,686	$1,384	$1,500	$31,500	$2,814
22	1,366.83	30,070.26	1,450	1,500	33,000	2,929.74
23	1,370.44	31,520.12	1,550	1,500	34,500	2,979.88
24	1,377.92	33,070.08	1,665	1,500	36,000	2,929.92
25	1,389.41	34,735.25	1,775	1,500	37,500	2,764.75
26	1,404.24	36,510.24	—	—	39,000	2,489.76

Q = Quantity of hang gliders produced per week
A/C = Average cost (total cost divided by quantity produced)
TC = Total cost
MC = Marginal cost (rounded to the nearest dollar)
MR = Marginal revenue (always equals market price when market is purely competitive)
TR = Total revenue
Profit = Total profit earned at each level of output

c. Competitive Markets in Long-Run Equilibrium

To complete our understanding of purely competitive markets, we look now at how suppliers react to changes in demand and overall costs. This discussion shows the difference between markets in short-run and long-run equilibrium, and why economists say that competitive markets in long-run equilibrium are more efficient than those in short-run equilibrium.

For this purpose we return to our hang glider shop. We saw that with the market price of hang gliders at $1,500, our profit-maximizing level of output was 23 gliders a week. Let us now assume (as we did when we drew the industry's supply curve) that the hang glider market is served by 100 manufacturers, and let us also assume that each of those 99 other suppliers has a plant of the same size and cost structure as your own.

Assume again that hang gliding enjoys a surge in popularity, and all 100 suppliers are swamped with orders. How will you and the other manufacturers respond to this increase in demand?

One response, of course, is to put on an extra shift and increase production to meet the new demand. But look again at Table A.4; at the present market price, an increase in production will reduce your profits. You and your competitors (who have the same cost structure as SD) will not increase production, therefore, until the competing buyers bid up the price.

Assume that the market price is bid up to $1,700. Now you and the other suppliers have a choice to make. You may continue producing at your present rate, or increase output to a new, profit-maximizing level.

The obvious way to make this choice is to compare marginal cost with marginal revenue at each likely level of higher output. Because the market is competitive, our marginal revenue is the market price—that is, the *new* market price of $1,700.

We turn again to Table A.4. At $1,700 we can increase production to 25 gliders a week and incur a marginal cost of $1,665. Since that number is lower than the marginal revenue of $1,700, expansion of production to 25 gliders is worthwhile. (Expansion to 26 gliders or more, however, will cause marginal cost to exceed $1,700 and will diminish our profits.)

We have seen, in this example, how a firm reaches short-run equilibrium in the hang glider market after the old equilibrium is upset by a change in demand [14]. While the market is in short-run equilibrium, the firms are enjoying *economic profit*—that is, they are earning more than the minimum return needed to cover their expenses and their cost of capital. If we look at the total profit column of Table A.4, we see a total profit at 23 gliders a week (the old profit-maximizing output) and a total profit at 25 gliders a week (the new profit-maximizing output). The difference between the latter number and the former number is economic profit.

Short-run equilibrium, however, does not last. If firms continue to earn economic profits, then new capital will be attracted to the industry: existing firms will expand or new firms will enter the business. Eventually, the increase in supply will depress the price of hang gliders. (Look again at Figure A.5, and picture an imaginary supply curve to the right of the old one, crossing the demand curve at a lower point on the price axis.) As the price declines, each firm will reduce production until it reaches the new, highest output at which marginal cost is lower than price. The price will continue to fall until it reaches $1,500 again, at which point the suppliers no longer can earn economic profits and the flow of new capital into the industry will cease. Now the suppliers are earning only the normal return on capital again [15].

As we discussed earlier, the reverse of this process occurs when demand declines. Suppliers experience losses and capital begins to leave the market: eventually, through contraction of the existing suppliers' capacity or the departure of some suppliers from the market altogether, supply returns to a level at which it can command a profit-maximizing price. Long-run equilibrium is restored when suppliers are making a normal profit once again.

Changes in cost, like changes in demand, can disturb the equilibrium of competitive markets. Such changes are of two kinds: changes in the cost structure of the entire industry, and changes in cost that affect only a single supplier.

First, suppose the hang glider market is faced with an industry-wide increase in the cost of one of its major inputs—aluminum tubing, for example. Unlike the change in demand we just discussed, a general increase in cost will not immediately raise the price of hang gliders. The hang glider suppliers cannot raise their prices in the short run (this is a purely competitive market), and as long as industry output is too high to command a price that will cover costs, the best firms can do is to reduce their output to the point at which they suffer the smallest losses. In the long run, firms will leave the industry until supply falls to the point at which prices equal the new, higher cost of making gliders.

An industry-wide decline in cost will have the opposite effect. The price cannot fall in the short run, so for a time firms will earn economic profit (just as they did, in the short run, when demand increased). In the long run, new firms or facilities will be added to the industry, supply will rise and the price will fall to the level of the new, lower cost of making gliders. In the long run, hang glider suppliers will earn only normal profits.

It is also possible for a single supplier to reduce its costs below the level typical for its industry. For example, SD Gliders may discover a cheaper (but equally functional) fabric for covering its glider frames. If the other firms do not catch on and switch to the cheaper fabric immediately, and if the market is in long-run equilibrium (so that price just equals the unit cost of the firms not yet using the new fabric), then SD will earn monopoly profits for some period of time. (The innovator may increase production, but because it is only one of 100 suppliers the overall market supply will not rise enough to reduce the market price.)

Of course, other firms are likely to emulate SD's idea eventually, and as more firms lower their costs and increase their production the overall market will begin to grow appreciably. The price will fall to a level below the unit cost of the firms that still use the more expensive fabric, and eventually everyone will adopt the new material or leave the industry.

2. How Competitive Firms Become Efficient

We now know how perfectly competitive firms achieve allocative, or pricing, efficiency. But all this guarantees is that consumers will pay no more than the costs of production. What if those costs are higher than they need to be? If firms are inefficient, then even prices set equal to cost will be too high. How does competition address the problem of productive efficiency?

The beginnings of an answer can be found in our hang glider shop. Remember that as long as total cost at a given level of production is lower than total revenue, expanded production brings more profit. Note also that our total cost, at any level of production, can be converted into cost per unit (average cost) by dividing the number of units produced into the total cost [16].

Average cost is the prime indicator of a firm's efficiency. If we compare two firms, each making 21 hang gliders a week, and we find that one is spending is $1,400 per hang glider per week while the other is spending $1,366, we know that the latter firm is more efficient. If we compare the average cost of the same two firms at 22 gliders a week and find that one is spending $1,370.44 per glider per week while the other is spending $1,380, we not only know that the first firm is more efficient—we also know that it will make more profit. This is because the first firm's marginal cost of going from 21 to 22 gliders is less than the market price, while the second company's marginal cost is more than the market price. The more efficient company can make the 22nd glider and earn the profit that results; the less efficient company cannot.

So companies in competition will benefit by reducing their average cost to a minimum. But how does a producer accomplish this result?

In the short run, the trick is for the firm to find the level of *output* at which its average cost is lowest; in the long run, the trick is for the firm to find the *size* (in physical plant, equipment, and other capital assets) at which suppliers of its product are most efficient. The greatest efficiencies are achieved by firms of optimal size operating at their level of output.

To understand how firms find the size and output levels that make them most efficient, we need to understand the elements that make up a firm's average, or unit, cost.

Minimizing Costs in the Short Run

Goods and services cannot be produced without incurring *costs*—land, labor, equipment, depreciation, capital, and the like. Costs may be related to tangible, physical inputs, or may be intangibles (such as depreciation).

A firm's *total costs* are the sum of its *fixed costs* plus its *variable costs*. Fixed costs remain constant regardless of output: they include executive salaries, depreciation, the normal return on capital, rent, property taxes, and the like. Variable costs change as the volume of production changes and can be eliminated by suspending production altogether. They include manual labor, materials, electric power, and others.

Note that even these fixed costs are truly "fixed" only in the short run. In the long run plants can be expanded, management can be restructured, and other measures can be taken that alter the fixed-cost picture substantially. This is why we distinguish short-run from long-run cost structures.

A firm's average costs also combine fixed and variable components. A firm's fixed cost per unit, called *average fixed cost* (AFC), always declines as production increases. (This makes sense: since our fixed costs do not vary, dividing their total by a larger number always yields a smaller average.)

A firm's *variable* cost per unit, called *average variable cost* (AVC), also declines as production increases—but only to a point. For any firm, increases in production eventually are accompanied by a *rise* in average variable cost.

To show why this is so, imagine that you meet with your shop manager one morning and tell him that you plan to increase SD's glider production to 40 gliders a week. You want to know how much more that level of production will cost. Your manager does some research and reports the following:

- Labor costs will increase. The additional production is too low to justify a second shift (the manager cannot find enough workers with the necessary skills in the short term anyhow); but he will ask the work force to put in two hours overtime each day, and he must pay each worker time-and-a-half for those overtime hours. (The manager also reminds you that workers on overtime get tired and work more slowly—in other words, you will be paying them more just when their productivity is lowest.)
- The company's usual aluminum and fabric suppliers cannot accommodate the additional demand for materials, but he can get them from suppliers in the next state. His materials budget will increase somewhat, however, because of higher transportation costs.
- Keeping the shop open two hours later every day will increase power requirements, but the rate per hour will not change.

Notice that the variable costs have not simply risen in proportion to the additional output: for two of the three variable cost items, the cost per glider actually increases.

This is an example of the phenomenon of diminishing returns. The specifics vary from industry to industry and from time to time, but any firm reaches a point at which increasing output leads to an increase in cost per unit, because the additional units of input (labor, materials, and so forth) yield a lower return. This is why economists refer to average variable cost as a U-shaped curve.

The same phenomenon applies to total average cost. Even after average variable costs begin to rise, average cost may continue to decline because of the continuing decline in the fixed-cost component of average cost; but eventually, at some level of production, the rise in average variable cost will overtake and exceed the decline in fixed cost. This is why, as we saw in Table A.4, average total cost grows as output increases.

Figure A.6 depicts on a graph the typical relationship among average fixed cost, average variable cost, and total cost as output grows [17].

The most efficient level of output, for any firm, is the output at which total average cost is lowest.

Minimizing Costs in the Long Run

We have just found SD Glider's most efficient level of production attainable *in the short run*. If the increased level of demand persists, the company can do better in the long run by expanding its plant.

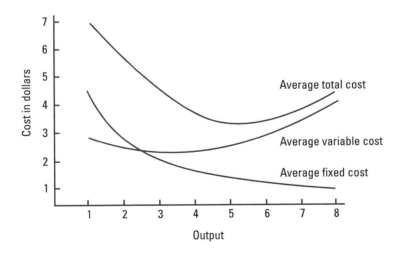

Figure A.6 Per-unit fixed, variable, and total cost. Even after average variable cost begins to rise, average total cost continues to decline until the increase in average variable cost outweighs the continuing decline of average fixed cost. At any given output, average total cost is the sum of the average fixed cost and average variable cost at that level of output.

Suppose you decide that you can sell 40 gliders a week for at least the next ten years. This is plenty of time to expand your shop, and a larger shop will accommodate more workers at one time than your present operation. Since the larger work force eliminates the need for overtime, your cost per worker will decline if you build the larger facility.

How large should your shop become? Each possible shop size will have its own average cost curve. If you look at Figure A.7, you will see the average cost curves for plants of three different sizes. If you are certain that production will hold steady at 40 gliders a week, size A is obviously your best choice. If you expect output to grow somewhat beyond that level, you may want to build size B, which is not optimal in the short or intermediate term but will be optimal after demand grows beyond 40 gliders per week.

The reader also may have noted that the lowest points on the cost curves of each size plant can be connected. If we draw that curve (shown on Figure A.7 as the long-run average cost curve) we find that the minimum-cost points on the short-run curves are not all at the same height: some plant sizes are more efficient than others. In fact, as the chart shows, increases in plant size improve efficiency to a point, after which the efficiencies are lost.

This U-shaped long-run average cost curve reflects the phenomenon known as *economies of scale*, and also reflects the limits of that phenomenon.

As Figure A.7 suggests, some plant sizes are too small to be optimally efficient. This reflects the fact that large-scale production is, to a point, more efficient than small-scale production. Suppose, for example, that someone invents a robot glider maker that produces gliders at a fraction of the

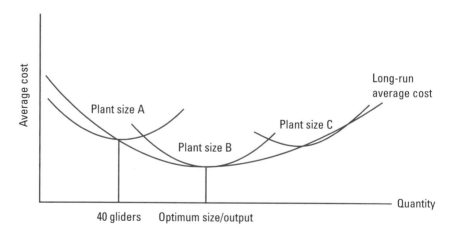

Figure A.7 Average cost curves of three hang glider shops.

average cost achieved by hand assembly. If a firm's output is too small to use one of those machines, it is condemned to be less efficient than those firms that do use it. The larger firm is taking advantage of the economies of its scale, and the smaller firm is suffering from the diseconomies of its smaller scale.

If larger firms enjoy economies of scale, then why isn't every market served by a single, huge supplier? The reason is that for most types of business, a point is reached at which greater size becomes a detriment. Some of the reasons are likely to be administrative or even cultural. As businesses become large and far-flung, the administrative costs of coordinating the activities of their personnel and subdivisions introduce inefficiencies. And as corporations acquire a larger managerial class, bureaucratic attitudes may come to outweigh entrepreneurial ones. (We all know of firms that have been overtaken by smaller competitors with owner-managers who controlled the operations of their firms personally.)

Optimal scale is very much a matter of the economics of the particular industry. Steel companies must achieve a substantial size before they become efficient; your local service station has no reason to aspire to the same scale. And some industries, as we shall see, are thought to have such a huge optimal size that they are believed to be natural monopolies.

3. Summing Up: The Character of Pure Competition

We now can reach some useful conclusions about pure competitive markets in short-run and long-run equilibrium.

Now we can summarize, with somewhat more precision than before, how competition enhances consumer welfare. Specifically, the competitive market in long-run equilibrium has the following features:

- Price tends to equal marginal cost [18]. This reinforces allocative efficiency: where prices of groups of related products are equal to their respective marginal costs of production, individuals and firms will make choices that promote the maximum efficiency in the society's use of its resources.

- Price also equals average cost. No firm earns excess profits, so price is not excessive in relation to costs and customers are not exploited.

- Average cost tends to be as low as suppliers can make it. Firms are of optimal size and produce at the minimum point on their short-run cost curve.

B. What Is a Monopoly Market?

Monopolies are firms with market power, which is defined as the ability to control price and output in the market for the firm's product. Monopolists, like firms in competitive markets, maximize their profits by setting their output at the point at which marginal cost equals marginal revenue. The difference between the monopolist's output decisions and those of competitive firms, however, is that the monopolist can choose to charge a higher price than the firm facing competition, and can induce customers to pay the higher price by restricting its output.

To show why this is so, we return to SD Gliders; but now we assume that your shop has no competition. You supply the entire hang glider market.

You still want to find the level of production that will maximize your profits, but you face a very different situation from the one presented by the competitive market. You will recall that under competition, you could not raise your price above the market price: since you could not materially affect the total supply of hang gliders, since other suppliers' gliders were perfect substitutes for yours and all consumers were perfectly informed about other suppliers' prices, raising your price would prevent you from selling a single glider.

Now that you are the sole supplier, however, raising your price will not keep you from selling any gliders at all: it only means that you will sell fewer gliders. And the number you will sell at any possible price is ascertainable: all you need to know is the demand curve for the hang glider market. For any possible price, the demand curve will tell you how many gliders will be purchased at that price [19].

So how do you find your profit-maximizing output in such a market?

You find it the same way you found it when you faced perfect competition: you determine the output at which marginal cost most closely approximates, but does not exceed, marginal revenue. The difference between competition and monopoly is that for a monopolist, the marginal revenue earned by choosing any given level of output is not the price (as it is under competition); it is lower than the price. This is how monopolists earn economic profit: marginal cost equals marginal revenue, but both marginal cost and marginal revenue are less than price. Monopolists price above marginal cost.

Table A.5 explains this crucial feature of monopoly. Like Table A.4, this table shows some relationships among SD Glider's costs and revenues at different levels of production. Unlike Table A.4, however, Table A.5 does

Table A.5
Monopoly Cost and Revenue Schedule for SD Gliders

Price (A/R)	Q	TR	TC	MR	MC	Total Profit
$1,700	21	$35,700	$28,686	−500	1,384	$7,014
1,600	22	35,200	30,070	−700	1,450	5,130
1,500	23	34,500	31,520	−900	1,550	2,980
1,400	24	33,600	33,070	−1,100	1,665	530
1,300	25	32,500	34,735	—	—	−2,235

not hold price constant at $1,500; instead, price changes with each level of output [20].

Notice that each time we choose to go build an additional hang glider, the marginal revenue we earn at that level of production is less than the market price. (Under competition, as Table A.4 shows, each additional hang glider brought marginal revenue equal to the market price.) The reason for this disparity is simple: while in competition we could sell as few or as many hang gliders as we wanted at $1,500, now changes in output cause changes in price. To find the marginal revenue associated with each increase in production, therefore, we now must subtract the reduction in revenue on each unit (caused by the decrease in price) from the revenue brought in by the last unit. As Table A.5 shows, the result of increased production under these circumstances can be a negative marginal revenue number.

Obviously, SD's profit-maximizing output under monopoly conditions is lower than any level of output shown in Table A.5. We must keep reducing output until marginal revenue turns positive, and then continue to reduce output until marginal revenue equals or exceeds marginal cost. And when we find that profit-maximizing output, both marginal cost and marginal revenue will be lower than price.

This analysis also shows us why monopolists, contrary to popular belief, do not simply charge all the traffic will bear. Once the output is reached at which marginal cost most closely approximates (but does not exceed) marginal revenue, the monopolist will not reduce output and increase price any further. To do so would sacrifice revenue (and profits) by producing at a level at which marginal cost exceeds marginal revenue.

From all of this we can draw some conclusions about the differences between competitive and monopolistic markets. Specifically:

1. The price charged by a monopolist is higher than the price set by supply and demand in a competitive market. The monopolist can raise the price above the level at which marginal cost equals price: a seller in a competitive market could not do that because its sales would drop to zero.

2. Economic profit can be a permanent, rather than a short-term, condition in a monopolistic market. (In competition market, economic profit exists only in the short run, when demand outruns productive capacity.)

3. In monopoly, price is always higher than marginal cost.

Does monopoly harm consumer welfare? We have the necessary tools to show that it does.

Monopolists are able to charge a higher price, make higher profits, and produce less of the product they control. Consumers who pay the higher price have less to spend than they otherwise would have had; their real income is reduced. Total consumer satisfaction is reduced because consumers pay more for (and get less of) the monopolized product, and have less to spend on other things they would like to have.

To show how this reduces consumer welfare, imagine that you are building a house and putting in wiring. You need electrical conduit. Plastic conduit works just as well as metal conduit for your purpose: plastic conduit costs $.50 a yard to make and metal conduit costs $.75 a yard to make. But the plastic conduit market is monopolized and plastic conduit sells for $.75 a yard; the metal conduit market is competitive and metal conduit also sells for $.75 a yard.

If you had bought plastic conduit at its marginal cost, you would have money left over to build a bigger house, add another bathroom, or put some money in your child's college fund. Because the plastic conduit market is monopolized, you lost that chance. And because demand for metal pipe will be artificially high (it costs the same, so why not have it?) and because metal conduit is so costly to make, the society's scarce resources will be wastefully allocated to a product that for most purposes is needlessly costly for the job.

Before we move on to particular controversies in telecommunications law, let us address one more subject—natural monopoly.

A natural monopoly market is one in which, because of the technology of production, a firm of optimal size is big enough to serve the entire market. This means that if several competitors tried to serve the market, their average cost would be higher than a single supplier could achieve.

Figure A.8 shows a natural monopoly market. Firm 1 is large enough to serve one-third of the market, but only at a lower output and a higher price than Firm 2, which is large enough to serve one-half of the market. If three firms as large as Firm 1 were competing, therefore, two of them could expand and split the market evenly at a lower price. But that situation would not be durable: a supplier as large as Firm 3 could serve the entire market at a higher output and lower price than Firm 1 *or* 2, so one firm would expand to the size of Firm 3 and occupy the entire market, driving the other two suppliers out of business. The surviving firm then would be in a position to make a monopolistic profit.

In this case, it seems that we must tolerate competition at artificially high costs, or tolerate a monopolist who has lower costs but will maximize his profits by declining to pass that cost advantage on to consumers [21]. But there is, of course, a third way: we can make the monopolist tolerable by regulating its earnings.

II. Specific Economic Issues in Telecommunications Law

Now we can use the tools we just acquired—and a few more we shall pick up as we go along—to illuminate some particular legal and regulatory issues.

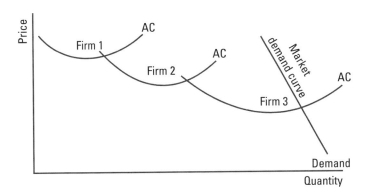

Figure A.8 A natural monopoly market. Firm 3 is optimally sized for its industry and large enough to serve the entire market.

Each of these questions is the subject of a vast literature, much of it far more technical than anything we attempt here: but the discussion that follows should give the student a useable introduction to the economic dimension of these problems.

A. The Puzzle of Natural Monopoly

Natural monopoly is an economic concept with legal implications. In this section we explain how natural monopoly markets can be identified, and how natural monopoly arguments have been applied to one telecommunications market in particular.

1. Identifying a Natural Monopoly Market

We have defined a natural monopoly market as one in which the most efficient firm is of a size sufficient to supply the entire demand for its services. Put another way, if a single firm supplying the entire market still is on the declining portion of its average cost curve, then any attempt to compete with that firm would be pointless [22].

Whether a particular market is naturally monopolistic depends heavily on the specific cost structure of production in that market. That cost structure, in turn, has much to do with the technology of production.

Suppose, for example, that telephone subscribers in the towns of Lincoln and Davis make frequent long-distance calls between the two cities. Their demand curve shows that they will make 1,000 such calls an hour at $.20 a call, 2,000 calls an hour at $.15 a call, 3,000 calls an hour at $.10 a call and 4,000 calls an hour at $.05 a call. Now suppose two firms put up microwave towers in both cities at substantial initial investment. In each city each firm buys land, builds a tower, installs an antenna and radio transceiver, makes an access road, and installs a climate-controlled equipment building.

If each firm can carry 1,000 calls an hour at a marginal cost of $.15 a call, then it appears that they each can supply half the market with no trouble. (Our demand curve shows that consumers will buy 2,000 calls an hour at $.15 a call.)

But what if each firm is still on the decreasing side of its average cost curve at the 1,000-call point? If the technology is such that no additional costs are incurred to carry more calls then there is no reason to stop here.

In fact, what if you learned that either of these firms could carry 4,000 calls a month with no additional investment, and that the marginal cost of one firm carrying those calls would be $.05? Then there is no reason for competition: the firm supplying the entire market is more efficient than two firms splitting the market.

Suppose, on the other hand, that the most efficient capacity of a microwave tower is 1,000 calls, and that after that output is reached another tower must be built, with its transceiver, access roads, and the like. If this is the case, then two firms *can* split the Lincoln-Davis long-distance market with no sacrifice of efficiency [23].

We can complicate the picture a bit more without changing the essential analysis. Assume that increasing a tower's capacity from 1,000 to 4,000 calls does require higher capacity radio equipment, but will not require the far greater expense of a whole new tower [24]. Is a single firm still more efficient? Yes, because the radio equipment, while causing a momentary increase in average cost, does not return average cost to the level that would result from construction of a new tower.

As this example suggests, arguments about natural monopoly are useless unless they are based on reliable cost data; and those data must be industry-specific, service-specific, and current [25].

2. Legal Implications of Natural Monopoly

Natural monopoly becomes an issue in telecommunications in at least two ways. First (as we have discussed) natural monopoly is an important predicate for granting or preserving an exclusive franchise: where markets will be more efficient under competition, the economic rationale for the exclusive franchise disappears. Second, proof that a market is a natural monopoly may be a defense to a charge that problems experienced by new entrants are the result of the established carriers' abuse of monopoly power. If a market is a natural monopoly, then a single firm's dominance may be attributed to its economies of scale rather than predatory conduct.

Both arguments were made by the Bell system during the early years of long-distance competition. When the FCC first considered permitting private and common carrier microwave systems to carry long-distance traffic, Bell argued that new entry necessarily would make the market less efficient. Later, when Bell was accused of antitrust violations in its treatment of the new carriers, similar arguments were made to show that Bell's behavior was reasonable.

If the reader has been attentive to our discussion of natural monopoly, however, these arguments must seem strange. If a market can be served most efficiently by a single firm, then there is no need for regulators to preserve the monopoly—the market will do that. And if a market is a natural monopoly, then why would rational people invest their money in competing systems that are bound, inexorably, to fail [26]? The very presence of competition in a market would seem to argue strongly against the presence of a natural monopoly cost structure.

This objection would be unanswerable if all telecommunications services were priced at cost. As we have mentioned earlier, however, the rates of the established carriers have been distorted by subsidies. As the Bell system has argued in its disputes with MCI and other competing carriers, the averaged rate structure of Bell's long-distance services (along with other anomalies in Bell's rate structure) meant that Bell's message rates on high-density routes with low unit costs were set no lower than message rates on lower density routes on which Bell's economies of scale were less fully exploited. This meant that a less efficient competitor could underprice Bell and still make a profit by offering service only on the highest density routes (a practice called "cream-skimming"). It also meant that if Bell cut its rates on those same rates to meet the competition, it was not necessarily engaged in below-cost pricing: it might only be departing from an inefficient, averaged-rate structure and moving its rates on those routes closer to the associated costs.

Bell never convinced its critics that long-distance service carried over microwave is a natural monopoly. More recently, however, a whole new technology has overtaken the long-distance business—and this time some experts believe that long-distance service is a natural monopoly at last [27].

The new factor is fiber-optic technology. A long-distance fiber optic network requires billions of dollars in initial investment, after which the volume of traffic that can be carried is almost infinitely expandable at negligible additional cost. In fact, the fiber optic installation necessary to carry the first lightwave message also can carry levels of capacity that would have a microwave network adding new transceivers and towers many times over. These facts give the dominant interexchange carrier—AT&T—an apparently irresistible tendency to eliminate its competitors [28].

So why has this not happened? Why are MCI, Sprint, and other competitors still in business? One explanation is that AT&T's competitors are sheltered (as they were said to be sheltered in the predivestiture era) by distortions in access pricing [29]. Another is that AT&T, fearing a new wave of antitrust suits, deliberately keeps its competitors alive by following their lead in pricing its services [30].

B. Does Regulation of Rates and Earnings Achieve Economic Efficiency?

Once we decide we must tolerate a telecommunications company as a natural monopoly, we want to restrain our single supplier from behaving like a monopoly. Ideally, we want our carrier to give its ratepayers the same benefits they would enjoy if the carrier faced a market in pure competition.

But pure competition, as we have seen, rigorously drives firms in the direction of pricing efficiency and productive efficiency. Do the command-and-control methods of regulation replicate these efficiencies?

In fact, both of the principal methods of controlling the rates and earnings of dominant carriers—rate-of-return regulation and price-cap regulation—are compromises. One is fairly effective at preventing carriers from earning monopoly profits, but is less effective at achieving productive efficiency; while the other emphasizes productive efficiency at the potential expense of pricing efficiency. And both are flawed, in their real-world incarnations, by the persistence of rate structures that do not consistently align rates with costs.

1. Traditional Regulation

As we discussed in Chapter 1, rate-of-return regulation controls a carrier's overall earnings but not its individual rates. It does this by assuring that the carrier earns a return no greater than its costs of capital. To put it in terms of our new economic knowledge, the regulators ensure that the carrier earns only normal profit and does not make economic profit.

But avoidance of monopoly profit is not always the same thing as pricing efficiency. Where a firm offers a number of services, that firm may earn only normal profits and still price some of those services above the associated costs. All that is required is for the firm to price other services below their associated costs, so that overall profits are held at the cost of capital.

Firms in competitive markets cannot successfully engage in such practices [31]. But regulated monopolies can, and regulators historically foster pricing inefficiency through rate averaging and other subsidies to basic, residential, and rural ratepayers. So while regulation of overall earnings may prevent economic profit, it does not do this the way competition does it (i.e., by forcing the supplier to price each product or service at the marginal cost of producing that good or service). Instead regulation exploits, for political ends, the unique ability of monopolists to practice price discrimination. The result is that consumers receive flawed price signals and overall allocative efficiency suffers.

In addition to its spotty record in promoting pricing efficiency, traditional regulation has only weak methods of promoting productive efficiency. These problems affect both the investment and the expenditure side of carriers' cost structures.

On the investment side, rate-of-return regulation may encourage the firm to underinvest or overinvest in capital assets. Firms in competition control their cost of capital in the same way they control other costs: they shop for it and ensure that they are paying no more than the current market price. Regulators, on the other hand, decide the cost of capital prospectively, on the basis of information that may be obsolete when the carrier actually goes to the capital markets. If the rate of return turns out to be lower than the cost of capital, then the company will be encouraged to overinvest, incurring unnecessary costs that will be passed along to ratepayers. If the rate of return turns out to be higher than the cost of capital, the company may underinvest, thereby failing to maintain and upgrade its network to meet customers' demands. This is the phenomenon usually referred to as the Averch-Johnson-Wellisz, or A-J-W, effect [32].

Another problem is that carriers experience no penalties for incurring excessive expenses unless those expenditures are disallowed by regulators [33]. It is unlikely that regulators will exercise the vigilance of a manager or entrepreneur whose livelihood depends on achieving minimum average cost in a world of competing firms with similar incentives.

2. Incentive Regulation

Incentive, or price-cap, regulation is intended primarily to improve carriers' productive efficiency and create cost savings that can be passed on to ratepayers. Under price caps, each service offered by a carrier has a maximum rate or range of rates: if the carrier can reduce its costs enough to earn more than the cost of its capital, it is allowed to retain some or all of that extra profit. As efficiency gains are realized from year to year, rates are expected to be capped at lower and lower levels.

Now that we know how economic efficiency is achieved, the danger of the incentive approach should be obvious. This system sets rates at a level sufficient for the firm to earn normal profits, then encourages the carrier to earn monopoly profits. To the extent carriers succeed under this system, rates exceed costs.

In fact, however, the system tries to permit monopoly profits only where a competitive economy also would allow them. As we have seen in the first part of this appendix, when markets are in long-run equilibrium all suppliers are pricing their services at cost and earning normal profits. If one supplier then finds a lower cost production method that has not yet occurred to its competitors, the innovative supplier earns monopoly profits until others emulate its method and the market returns to long-run equilibrium. These temporary profits can be viewed as a reasonable reward for the risks posed by innovation—particularly since productivity improvements eventually benefit all consumers of the product.

Price-cap regimes can be seen as an effort to approximate this process. The initial cap is set by the regulator at a level sufficient to earn normal profit. Those rates then are indexed downward to match normal, industry-wide productivity gains (thereby maintaining normal profit levels as industry-wide efficiencies improve). *Under this system, only carriers that become more efficient than the average may earn monopoly profits.* Combined with sharing, this system is a politically palatable, rough-and-ready method of imitating the way in which competitive markets reward innovation.

As devices for achieving pricing efficiency, incentive regimes (like traditional regulation) are hampered by the difficulty of acquiring accurate cost and productivity data [34], and by rate averaging and other built-in subsidies that persist in the capped rates. Both problems are likely to persist, with greater or lesser severity, so long as regulators are charged with controlling the rates or earnings of firms subject to their jurisdictions.

C. Efficient Pricing by Carriers

So far we have spoken of the characteristics of purely competitive markets and of purely monopolistic markets, and have surveyed the means by which various regulatory schemes try to encourage the virtues of the former and avoid the vices of the latter. One of the efficiency benchmarks suggested by pure competition was pricing efficiency, or the tendency of prices to equal cost.

But finding the most efficient price for a regulated utility can be more complicated than we have admitted so far. For one thing, a carrier that prices at marginal cost may not be able to recover its costs. For another, carriers provide many services and it often is difficult to decide which costs should be attributed to a particular service, for purposes of deciding what the rate for that service should be.

These problems are far too complex for us to explore exhaustively in this appendix. We can, however, explain the dilemmas they present and at least allude to some of the responses that economists and regulators have proposed.

1. Why Carriers May Not Be Able to Price at Marginal Cost

Our perfectly competitive market demonstrated that suppliers achieve pricing efficiency by achieving that output when marginal cost equals the market price. This is the price at which the firm is most efficient, and at which consumers pay no more than the cost of producing at that level of output.

Regulators, with their godlike power to dictate rates, would seem to have an easy task: all rates should be set equal to the marginal cost of producing at a level sufficient to meet market demand. How can it be more complicated than that?

It can be more complicated because utilities tend not to be of optimal size for producing at normal levels of demand. Specifically, utilities (whether they are natural monopolies or not) build in excess capacity sufficient to meet peak demand loads and provide for future increases in demand. The result is that at any given output within the normal range of demand, the company's average cost is likely to be higher than its marginal cost [35].

Figure A.9 shows these cost relationships and depicts the choices available to regulators under this circumstances. We already have rejected the notion of letting the carrier set its price at the monopoly level. Marginal cost reflects the costs incurred to produce at each level of output shown, but if prices are set at marginal cost the carrier will not recover all of its costs, including its costs of capital. And if the price is set at average cost the

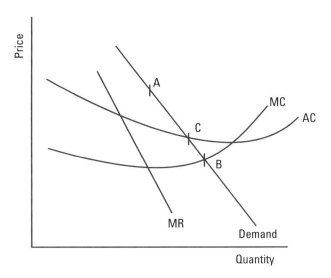

A = Monopoly price and output (output at which MC = MR)
B = Price and output at marginal cost
C = Price and output at average cost

Figure A.9 Pricing alternatives for a decreasing-cost carrier.

consumers will be paying for capacity they are not using—in effect, paying the cost of service not supplied.

Some commentators have suggested pricing services at marginal cost and making up the shortfall from general tax revenues. The typical regulator's solution is to price services at average cost. A third solution is to charge differential rates (i.e., charge some customers marginal cost and charge other customers [or the same customers at different times] average cost). The usual rationale is that if the marginal-cost customers use more service than they would have used if they were charged average cost, the resulting revenues might cover all of the firm's costs.

Certain two-tier rate schemes are quite common. For example, the long-distance companies historically have charged customers less to call at off-peak times than at peak calling times. If the off-peak rates are set at slightly above marginal cost but less than average cost, and customers do more off-peak calling than they would have at average cost, then the lower rates make a net contribution to the carrier's total costs.

Economists differ as to how such multitier rate schemes should be structured. Generally speaking, however, economists argue that the rate structure should exploit, as much as possible, the different demand elasticities of the carrier's customers. This means that customers who value the service more and are willing to buy more of it will pay more, while those whose demand is more price-elastic will be charged less to induce them to use the service more. If the favored customers are making a contribution to the common costs that the less elastic customers otherwise would bear by themselves, then the disfavored customers benefit from the lower rates charged to the favored customers. If the favored customers pay less than the cost of their service, however, they are enjoying an outright subsidy: and if they make no contribution to the carriers' common costs, then the scheme confers no benefit on the disfavored customers [36].

2. Allocating Costs Among Regulated Services

No matter what pricing method we adopt—marginal cost, average cost, or some other approach—it will be complicated by the fact that our carrier offers more than one service. Much of the carrier's property, equipment, and personnel will be used to provide more than one service, and there may be no obvious way to decide what part of those shared costs each service should bear.

Suppose, for example, that Peninsula Telephone has grown much larger and offers both local and long-distance (intraLATA) service. Some of Peninsula's assets (such as trunk lines connecting distant central offices) are used only to provide long-distance: the cost of those facilities will be recovered entirely through long-distance rates. Other facilities, however, are used to provide both local and long-distance service. How will the cost of those facilities be allocated between the two services?

To find some possible answers, consider just one facility—the switch located in Peninsula's original central office. The switch carries both local and long-distance calls.

One rational way to allocate the cost of the switch is to identify the marginal switching cost imposed by each local call and each long-distance call, and assign those marginal costs to the two services accordingly. You remember how marginal cost is calculated: we measure the total switching cost incurred to provide the number of long-distance calls demanded, less the cost incurred to provide one fewer; and we do the same calculation for the local calls.

Suppose, however, that the switch (like much telephone equipment) has considerable excess capacity, so that at any likely level of demand no calls are blocked. And suppose that each additional call (local or long-distance) imposes no additional, variable cost whatever. In this case, the marginal cost of each call is zero [37]. Yet the average cost of each call—the total cost of the switch divided by the number of calls—is probably quite substantial. How do we recover the total cost of the switch?

An intuitively appealing approach is to find the total cost of the switch and divide it between the two services according to some simple formula. So, for example, we might divide the number of calls (both local and long-distance) into the total cost of the switch; or we might divide the minutes spent on those calls into the total switch cost; or we might divide the switch costs between the services in the same proportion as those costs that can be directly attributed to each service. These are all examples of *fully distributed cost methods,* and they all are arbitrary in some degree.

They also may be inefficient. Suppose the elasticity of the long-distance customers' demand is far greater than that of the local customers, and that pricing long-distance calls at fully distributed cost discourages usage. What if we also found that if long-distance was priced at less than its fully distributed cost, usage would increase to such a degree that long-distance would end up contributing more to the total costs of the carrier than it would have contributed if priced at its fully distributed cost?

This shows why rigid fully distributed cost methodologies, to the extent they prevent carriers from exploiting different cost elasticities among services and customer groups, are not optimally efficient. As long as no customers are actually subsidized (by paying less than the marginal cost of their service), and as long as the less elastic customers pay no more for their service than they would have paid without the other customers' contribution, fully distributed cost pricing is not necessary to achieve fairness or efficiency.

Regulators, nonetheless, sometimes are reluctant to depart from fully distributed cost pricing, particularly where that method is thought to favor basic ratepayers. The FCC, for example, continues to require fully distributed cost allocations for jurisdictional separations and to account for the cost of basic and enhanced services.

While fully distributed cost methodologies may create inefficiencies when they are used to allocate costs among monopoly services, they are especially inappropriate when applied to services for which carriers face competition from unregulated suppliers. In the next section, we explain why this is so.

3. Allocating Costs When Carriers Face Competition

In Chapter 2 of this book we described your visit to Peninsula telephone's state commission to plead the case for lowering the price of off-premises switching (OPS) service—a service that was facing strong competition from PBX manufacturers. The commission had required you to price OPS, along with your other business-oriented services, at a steep premium that contributed a subsidy to basic telephone rates. You wanted to lower the price to a competitive level.

The commission agreed that some reduction was in order, and that you should be allowed to price OPS at a level that only recovered the associated costs. But the commission's idea of the costs OPS should recover differed from yours—they wanted OPS to recover some part of Peninsula's overhead, including investment that was not caused by the decision to offer OPS. The commission, in short, wanted OPS priced at fully distributed cost.

As you argued at the time, no owner or manager of an unregulated business sets prices this way. When your competitors decide whether to offer a new product or service, they compare the revenues they expect to earn with the costs (fixed and variable) they will incur to offer the service. If the revenues earned will exceed the costs incurred (including a reasonable return on investment) then the service is worth offering.

In making this analysis the business owner or manager does not ask whether the anticipated revenues will cover the cost of plant already in use. Those costs will be there whether the new service is offered or not.

When carriers are required to price competitive services at fully distributed cost, their prices are higher than those that would be charged by an efficient firm pricing at marginal cost. This disparity may result in the carrier being driven from the market by more efficient competitors, or it simply may create a price umbrella under which inefficient competitors can find shelter. Either result is bad for consumers.

Many economists argue that where carriers face effective competition for the sale of a service, they should be permitted to price at marginal cost (or at average variable cost, which is a surrogate more readily generated from conventional accounting data). In regulatory practice, the debate typically is between pricing at fully distributed cost and pricing at long-run incremental cost.

The long-run incremental cost (LRIC) of any product has been defined as "total company cost minus what the total cost of the company would be in the absence of production of (the product), all divided by the quantity of (the product) being produced" [38]. To identify the LRIC price, you identify all of the costs—both fixed and variable—of adding your new product or service, and divide by the output. You do not include any costs not caused by the new service. The resulting price may differ from the marginal (or average variable) cost of producing at that level of output, but it will recover total cost and will be lower than a price that allocated some portion of the company's embedded costs to the new service.

A number of commissions permit carriers to price competitive services on an LRIC basis, or at LRIC plus some contribution to overhead. Where commissions require competitive services to be priced at fully distributed cost, they may argue that the use of shared facilities creates economies of scope and scale that benefit all services provided through those facilities, and that consumers of the carriers' basic services should share in those efficiencies [39].

D. Predatory Pricing

When regulators scrutinize the relationships between a carrier's rates and the associated costs, they generally are trying to protect ratepayers. They want to ensure that ratepayers are not exploited by overall earnings that exceed overall costs; or want to decide whether earnings are high enough to be "shared" with ratepayers; or they want to ensure that basic ratepayers are not subsidizing other services.

When antitrust courts scrutinize the relationships between a carrier's rates and the associated costs, the emphasis is on protection of competition [40]. The concern is that dominant carriers may price their competitive services below cost, drive new entrants from the market, then take advantage of the absence of competition to raise prices to monopoly levels. Where a carrier offers both monopoly and competitive services, and particularly where the monopoly services are subject to rate-of-return regulation, the alleged method of successful below-cost pricing is likely to be the same one against which regulators guard on behalf of basic ratepayers—cross-subsidization [41].

The leading case on predatory pricing by telephone companies is the decision of the U.S. Court of Appeals for the Seventh Circuit in *MCI v. AT&T*. A brief review of that decision is useful, both for the facts and for the Court's reasoning.

MCI's lawsuit dated from the time when MCI offered only private-line, long-distance service in competition with the unified Bell system. MCI had started by building microwave facilities connecting densely populated cities with substantial business telephone traffic between them. MCI's rates reflected only the costs of offering those services—costs that were relatively low, not only because of the economics of microwave, but also because the heavy usage of MCI's facilities between these urban centers ensured a relatively low per-message cost. The Bell system's private-line rates between those city-pairs were not so low: Bell's rates were averaged with the rates connecting lower density routes for which unit costs were higher, and also (as Bell claimed at trial) were burdened with the subsidies that all long-distance service then provided to local service.

Bell responded to MCI's proposed service by selectively lowering its own private-line rates. The so-called "hi-lo" tariff lowered Bell's rates for private-line service on high-density routes and raised the rates for those same services on lower density routes. Naturally enough, hi-lo resulted in more competitive Bell rates between the city-pairs MCI planned to serve [42].

MCI claimed that the hi-lo rates were predatory because they failed to cover Bell's fully distributed cost of serving the city-pairs on which discounted rates were offered. Bell argued that fully distributed cost was an inappropriate standard for proof of predatory pricing, and that the hi-lo rates were lawful because they covered the long-run incremental costs of the service. The jury found that fully distributed cost was the appropriate standard, and further found that the hi-lo rates failed to cover FDC and were predatory.

The Court of Appeals agreed with Bell that fully distributed costs are irrelevant to a claim of predatory pricing [43]. The Court reasoned that decisions to enter new markets are not made on the basis of "historical or embedded costs," but involve a comparison of anticipated revenues attributable to the new service with the "current and anticipated cost" incurred to provide the service [44]. Any price designed to recover embedded costs will create a price umbrella over inefficient competitors and "may thus misallocate resources and force consumers to pay for less production than competition would dictate" [45].

In the course of its opinion the Court made some interesting observations about the differences between the two contending standards in the case before it and the marginal-cost standard. The Court noted that both the FDC and LRIC methods are ways of pricing at the average total cost of a service. Unlike marginal-cost methods, which price at the addition to short-run costs caused by the last distinguishable unit of production, LRIC measures all the costs—fixed and variable—of providing a new service. As the Court points out (quoting Professor Kahn):

"[M]arginal cost, strictly speaking, refers to the additional cost of supplying a single, infinitesimally small additional unit, while 'incremental' refer[s] to the *average* additional cost of a finite and possibly a large change in production or sales" [46].

While the Court suggested that LRIC might be superior to marginal costs in highly capital-intensive industries, the opinion also shows considerable sympathy with the predatory pricing rules suggested by Professors Areeda and Turner. The Areeda-Turner rules, which have been adopted by a number of courts, would hold that prices below marginal cost are presumptively predatory; that prices above marginal cost but below long-run incremental cost are presumptively lawful; and that prices at or above long-run incremental cost are always lawful.

The *MCI* decision is logical and instructive, but of course it is not binding on the other Courts of Appeals or on the U.S. Supreme Court, which has expressed its skepticism about predatory pricing claims but has not adopted a cost standard for review of such claims.

The reader also should be aware that some commentators argue against *any* cost standard for predatory pricing claims. Some scholars argue that predatory pricing is so self-defeating that it never can be practiced successfully, and that cost-based standards (because of the possibility of error in the use of accounting data) pose a risk that harmless behavior will be punished. Other commentators claim that firms may drive competitors from the market through pricing and other strategies that do not include below-cost pricing.

E. Measuring a Carrier's Market Power

Both the FCC and the state commissions have deregulated, or at least loosened regulatory scrutiny of, various telecommunications services. Where the established carriers (especially the LECs) ask for pricing flexibility or other relief in a particular market, the Commission must decide whether that market is sufficiently *competitive* to make such relief in the public interest. If a market meets the relevant criteria of competitiveness, the Commission decides that the established carrier lacks market power (i.e., the ability to control price or output) in that market. Relaxed regulation then is appropriate.

We described the features of a perfectly competitive market in the first part of this appendix. Such a market, we saw, has many competing suppliers, perfectly informed consumers, a product that is purchased entirely on the basis of price, and a number of other qualities that defeat any effort of individual suppliers to determine price and output. If a commission found a real-world market that displayed all of these qualities, its decision would be simple: but real-world markets do not, of course, match the purity of the theoretical model.

Recognizing this, many economists have developed criteria for identifying markets that are reasonably, if not perfectly, competitive. One influential theory is that of the "contestable" market (i.e., a market in which new entry is relatively easy, prices adjust rapidly to changes in supply and demand, and established firms anticipate that monopoly pricing will cause new competitors to enter the market). Under contestable-market theory, even a market with a single supplier may be reasonably competitive if that supplier's behavior is sufficiently restrained by fear of new entry.

Some states have adopted express, statutory criteria to guide their commissions' efforts to identify competitive markets. These statutes generally emphasize the availability of substitute suppliers and services, the difficulty of entering the market, and the relative size of the established carriers and their competitors.

Oregon, for example, identifies the following standards:

1. The extent to which services are available from alternative providers in the relevant market;
2. The extent to which the services of alternate providers are functionally equivalent or substitutable at comparable rates, terms, and conditions;
3. Existing economic or regulatory barriers to entry;
4. Any other factors deemed relevant by the commission [47].

The Missouri statute creates three classes of services—noncompetitive, transitionally competitive, and competitive [48]. Transitionally competitive services are subject to less stringent regulation than noncompetitive services, and competitive are regulated least. The statute specifies a number of factors the Missouri Commission must consider in reviewing a carrier's petition to have a service classified as competitive or transitionally competitive.

Texas permits carriers to show that particular services are subject to "significant competitive challenge," and specifies twelve criteria for the Commission's use in ruling on those petitions.

The FCC also has engaged in careful market-power analysis to justify regulatory relief, notably in its forbearance decisions and its 1991 long-distance competition order.

Conclusion

Economics played only a minor role in the early years of telephone regulation. One reason for this was that in the absence of competition, many of the economic tasks that commissions now face (such as deciding whether competitive services cover their costs) did not arise; another is that the tools of economic analysis were themselves primitive and commanded slight respect.

Today, both the complexity of telecommunications law and the credibility of economic analysis have grown substantially. The courts and the commissions are increasingly receptive to economic arguments, and the standard of economic debate has risen to a level at which advocacy based on incomplete or imprecise economic learning no longer will serve.

This appendix has been a bare introduction to the subject. The serious student or practitioner will want to dig deeper, starting perhaps with Alfred Kahn's classic work, *The Economics of Regulation*.

Endnotes

[1] Competition, as we know, is a bruising process that takes a heavy toll on many of us in our capacities as workers or business owners.

[2] "Costs" in this context include a normal return on capital, or profit.

[3] In the real world, of course, this tends not to be true of hang gliders or most other products. But some highly utilitarian items come close: 4" anthracite coal, for example, is the same product no matter who produces it. Purchasers of such products tend to be interested only in price.

[4] This is a feature found in few markets outside of securities exchanges.

[5] Most products do not exist in a single place: the retail auto market, for example, includes all of the outlets through which cars are sold. And few markets involve a single buyer and a single seller: usually many buyers compete for the available goods and many producers compete for the business of the buyers.

[6] Economists describe this tendency by saying that demand usually is *elastic*. If we find a consumer who is willing to buy the same quantity of a product regardless of price, that consumer's demand for the product is said to be *inelastic*.

[7] We explain later how suppliers decide how much to produce at any particular market price.

[8] The supply curve will vary, depending on whether we are depicting the quantity a producer will supply in the short run or in the long run. (The long run is a time sufficient for the producer to expand the scale of its operation.)

[9] Remember that in this perfectly competitive market there are no brand preferences and all consumers are fully informed about the prices at which different suppliers are offering the product.

[10] We say more later about the relationship between efficiency and output. For now, it is enough to know that firms are more efficient at some levels of output than at others.

[11] The revenue associated with an additional unit of production is called the *marginal revenue*. Marginal revenue always equals market price when a firm faces perfect competition, but not (as we shall see) when the firm is a monopoly.

[12] The reader may have noticed that there is no pair of outputs in the table at which marginal cost equals marginal revenue. Does this call into question the familiar maxim that profits are maximized when marginal cost equals marginal revenue? No: it just means that hang gliders, like most products, cannot be divided indefinitely into smaller and smaller units. While profits theoretically are maximized when marginal cost equals marginal revenue, the best we usually can do in practice is to produce the largest output at which marginal cost does not exceed marginal revenue.

[13] Bear in mind that the marginal cost of making the 23rd glider is not a discrete sum expended to make that glider alone: it is the difference in *total* cost between making 23 gliders and making 22 gliders.

[14] The short run is a time too short for existing suppliers to expand their plants or for new firms to enter the market.

[15] Our example assumes no increase in costs. In fact, increased demand for hang gliders will cause increased demand for materials, labor and other "inputs" into hang glider manufacture, which may raise the cost of those inputs. If those markets do not also return to equilibrium, then hang glider prices may not return all the way to their former level.

[16] We see this in the first three columns of Table A.4.

[17] The precise shape of these curves will vary substantially from one industry to another.

[18] This is also true of competitive markets in the short run.

[19] The supplier in a purely competitive market is said to face a flat demand curve: there is only one price at which the product can be sold, but the supplier can sell as much of the product as it likes at that price. The monopolist, however, is said to face the industry demand curve, which is a sloping curve.

[20] To simplify the calculations, Table A.5 divides by 100 the number of hang gliders that will be purchased at each price shown in the marketwide demand curve (Figure A.1). This brings the demand numbers in line with the cost and output numbers given in Table A.4 for SD, which we assumed to represent 1/100 of the total market output.

[21] Figure A.9 depicts the pricing options available to a natural monopoly carrier (or to any carrier serving the entire market on the decreasing side of its average cost curve).

[22] We have treated natural monopoly as a product of economies of scale, but the reader should be aware that firms may achieve efficiencies through economies of scope—that is, cost advantages derived from using the same inputs to make more than one product or service. Economies of scope also are relevant in determining whether a market is most efficiently served by a single firm. *See*, e.g., Waverman, L., "U.S. Interexchange Competition," in R. Crandall and K. Flamm (editors), *Changing the Rules: Technological Change, International Competition, and Regulation in Communications* at 72-74 (1989).

[23] Keep in mind that we are excluding all but the transmission costs—not the two firms' administrative or other costs. All costs should be considered before we say definitively that two firms are as efficient as one.

[24] Microwave sites have been made more efficient in just this way. In the 1980s, for example, the Bell system introduced "overbuild" microwave radio equipment that increased the capacity of existing towers from 1,800 voice channels per radio frequency channel to 6,000 voice channels per radio frequency channel. *Engineering and Operations in the Bell System* at 355-56 (2nd Ed. 1983); *see also* Kahn, A., *The Economics of Regulation, Part I* at 125-26 (1988).

[25] We already have seen that optimal firm size and output vary dramatically from industry to industry (a dry cleaning shop achieves its greatest efficiency at a much smaller scale than an automobile plant does). Similarly, economies of scale are not necessarily the same for all products and services within an industry. (Consider a mill that makes steel and also fabricates pipe, using entirely different equipment for each product.) And technologies change over time, so that a given industry may be a natural monopoly at one time and not another. When we add to these complexities the difficulty of allocating costs to particular products in a multiproduct firm, the challenge of identifying true natural monopoly markets is apparent.

[26] The advent of microwave made new entry possible because it did the same job as copper cable at drastically lower cost. (Two microwave towers, transmitting over a "free" medium—the air—can cover 25 or so miles. To acquire the same capability through copper cable requires buying rights of way, digging trenches or putting up poles, and laying or stringing expensive cables over every laborious inch of that 25 miles.) While

this lower initial investment made competition thinkable, it did not guarantee its durability. The Bell system also was investing in microwave, and some believed that the economics of that technology would perpetuate—rather than end—natural monopoly in the long-distance market.

[27] *See*, e.g., Crandall, R., *After the Breakup* at 4 (1990); Huber, P., M. Kellogg, and J. Thorne, *The Geodesic Network II* at 3.2–3.5 and 3.33–3.35 (1992).

[28] If this is true, it is of course profoundly ironic. The divestiture of the Bell system was based on the notion that local service is a natural monopoly, while interexchange service lends itself to competition. Now the LECs are facing growing competition within the local exchange just as optical fiber is (arguably) turning long-distance into a natural monopoly.

[29] *See* The Geodesic Network II, *supra*, 3.29–3.30.

[30] *Id.* at 3.44.

[31] If a firm in competition charged any group of customers more than the marginal cost of their service, competitors would snatch those customers away.

[32] *See* Averch, H., and L., Johnson, "Behavior of the Firm under Regulatory Constraint," *American Econ. Review,* (Dec. 1962); *see also,* Wellisz, S., "Regulation of Natural Gas Pipeline Companies: An Economic Analysis," *Jour. Pol. Econ.* (Feb. 1963). Some economists have argued that this phenomenon, if it exists, is likely to be offset by regulatory lag. So, for example, if all reductions in cost resulted in immediate rate reductions, then carriers would have an incentive to invest excessively and inflate their costs. If cost reductions were not accompanied by immediate rate reductions, however, then carriers might have an offsetting incentive to reduce costs and enjoy the earnings that will be generated before the regulators catch up. Alfred Kahn argues that to the extent the A-J-W effect has some residual effect, it may be helpful as a spur to risk-taking and investment. A. Kahn, *The Economics of Regulation, supra,* Part II at 107.

[33] Of course, because expenses are recovered on a dollar-for-dollar basis, wasteful spending on salaries and equipment does not increase profits but neither does it decrease profits.

[34] Sharing mechanisms are partly a hedge against errors in capping rates and establishing productivity adjustments. "A regulatory structure which combines the price-cap indexing approach with a sharing mechanism can provide protection to both shareholders and ratepayers from the risks that the indexing method may over- or underestimate the revenue changes which are needed to keep the utility financially healthy—but not too healthy." *California Price Cap Decision, supra,* 107 PUR 4th at 88.

[35] When a firm is at its optimal size and output and the market is competitive and in long-run equilibrium, the marginal cost curve crosses the average cost curve at the average cost curve's lowest point. If a utility is producing on the decreasing side of its average cost curve, however, then marginal cost is lower than average cost.

[36] As we noted before, discriminatory strategies of this kind are a luxury of monopolists.

[37] We are speaking here of short-run marginal cost, which economists regard as the optimum cost for pricing. If we were speaking of long-run marginal cost we might find

[38] Baumol, W., *Quasi-Permanence of Price Reductions: A Policy for Prevention of Predatory Pricing,* 89 Yale L. J. 1, n. 26 at 9 (1979).

[39] *See* discussion at p.25, *supra.*

[40] Commissions sometimes have the same concern. The FCC, for example, examined AT&T's private-line tariffs to determine if AT&T was trying to drive its competitors out of business. And as we mentioned earlier, the FCC is expressly empowered to enforce the antitrust laws.

[41] As we noted earlier, the courts generally have grown more skeptical of predatory pricing claims. *See* discussion at p. 65, *supra* and authorities cited therein. The U.S. Supreme Court has endorsed the views of those commentators who argue that "predatory pricing schemes are rarely tried, and even more rarely successful." *Matsushita Elec. Ind. Co. v. Zenith Radio Corp.,* 475 U.S. 574, 589 m(1986). Where cross-subsidization is alleged, however, the charge may be taken more seriously.

[42] Bell also had in place a bulk-discounted private-line service called Telpak, which had been instituted in response to an earlier FCC decision permitting large users to build and operate private microwave systems. MCI claimed that the Telpak rates also were predatory, but the jury found for AT&T on this charge.

[43] The Court of Appeals also found that the trial court had erred when it let the *jury* decide which standard—FDC or LRIC—was correct.

[44] 708 F.2d at 1116-17.

[45] *Id.* at 1117.

[46] Quoting Kahn, A., *The Economics of Regulation* 66 (1970) (emphasis in original).

[47] Or. Rev. Stat. Sections 759.020, 759.030 (1991). I am indebted for this discussion of statutory standards to the excellent article on this subject by Larson, A. C., "An Economic Guide to Competitive Standards in Telecommunications Regulation," 1 *Comm-Law Conspectus,* 31 (1993).

[48] Mo. Rev. Stat. Section 392.361 *et seq.* (1987).

Appendix B
Selected Sections of the Communications Act of 1934, as Amended by the Telecommunications Act of 1996

Title 47. Telegraphs, Telephones, and Radiotelegraphs

§ 151. Purposes of Chapter; Federal Communications Commission Created

For the purpose of regulating interstate and foreign commerce in communication by wire and radio so as to make available, so far as possible, to all the people of the United States, without discrimination on the basis of race, color, religion, national origin, or sex, a rapid, efficient, nationwide, and world-wide wire and radio communication service with adequate facilities at reasonable charges, for the purpose of national defense, for the purpose of promoting safety of life and property through the use of wire and radio communication, and for the purpose of securing a more effective execution of this policy by centralizing authority heretofore granted by law to several agencies and by granting additional authority with respect to interstate and foreign commerce in wire and radio communication, there is hereby created a commission to be known as the "Federal Communications Commission," which shall be constituted as hereinafter provided, and which shall execute and enforce the provisions of this Act.

§ 152. Application of Chapter

(a) The provisions of this Act shall apply to all interstate and foreign communication by wire or radio and all interstate and foreign transmission of energy by radio, which originates and/or is received within the United States, and to all persons engaged within the United States in such communication or such transmission of energy by radio, and to the licensing and regulating of all radio stations as hereinafter provided; but it shall not apply to persons engaged in wire or radio communication or transmission in [the Philippine Islands or] the Canal Zone, or to wire or radio communication or transmission wholly within [the Philippine Islands or] the Canal Zone. The provisions of this Act shall apply with respect to cable service, to all persons engaged within the United States in providing such service, and to the facilities of cable operators which relate to such service, as provided in title VI.

(b) Except as provided in sections 223 through 227, inclusive, and section 332, and subject to the provisions of section 301 and Title VI, nothing in this Act shall be construed to apply or to give the Commission jurisdiction with respect to (1) charges, classifications, practices, services, facilities, or regulations for or in connection with intrastate communication service by wire or radio of any carrier, or (2) any carrier engaged in interstate or foreign communication solely through physical connection with the facilities of another carrier not directly or indirectly controlling or controlled by, or under direct or indirect common control with such carrier, or (3) any carrier engaged in interstate or foreign communication solely through connection by radio or by wire and radio, with facilities, located in an adjoining State or in Canada or Mexico (where they adjoin the State in which the carrier is doing business), of another carrier not directly or indirectly controlling or controlled by, or under direct or indirect common control with such carrier, or (4) any carrier to which clause (2) or clause (3) would be applicable except for furnishing interstate mobile radio communication service or radio communication service to mobile stations on land vehicles in Canada or Mexico; except that sections 201 through 205 of this Act, both inclusive, shall, except as otherwise provided therein, apply to carriers described in clauses (2), (3), and (4).

§ 153. Definitions

For the purposes of this Act, unless the context otherwise requires—

(1) Affiliate. The term "affiliate" means a person that (directly or indirectly) owns or controls, is owned or controlled by, or is under common ownership or control with, another person. For purposes of this paragraph, the term "own" means to own an equity interest (or the equivalent thereof) of more than 10 percent.

(2) Amateur station. The term "amateur station" means a radio station operated by a duly authorized person interested in radio technique solely with a personal aim and without pecuniary interest.

(3) AT&T Consent Decree. The term "AT&T Consent Decree" means the order entered August 24, 1982, in the antitrust action styled United States v. Western Electric, Civil Action No. 82-0192, in the United States District Court for the District of Columbia, and includes any judgment or order with respect to such action entered on or after August 24, 1982.

(4) Bell operating company. The term "Bell operating company"—

(A) means any of the following companies: Bell Telephone Company of Nevada, Illinois Bell Telephone Company, Indiana Bell Telephone Company, Incorporated, Michigan Bell Telephone Company, New England Telephone and Telegraph Company, New Jersey Bell Telephone Company, New York Telephone Company, U S West Communications Company, South Central Bell Telephone Company, Southern Bell Telephone and Telegraph Company, Southwestern Bell Telephone Company, The Bell Telephone Company of Pennsylvania, The Chesapeake and Potomac Telephone Company, The Chesapeake and Potomac Telephone Company of Maryland, The Chesapeake and Potomac Telephone Company of Virginia, The Chesapeake and Potomac Telephone Company of West Virginia, The Diamond State Telephone Company, The Ohio Bell Telephone Company, The Pacific Telephone and Telegraph Company, or Wisconsin Telephone Company; and

(B) includes any successor or assign of any such company that provides wireline telephone exchange service; but

(C) does not include an affiliate of any such company, other than an affiliate described in subparagraph (A) or (B).

(5) Broadcast station. The term "broadcast station", "broadcasting station", or "radio broadcast station" means a radio station equipped to engage in broadcasting as herein defined.

(6) Broadcasting. The term "broadcasting" means the dissemination of radio communications intended to be received by the public, directly or by the intermediary of relay stations.

(7) Cable service. The term "cable service" has the meaning given such term in section 602.

(8) Cable system. The term "cable system" has the meaning given such term in section 602.

(9) Chain broadcasting. The term "chain broadcasting" means simultaneous broadcasting of an identical program by two or more connected stations.

(10) Common carrier. The term "common carrier" or "carrier" means any person engaged as a common carrier for hire, in interstate or foreign communication by wire or radio or in interstate or foreign radio transmission of energy, except where reference is made to common carriers not subject to this Act; but a person engaged in radio broadcasting shall not, insofar as such person is so engaged, be deemed a common carrier.

(11) Connecting carrier. The term "connecting carrier" means a carrier described in clauses (2), (3), or (4) of section 2(b).

(12) Construction permit. The term "construction permit" or "permit for construction" means that instrument of authorization required by this Act or the rules and regulations of the Commission made pursuant to this Act for the construction of a station, or the installation of apparatus, for the transmission of energy, or communications, or signals by radio, by whatever name the instrument may be designated by the Commission.

(13) Corporation. The term "corporation" includes any corporation, joint-stock company, or association.

(14) Customer premises equipment. The term "customer premises equipment" means equipment employed on the premises of a person (other than a carrier) to originate, route, or terminate telecommunications.

(15) Dialing parity. The term "dialing parity" means that a person that is not an affiliate of a local exchange carrier is able to provide telecommunications services in such a manner that customers have the ability to route automatically, without the use of any access code, their telecommunications to the telecommunications services provider of the customer's designation from among 2 or more telecommunications services providers (including such local exchange carrier).

(16) Exchange access. The term "exchange access" means the offering of access to telephone exchange services or facilities for the purpose of the origination or termination of telephone toll services.

(17) Foreign communication. The term "foreign communication" or "foreign transmission" means communication or transmission from or to any place in the United States to or from a foreign country, or between a

station in the United States and a mobile station located outside the United States.

(18) Great Lakes Agreement. The term "Great Lakes Agreement" means the Agreement for the Promotion of Safety on the Great Lakes by Means of Radio in force and the regulations referred to therein.

(19) Harbor. The term "harbor" or "port" means any place to which ships may resort for shelter or to load or unload passengers or goods, or to obtain fuel, water, or supplies. This term shall apply to such places whether proclaimed public or not and whether natural or artificial.

(20) Information service. The term "information service" means the offering of a capability for generating, acquiring, storing, transforming, processing, retrieving, utilizing, or making available information via telecommunications, and includes electronic publishing, but does not include any use of any such capability for the management, control, or operation of a telecommunications system or the management of a telecommunications service.

(21) InterLATA service. The term "interLATA service" means telecommunications between a point located in a local access and transport area and a point located outside such area.

(22) Interstate communication. The term "interstate communication" or "interstate transmission" means communication or transmission (A) from any State, Territory, or possession of the United States (other than the [Philippine Islands and] the Canal Zone), or the District of Columbia, to any other State, Territory, or possession of the United States (other than [the Philippine Islands and] the Canal Zone), or the District of Columbia, (B) from or to the United States to or from [the Philippine Islands or] the Canal Zone, insofar as such communication or transmission takes place within the United States, or (C) between points within the United States but through a foreign country; but shall not, with respect to the provisions of title II of this Act (other than section 223 thereof include wire or radio communication between points in the same State, Territory, or possession of the United States, or the District of Columbia, through any place outside thereof, if such communication is regulated by a State commission.

(23) Land station. The term "land station" means a station, other than a mobile station, used for radio communication with mobile stations.

(24) Licensee. The term "licensee" means the holder of a radio station license granted or continued in force under authority of this Act.

(25) Local access and transport area. The term "local access and transport area" or "LATA" means a contiguous geographic area—

(A) established before the date of enactment of the Telecommunications Act of 1996 by a Bell operating company such that no exchange area includes points within more than 1 metropolitan statistical area, consolidated metropolitan statistical area, or State, except as expressly permitted under the AT&T Consent Decree; or

(B) established or modified by a Bell operating company after such date of enactment and approved by the Commission.

(26) Local exchange carrier. The term "local exchange carrier" means any person that is engaged in the provision of telephone exchange service or exchange access. Such term does not include a person insofar as such person is engaged in the provision of a commercial mobile service under section 332(c), except to the extent that the Commission finds that such service should be included in the definition of such term.

(27) Mobile service. The term "mobile service" means a radio communication service carried on between mobile stations or receivers and land stations, and by mobile stations communicating among themselves, and includes (A) both one-way and two-way radio communication services, (B) a mobile service which provides a regularly interacting group of base, mobile, portable, and associated control and relay stations (whether licensed on an individual, cooperative, or multiple basis) for private one-way or two-way land mobile radio communications by eligible users over designated areas of operation, and (C) any service for which a license is required in a personal communications service established pursuant to the proceeding entitled "Amendment to the Commission's Rules to Establish New Personal Communications Services" (GEN Docket No. 90-314; ET Docket No. 92-100), or any successor proceeding.

(28) Mobile station. The term "mobile station" means a radiocommunication station capable of being moved and which ordinarily does move.

(29) Network element. The term "network element" means a facility or equipment used in the provision of a telecommunications service. Such term also includes features, functions, and capabilities that are provided by means of such facility or equipment, including subscriber numbers, databases, signaling systems, and information sufficient for billing and collection or used in the transmission, routing, or other provision of a telecommunications service.

(30) Number portability. The term "number portability" means the ability of users of telecommunications services to retain, at the same location, existing telecommunications numbers without impairment of quality, reliability, or convenience when switching from one telecommunications carrier to another.

(31) Operator.

(A) "Operator" on a ship of the United States means, for the purpose of parts II and III of title III of this Act a person holding a radio operator's license of the proper class as prescribed and issued by the Commission.

(B) "Operator" on a foreign ship means, for the purpose of part II of title III of this Act, a person holding a certificate as such of the proper class complying with the provisions of the radio regulations annexed to the International Telecommunication Convention in force, or complying with an agreement or treaty between the United States and the country in which the ship is registered.

(32) Person. The term "person" includes an individual, partnership, association, joint-stock company, trust, or corporation.

(33) Radio communication. The term "radio communication" or "communication by radio" means the transmission by radio of writing, signs, signals, pictures, and sounds of all kinds, including all instrumentalities, facilities, apparatus, and services (among other things, the receipt, forwarding, and delivery of communications) incidental to such transmission.

(34) Radio officer.

(A) "Radio officer" on a ship of the United States means, for the purpose of part II of title III of this Act, a person holding at least a first or second class radiotelegraph operator's license as prescribed and issued by the Commission. When such person is employed to operate a radiotelegraph station aboard a ship of the United States, he is also required to be licensed as a "radio officer" in accordance with the Act of May 12, 1948.

(B) "Radio officer" on a foreign ship means, for the purpose of part II of title III of this Act, a person holding at least a first or second class radiotelegraph operator's certificate complying with the provisions of the radio regulations annexed to the International Telecommunication Convention in force.

(35) Radio station. The term "radio station" or "station" means a station equipped to engage in radio communication or radio transmission of energy.

(36) Radiotelegraph auto alarm. The term "radiotelegraph auto alarm" on a ship of the United States subject to the provisions of part II of title III of this Act means an automatic alarm receiving apparatus which responds to the radiotelegraph alarm signal and has been approved by the Commission. "Radiotelegraph auto alarm" on a foreign ship means an automatic alarm receiving apparatus which responds to the radiotelegraph alarm signal and has been approved by the government of the country in which the ship is registered: Provided, That the United States and the country in which the ship is registered are parties to the same treaty, convention, or agreement

prescribing the requirements for such apparatus. Nothing in this Act or in any other provision of law shall be construed to require the recognition of a radiotelegraph auto alarm as complying with part II of title III of this Act, on a foreign ship subject to such part, where the country in which the ship is registered and the United States are not parties to the same treaty, convention, or agreement prescribing the requirements for such apparatus.

(37) Rural telephone company. The term "rural telephone company" means a local exchange carrier operating entity to the extent that such entity—

(A) provides common carrier service to any local exchange carrier study area that does not include either—

(i) any incorporated place of 10,000 inhabitants or more, or any part thereof, based on the most recently available population statistics of the Bureau of the Census; or

(ii) any territory, incorporated or unincorporated, included in an urbanized area, as defined by the Bureau of the Census as of August 10, 1993;

(B) provides telephone exchange service, including exchange access, to fewer than 50,000 access lines;

(C) provides telephone exchange service to any local exchange carrier study area with fewer than 100,000 access lines; or

(D) has less than 15 percent of its access lines in communities of more than 50,000 on the date of enactment of the Telecommunications Act of 1996.

(38) Safety convention. The term "safety convention" means the International Convention for the Safety of Life at Sea in force and the regulations referred to therein.

(39) Ship.

(A) "Ship" or "vessel" includes every description of watercraft or other artificial contrivance, except aircraft, used or capable of being used as a means of transportation on water, whether or not it is actually afloat.

(B) A ship shall be considered a passenger ship if it carries or is licensed or certificated to carry more than twelve passengers.

(C) A cargo ship means any ship not a passenger ship.

(D) A passenger is any person carried on board a ship or vessel except (1) the officers and crew actually employed to man and operate the ship, (2) persons employed to carry on the business of the ship, and (3) persons on board a ship when they are carried, either because of the obligation laid upon the master to carry shipwrecked, distressed, or other persons in like or similar situations or by reason of any circumstance over which neither the master, the owner, nor the charterer (if any) has control.

(E) "Nuclear ship" means a ship provided with a nuclear power plant.

(40) State. The term "State" includes the District of Columbia and the Territories and possessions.

(41) State commission. The term "State commission" means the commission, board, or official (by whatever name designated) which under the laws of any State has regulatory jurisdiction with respect to intrastate operations of carriers.

(42) Station license. The term "station license," "radio station license," or "license" means that instrument of authorization required by this Act or the rules and regulations of the Commission made pursuant to this Act, for the use or operation of apparatus for transmission of energy, or communications, or signals by radio, by whatever name the instrument may be designated by the Commission.

(43) Telecommunications. The term "telecommunications" means the transmission, between or among points specified by the user, of information of the user's choosing, without change in the form or content of the information as sent and received.

(44) Telecommunications carrier. The term "telecommunications carrier" means any provider of telecommunications services, except that such term does not include aggregators of telecommunications services (as defined in section 226). A telecommunications carrier shall be treated as a common carrier under this Act only to the extent that it is engaged in providing telecommunications services, except that the Commission shall determine whether the provision of fixed and mobile satellite service shall be treated as common carriage.

(45) Telecommunications equipment. The term "telecommunications equipment" means equipment, other than customer premises equipment, used by a carrier to provide telecommunications services, and includes software integral to such equipment (including upgrades).

(46) Telecommunications service. The term "telecommunications service" means the offering of telecommunications for a fee directly to the public, or to such classes of users as to be effectively available directly to the public, regardless of the facilities used.

(47) Telephone exchange service. The term "telephone exchange service" means (A) service within a telephone exchange, or within a connected system of telephone exchanges within the same exchange area operated to furnish to subscribers intercommunicating service of the character ordinarily furnished by a single exchange, and which is covered by the exchange service charge, or (B) comparable service provided through a

system of switches, transmission equipment, or other facilities (or combination thereof) by which a subscriber can originate and terminate a telecommunications service.

(48) Telephone toll service. The term "telephone toll service" means telephone service between stations in different exchange areas for which there is made a separate charge not included in contracts with subscribers for exchange service.

(49) Television service.

(A) Analog television service. The term "analog television service" means television service provided pursuant to the transmission standards prescribed by the Commission in section 73.682(a) of its regulations (47 C.F.R. 73.682(a)).

(B) Digital television service. The term "digital television service" means television service provided pursuant to the transmission standards prescribed by the Commission in section 73.682(d) of its regulations (47 C.F.R. 73.682(d)).

(50) Transmission of energy by radio. The term "transmission of energy by radio" or "radio transmission of energy" includes both such transmission and all instrumentalities, facilities, and services incidental to such transmission.

(51) United States. The term "United States" means the several States and Territories, the District of Columbia, and the possessions of the United States, but does not include [the Philippine Islands or] the Canal Zone.

(52) Wire communication. The term "wire communication or "communication by wire" means the transmission of writing, signs, signals, pictures, and sounds of all kinds by aid of wire, cable, or other like connection between the points of origin and reception of such transmission, including all instrumentalities, facilities, apparatus, and services (among other things, the receipt, forwarding, and delivery of communications) incidental to such transmission.

§ 154. Federal Communications Commission

(a) Number of commissioners; appointment. The Federal Communications Commission (in this Act referred to as the "Commission") shall be composed of five commissioners appointed by the President, by and with the advice and consent of the Senate, one of whom the President shall designate as chairman.

(b) Qualifications.

(1) Each member of the Commission shall be a citizen of the United States.

(2) (A) No member of the Commission or person employed by the Commission shall—

(i) be financially interested in any company or other entity engaged in the manufacture or sale of telecommunications equipment which is subject to regulation by the Commission;

(ii) be financially interested in any company or other entity engaged in the business of communication by wire or radio or in the use of the electromagnetic spectrum;

(iii) be financially interested in any company or other entity which controls any company or other entity specified in clause (i) or clause (ii), or which derives a significant portion of its total income from ownership of stocks, bonds, or other securities of any such company or other entity; or

(iv) be employed by, hold any official relation to, or own any stocks, bonds, or other securities of, any person significantly regulated by the Commission under this Act; except that the prohibitions established in this subparagraph shall apply only to financial interests in any company or other entity which has a significant interest in communications, manufacturing, or sales activities which are subject to regulation by the Commission.

(B) (i) The Commission shall have authority to waive, from time to time, the application of the prohibitions established in subparagraph (A) to persons employed by the Commission if the Commission determines that the financial interests of a person which are involved in a particular case are minimal, except that such waiver authority shall be subject to the provisions of section 208 of title 18, United States Code. The waiver authority established in this subparagraph shall not apply with respect to members of the Commission.

(ii) In any case in which the Commission exercises the waiver authority established in this subparagraph, the Commission shall publish notice of such action in the Federal Register and shall furnish notice of such action to the appropriate committees of each House of the Congress. Each such notice shall include information regarding the identity of the person receiving the waiver, the position held by such person, and the nature of the financial interests which are the subject of the waiver.

(3) The Commission, in determining whether a company or other entity has a significant interest in communications, manufacturing, or sales

activities which are subject to regulation by the Commission, shall consider (without excluding other relevant factors)—

(A) the revenues, investments, profits, and managerial efforts directed to the related communications, manufacturing, or sales activities of the company or other entity involved, as compared to the other aspects of the business of such company or other entity;

(B) the extent to which the Commission regulates and oversees the activities of such company or other entity;

(C) the degree to which the economic interests of such company or other entity may be affected by any action of the Commission; and

(D) the perceptions held by the public regarding the business activities of such company or other entity.

(4) Members of the Commission shall not engage in any other business, vocation, profession, or employment while serving as such members.

(5) The maximum number of commissioners who may be members of the same political party shall be a number equal to the least number of commissioners which constitute a majority of the full membership of the Commission.

(c) Terms of office; vacancies. [Commissioners] shall be appointed for terms of five years and until their successors are appointed and have been confirmed and taken the oath of office, except that they shall not continue to serve beyond the expiration of the next session of Congress subsequent to the expiration of said fixed term of office; except that any person chosen to fill a vacancy shall be appointed only for the unexpired term of the Commissioner whom he succeeds. No vacancy in the Commission shall impair the right of the remaining commissioners to exercise all the powers of the Commission.

(d) Compensation of Commission members. Each Commissioner shall receive an annual salary at the annual rate payable from time to time for level IV of the Executive Schedule, payable in monthly installments. The Chairman of the Commission, during the period of his service as Chairman, shall receive an annual salary at the annual rate payable from time to time for level III of the Executive Schedule.

(e) Principal office; special sessions. The principal office of the Commission shall be in the District of Columbia, where its general sessions shall be held; but whenever the convenience of the public or of the parties may be promoted or delay or expense prevented thereby, the Commission may hold special sessions in any part of the United States.

(f) Employees and assistants; compensation of members of Field Engineering and Monitoring Bureau; use of amateur volunteers for certain purposes; commercial radio operator examinations.

    (1) The Commission shall have authority, subject to the provisions of the civil-service laws and the Classification Act of 1949, as amended, to appoint such officers, engineers, accountants, attorneys, inspectors, examiners, and other employees as are necessary in the exercise of its functions.

    (2) Without regard to the civil-service laws, but subject to the Classification Act of 1949, each commissioner may appoint three professional assistants and a secretary, each of whom shall perform such duties as such commissioner shall direct. In addition, the chairman of the Commission may appoint, without regard to the civil-service laws, but subject to the Classification Act of 1949, an administrative assistant who shall perform such duties as the chairman shall direct.

    (3) The Commission shall fix a reasonable rate of extra compensation for overtime services of engineers in charge and radio engineers of the Field Engineering and Monitoring Bureau of the Federal Communications Commission, who may be required to remain on duty between the hours of 5 o'clock post meridian and 8 o'clock ante meridian or on Sundays or holidays to perform services in connection with the inspection of ship radio equipment and apparatus for the purposes of part II of title III of this Act or the Great Lakes Agreement, on the basis of one-half day's additional pay for each two hours or fraction thereof of at least one hour that the overtime extends beyond 5 o'clock post meridian (but not to exceed two and one-half days' pay for the full period from 5 o'clock post meridian to 8 o'clock ante meridian) and two additional days' pay for Sunday or holiday duty. The said extra compensation for overtime services shall be paid by the master, owner, or agent of such vessel to the local United States collector of customs [Secretary of the Treasury] or his representative, who shall deposit such collection into the Treasury of the United States to an appropriately designated receipt account: Provided, That the amounts of such collections received by the said collector of customs or his representatives shall be covered into the Treasury as miscellaneous receipts; and the payments of such extra compensation to the several employees entitled thereto shall be made from the annual appropriations for salaries and expenses of the Commission: Provided further, That to the extent that the annual appropriations which are hereby authorized to be made from the general fund of the Treasury are insufficient, there

are hereby authorized to be appropriated from the general fund of the Treasury such additional amounts as may be necessary to the extent that the amounts of such receipts are in excess of the amounts appropriated: Provided further, That such extra compensation shall be paid if such field employees have been ordered to report for duty and have so reported whether the actual inspection of the radio equipment or apparatus takes place or not: And provided further, That in those ports where customary working hours are other than those hereinabove mentioned, the engineers in charge are vested with authority to regulate the hours of such employees so as to agree with prevailing working hours in said ports where inspections are to be made, but nothing contained in this proviso shall be construed in any manner to alter the length of a working day for the engineers in charge and radio engineers or the overtime pay herein fixed: and Provided further, That, in the alternative, an entity designated by the Commission may make the inspections referred to in this paragraph.

(4) (A) The Commission, for purposes of preparing or administering any examination for an amateur station operator license, may accept and employ the voluntary and uncompensated services of any individual who holds an amateur station operation license of a higher class than the class of license for which the examination is being prepared or administered. In the case of examinations for the highest class of amateur station operator license, the Commission may accept and employ such services of any individual who holds such class of license.

(B) (i) The Commission, for purposes of monitoring violations of any provision of this Act (and of any regulation prescribed by the Commission under this Act) relating to the amateur radio service, may—

(I) recruit and train any individual licensed by the Commission to operate an amateur station; and

(II) accept and employ the voluntary and uncompensated services of such individual.

(ii) The Commission, for purposes of recruiting and training individuals under clause (i) and for purposes of screening, annotating, and summarizing violation reports referred under clause (i), may accept and employ the voluntary and uncompensated services of any amateur station operator organization.

(iii) The functions of individuals recruited and trained under this subparagraph shall be limited to—

(I) the detection of improper amateur radio transmissions;

(II) the conveyance to Commission personnel of information which is essential to the enforcement of this Act (or regulations prescribed by the Commission under this Act) relating to the amateur radio service; and

(III) issuing advisory notices, under the general direction of the Commission, to persons who apparently have violated any provision of this Act (or regulations prescribed by the Commission under this Act) relating to the amateur radio service.

Nothing in this clause shall be construed to grant individuals recruited and trained under this subparagraph any authority to issue sanctions to violators or to take any enforcement action other than any action which the Commission may prescribe by rule.

(C) (i) The Commission, for purposes of monitoring violations of any provision of this Act (and of any regulation prescribed by the Commission under this Act) relating to the citizens band radio service, may—

(I) recruit and train any citizens band radio operator; and

(II) accept and employ the voluntary and uncompensated services of such operator.

(ii) The Commission, for purposes of recruiting and training individuals under clause (i) and for purposes of screening, annotating, and summarizing violation reports referred under clause (i), may accept and employ the voluntary and uncompensated services of any citizens band radio operator organization. The Commission, in accepting and employing services of individuals under this subparagraph, shall seek to achieve a broad representation of individuals and organizations interested in citizens band radio operation.

(iii) The functions of individuals recruited and trained under this subparagraph shall be limited to—

(I) the detection of improper citizens band radio transmissions;

(II) the conveyance to Commission personnel of information which is essential to the enforcement of this Act (or regulations prescribed by the Commission under this Act) relating to the citizens band radio service; and

(III) issuing advisory notices, under the general direction of the Commission, to persons who apparently have violated any provision of this Act (or regulations prescribed by the Commission under this Act) relating to the citizens band radio service.

Nothing in this clause shall be construed to grant individuals recruited and trained under this subparagraph any authority to issue sanctions to violators or to take any enforcement action other than any action which the Commission may prescribe by rule.

(D) The Commission shall have the authority to endorse certification of individuals to perform transmitter installation, operation, maintenance, and repair duties in the private land mobile services and fixed services (as defined by the Commission by rule) if such certification programs are conducted by organizations or committees which are representative of the users in those services and which consist of individuals who are not officers or employees of the Federal Government.

(E) The authority of the Commission established in this paragraph shall not be subject to or affected by the provisions of part III of title 5, United States Code, or section 3679(b) of the Revised Statutes.

(F) Any person who provides services under this paragraph shall not be considered, by reason of having provided such services, a Federal employee.

(G) The Commission, in accepting and employing services of individuals under subparagraphs (A) and (B), shall seek to achieve a broad representation of individuals and organizations interested in amateur station operation.

(H) The Commission may establish rules of conduct and other regulations governing the service of individuals under this paragraph.

(I) With respect to the acceptance of voluntary uncompensated services for the preparation, processing, or administration of examinations for amateur station operator licenses pursuant to subparagraph (A) of this paragraph, individuals, or organizations which provide or coordinate such authorized volunteer services may recover from examinees reimbursement for out-of-pocket costs.

(J) [Redesignated]

(5)(A) The Commission, for purposes of preparing and administering any examination for a commercial radio operator license or endorsement, may accept and employ the services of persons that the Commission determines to be qualified. Any person so employed may not receive compensation for such services, but may recover from examinees such fees as the Commission permits, considering such factors as public service and cost estimates submitted by such person.

(B) The Commission may prescribe regulations to select, oversee, sanction, and dismiss any person authorized under this paragraph to be employed by the Commission.

(C) Any person who provides services under this paragraph or who provides goods in connection with such services shall not, by reason of having provided such service or goods, be considered a Federal or special government employee.

(g) Expenditures

(1) The Commission may make such expenditures (including expenditures for rent and personal services at the seat of government and elsewhere, for office supplies, law books, periodicals, and books of reference, for printing

and binding, for land for use as sites for radio monitoring stations and related facilities, including living quarters where necessary in remote areas, for the construction of such stations and facilities, and for the improvement, furnishing, equipping, and repairing of such stations and facilities and of laboratories and other related facilities (including construction of minor subsidiary buildings and structures not exceeding $25,000 in any one instance) used in connection with technical research activities, as may be necessary for the execution of the functions vested in the Commission and as may be appropriated for by the Congress in accordance with the authorizations ofappropriations established in section 6. All expenditures of the Commission, including all necessary expenses for transportation incurred by the commissioners or by their employees, under their orders, in making any investigation or upon any official business in any other places than in the city of Washington, shall be allowed and paid on the presentation of itemized vouchers therefor approved by the chairman of the Commission or by such other member or officer thereof as may be designated by the Commission for that purpose.

(2) (A) If—

(i) the necessary expenses specified in the last sentence of paragraph (1) have been incurred for the purpose of enabling commissioners or employees of the Commission to attend and participate in any convention, conference, or meeting;

(ii) such attendance and participation are in furtherance of the functions of the Commission; and

(iii) such attendance and participation are requested by the person sponsoring such convention, conference, or meeting; then the Commission shall have authority to accept direct reimbursement from such sponsor for such necessary expenses.

(B) The total amount of unreimbursed expenditures made by the Commission for travel for any fiscal year, together with the total amount of reimbursements which the Commission accepts under subparagraph (A) for such fiscal year, shall not exceed the level of travel expenses appropriated to the Commission for such fiscal year.

(C) The Commission shall submit to the appropriate committees of the Congress, and publish in the Federal Register, quarterly reports specifying reimbursements which the Commission has accepted under this paragraph.

(D) The provisions of this paragraph shall cease to have any force or effect at the end of fiscal year 1994.

(E) Funds which are received by the Commission as reimbursements under the provisions of this paragraph after the close of a fiscal year shall remain available for obligations.

(3)

(A) Notwithstanding any other provision of law, in furtherance of its functions the Commission is authorized to accept, hold, administer, and use unconditional gifts, donations, and bequests of real, personal, and other property (including voluntary and uncompensated services, as authorized by section 3109 of title 5, United States Code).

(B) The Commission, for purposes of providing radio club and military-recreational call signs, may utilize the voluntary, uncompensated, and unreimbursed services of amateur radio organizations authorized by the Commission that have tax-exempt status under section 501(c)(3) of the Internal Revenue Code of 1986.

(C) For the purpose of Federal law on income taxes, estate taxes, and gift taxes, property or services accepted under the authority of subparagraph (A) shall be deemed to be a gift, bequest, or devise to the United States.

(D) The Commission shall promulgate regulations to carry out the provisions of this paragraph. Such regulations shall include provisions to preclude the acceptance of any gift, bequest, or donation that would create a conflict of interest or the appearance of a conflict of interest.

(h) Quorum; seal. Three members of the Commission shall constitute a quorum thereof. The Commission shall have an official seal which shall be judicially noticed.

(i) Duties and powers. The Commission may perform any and all acts, make such rules and regulations, and issue such orders, not inconsistent with this Act, as may be necessary in the execution of its functions.

(j) Conduct of proceedings; hearings. The Commission may conduct its proceedings in such manner as will best conduce to the proper dispatch of business and to the ends of justice. No commissioner shall participate in any hearing or proceeding in which he has a pecuniary interest. Any party may appear before the Commission and be heard in person or by attorney. Every vote and official act of the Commission shall be entered of record, and its proceedings shall be public upon the request of any party interested. The

Commission is authorized to withhold publication of records or proceedings containing secret information affecting the national defense.

(k) Annual reports to Congress. The Commission shall make an annual report to Congress, copies of which shall be distributed as are other reports transmitted to Congress. Such reports shall contain—
    (1) such information and data collected by the Commission as may be considered of value in the determination of questions connected with the regulation of interstate and foreign wire and radio communication and radio transmission of energy;
    (2) such information and data concerning the functioning of the Commission as will be of value to Congress in appraising the amount and character of the work and accomplishments of the Commission and the adequacy of its staff and equipment;
    (3) an itemized statement of all funds expended during the preceeding year by the Commission, of the sources of such funds, and of the authority in this Act or elsewhere under which such expenditures were made; and
    (4) specific recommendations to Congress as to additional legislation which the Commission deems necessary or desirable, including all legislative proposals submitted for approval to the Director of the Office of Management and Budget.
    (5) [Redesignated]

(l) Record of reports. All reports of investigations made by the Commission shall be entered of record, and a copy thereof shall be furnished to the party who may have complained, and to any common carrier or licensee that may have been complained of.

(m) Publication of reports; admissibility as evidence. The Commission shall provide for the publication of its reports and decisions in such form and manner as may be best adapted for public information and use, and such authorized publications shall be competent evidence of the reports and decisions of the Commission therein contained in all courts of the United States and of the several States without any further proof or authentication thereof.

(n) Compensation of appointees. Rates of compensation of persons appointed under this section shall be subject to the reduction applicable to officers and employees of the Federal Government generally.

(o) Use of communications in safety of life and property. For the purpose of obtaining maximum effectiveness from the use of radio and wire communications in connection with safety of life and property, the Commission shall investigate and study all phases of the problem and the best methods of obtaining the cooperation and coordination of these systems.

§ 155. Commission

(a) Chairman; duties; vacancy. The member of the Commission designated by the president as chairman shall be the chief executive officer of the Commission. It shall be his duty to preside at all meetings and sessions of the Commission, to represent the Commission in all matters relating to legislation and legislative reports, except that any commissioner may present his own or minority views or supplemental reports, to represent the Commission in all matters requiring conferences or communications with other governmental officers, departments or agencies, and generally to coordinate and organize the work of the Commission in such manner as to promote prompt and efficient disposition of all matters within the jurisdiction of the Commission. In the case of a vacancy in the office of the chairman of the Commission, or the absence or inability of the chairman to serve, the Commission may temporarily designate one of its members to act as chairman until the cause or circumstance requiring such designation shall have been eliminated or corrected.

(b) Organization of staff. From time to time [thereafter] as the Commission may find necessary, the Commission shall organize its staff into (1) integrated bureaus, to function on the basis of the Commission's principal workload operations, and (2) such other divisional organizations as the Commission may deem necessary. Each such integrated bureau shall include such legal, engineering, accounting, administrative, clerical, and other personnel as the Commission may determine to be necessary to perform its functions.

(c) Delegation of functions; exceptions to initial orders; force, effect and enforcement of orders; administrative and judicial review; qualifications and compensation of delegates; assignment of cases; separation of review and investigative or prosecuting functions; secretary; seal.

(1) When necessary to the proper functioning of the Commission and the prompt and orderly conduct of its business, the Commission may, by published rule or by order, delegate any of its functions (except

functions granted to the Commission by this paragraph and by paragraphs (4), (5), and (6) of this subsection and except any action referred to in sections 204(a)(2), 208(b), and 405(b)) to a panel of commissioners, an individual commissioner, and employee board, or an individual employee, including functions with respect to hearing, determining, ordering, certifying, reporting, or otherwise acting as to any work, business, or matter; except that in delegating review functions to employees in cases of adjudication (as defined in the Administrative Procedure Act; the delegation in any such case may be made only to an employee board consisting of two or more employees referred to in paragraph (8). Any such rule or order may be adopted, amended, or rescinded only by a vote of a majority of the members of the Commission then holding office. Except for cases involving the authorization of service in the instructional television fixed service, or as otherwise provided in this Act, nothing in this paragraph shall authorize the Commission to provide for the conduct, by any person or persons other than persons referred to in paragraph (2) or (3) of section 556(b) of title 5, United States Code, of any hearing to which such section applies.

(2) As used in this subsection (d) the term "order, decision, report, or action" does not include an initial, tentative, or recommended decision to which exceptions may be filed as provided in section 409(b).

(3) Any order, decision, report, or action made or taken pursuant to any such delegation, unless reviewed as provided in paragraph (4), shall have the same force and effect, and shall be made, evidenced, and enforced in the same manner, as orders, decisions, reports, or other actions of the Commission.

(4) Any person aggrieved by any such order, decision, report or action may file an application for review by the Commission within such time and in such manner as the Commission shall prescribe, and every such application shall be passed upon by the Commission. The Commission, on its own initiative, may review in whole or in part, at such time and in such manner as it shall determine, any order, decision, report, or action made or taken pursuant to any delegation under paragraph (1).

(5) In passing upon applications for review, the Commission may grant, in whole or in part, or deny such applications without specifying any reasons therefor. No such application for review shall rely on questions of fact or law upon which the panel of commissioners, individual commissioner, employee board, or individual employee has been afforded no opportunity to pass.

(6) If the Commission grants the application for review, it may affirm, modify, or set aside the order, decision, report, or action, or it may

order a rehearing upon such order, decision, report, or action in accordance with section 405.

(7) The filing of an application for review under this subsection shall be a condition precedent to judicial review of any order, decision, report, or action made or taken pursuant to a delegation under paragraph (1). The time within which a petition for review must be filed in a proceeding to which section 402(a) applies, or within which an appeal must be taken under section 402(b), shall be computed from the date upon which public notice is given of orders disposing of all applications for review filed in any case.

(8) The employees to whom the Commission may delegate review functions in any case of adjudication (as defined in the Administrative Procedure Act) shall be qualified, by reason of their training, experience, and competence, to perform such review functions, and shall perform no duties inconsistent with such review functions. Such employees shall be in a grade classification or salary level commensurate with their important duties, and in no event less than the grade classification or salary level of the employee or employees whose actions are to be reviewed. In the performance of such review functions such employees shall be assigned to cases in rotation so far as practicable and shall not be responsible to or subject to the supervision or direction of any officer, employee, or agent engaged in the performance of investigative or prosecuting functions for any agency.

(9) The secretary and seal of the Commission shall be the secretary and seal of each panel of the Commission, each individual commissioner, and each employee board or individual employee exercising functions delegated pursuant to paragraph (1) of this subsection.

(d) Meetings. Meetings of the Commission shall be held at regular intervals, not less frequently than once each calendar month, at which times the functioning of the Commission and the handling of its work load shall be reviewed and such orders shall be entered and other action taken as may be necessary or appropriate to expedite the prompt and orderly conduct of the business of the Commission with the objective of rendering a final decision (1) within three months from the date of filing in all original application, renewal, and transfer cases in which it will not be necessary to hold a hearing and (2) within six months from the final date of the hearing in all hearing cases.

(e) Managing Director; appointment, functions, pay. The Commission shall have a Managing Director who shall be appointed by the Chairman subject to the approval of the Commission. The Managing Director, under the supervision and direction of the Chairman, shall perform such administrative and executive functions as the Chairman shall delegate. The Managing Director shall be paid at a rate equal to the rate then payable for level V of the Executive Schedule.

§ 201. Service and Charges

(a) It shall be the duty of every common carrier engaged in interstate or foreign communication by wire or radio to furnish such communication service upon reasonable request therefor; and, in accordance with the orders of the Commission, in cases where the Commission, after opportunity for hearing, finds such action necessary or desirable in the public interest, to establish physical connections with other carriers, to establish through routes and charges applicable thereto and the divisions of such charges, and to establish and provide facilities and regulations for operating such through routes.

(b) All charges, practices, classifications, and regulations for and in connection with such communication service, shall be just and reasonable, and any such charge, practice, classification, or regulation that is unjust or unreasonable is hereby declared to be unlawful: Provided, That communications by wire or radio subject to this Act may be classified into day, night, repeated, unrepeated, letter, commercial, press, Government, and such other classes as the Commission may decide to be just and reasonable, and different charges may be made for the different classes of communications: Provided further, That nothing in this or in any other provision of law shall be construed to prevent a common carrier subject to this Act from entering into or operating under any contract with any common carrier not subject to this Act, for the exchange of their services, if the Commission is of the opinion that such contract is not contrary to the public interest: Provided further, That nothing in this Act or in any other provision of law shall prevent a common carrier subject to this Act from furnishing reports of positions of ships at sea to newspapers of general circulation, either at a nominal charge or without charge, provided the name of such common carrier is displayed along with such ship position reports. The Commission may prescribe such rules and regulations as may be necessary in the public interest to carry out the provisions of this Act.

§ 202. Discriminations and Preferences

(a) Charges, services, etc. It shall be unlawful for any common carrier to make any unjust or unreasonable discrimination in charges, practices, classifications, regulations, facilities, or services for or in connection with like communication service, directly or indirectly, by any means or device, or to make or give any undue or unreasonable preference or advantage to any particular person, class of persons, or locality, or to subject any particular person, class of persons, or locality to any undue or unreasonable prejudice or disadvantage.

(b) Charges or services included. Charges or services, whenever referred to in this Act, include charges for, or services in connection with, the use of common carrier lines of communication, whether derived from wire or radio facilities, in chain broadcasting or incidental to radio communication of any kind.

(c) Penalty. Any carrier who knowingly violates the provisions of this section shall forfeit to the United States the sum of $6,000 for each such offense and $300 for each and every day of the continuance of such offense.

§ 203. Schedules of Charges

(a) Filing; public display. Every common carrier, except connecting carriers, shall, within such reasonable time as the Commission shall designate, file with the Commission and print and keep open for public inspection schedules showing all charges for itself and its connecting carriers for interstate and foreign wire or radio communication between the different points on its own system, and between points on its own system and points on the system of its connecting carriers or points on the system of any other carrier subject to this Act when a through route has been established, whether such charges are joint or separate, and showing the classifications, practices, and regulations affecting such charges. Such schedules shall contain such other information, and be printed in such form, and be posted and kept open for public inspection in such places, as the Commission may by regulation require, and each such schedule shall give notice of its effective date; and such common carrier shall furnish such schedules to each of its connecting carriers, and such connecting carriers shall keep such schedules open for inspection in such public places as the Commission may require.

(b) Changes in schedule; discretion of Commission to modify requirements.
    (1) No change shall be made in the charges, classifications, regulations, or practices which have been so filed and published except after one hundred and twenty days notice to the Commission and to the public, which

shall be published in such form and contain such information as the Commission may by regulations prescribe.

(2) The Commission may, in its discretion and for good cause shown, modify any requirement made by or under the authority of this section either in particular instances or by general order applicable to special circumstances or conditions except that the Commission may not require the notice period specified in paragraph (1) to be more than one hundred and twenty days.

(c) Overcharges and rebates. No carrier, unless otherwise provided by or under authority of this Act, shall engage, or participate in such communication unless schedules have been filed and published in accordance with the provisions of this Act and with the regulations made thereunder; and no carrier shall (1) charge, demand, collect, or receive a greater or less or different compensation for such communication, or for any service in connection therewith, between the points named in any such schedule than the charges specified in the schedule then in effect, or (2) refund or remit by any means or device any portion of the charges so specified, or (3) extend to any person any privileges or facilities in such communication, or employ or enforce any classifications, regulations, or practices affecting such charges, except as specified in such schedule.

(d) Rejection or refusal. The Commission may reject and refuse to file any schedule entered for filing which does not provide and give lawful notice of its effective date. Any schedule so rejected by the Commission shall be void and its use shall be unlawful.

(e) Penalty for violations. In case of failure or refusal on the part of any carrier to comply with the provisions of this section or of any regulation or order made by the Commission thereunder, such carrier shall forfeit to the United States the sum of $6,000 for each such offense, and $300 for each and every day of the continuance of such offense.

§ 205. Commission Authorized to Prescribe Just and Reasonable Charges; Penalties for Violations

(a) Whenever, after full opportunity for hearing, upon a complaint or under an order for investigation and hearing made by the Commission on its own initiative, the Commission shall be of opinion that any charge, classification, regulation, or practice of any carrier or carriers is or will be in violation of any of the provisions of this Act, the Commission is authorized and empowered to determine and prescribe what will be the just and reasonable charge or the maximum or minimum, or maximum and minimum, charge or charges to be thereafter observed, and what classification, regulation, or practice is or will be just, fair, and reasonable, to be thereafter followed, and to make an order that the carrier or carriers shall cease and desist from such violation to the extent that the Commission finds that the same does or will exist, and shall not thereafter publish, demand, or collect any charge other than the charge so prescribed, or in excess of the maximum or less than the minimum so prescribed, as the case may be, and shall adopt the classification and shall conform to and observe the regulation or practice so prescribed.

(b) Any carrier, any officer, representative, or agent of a carrier, or any receiver, trustee, lessee, or agent of either of them, who knowingly fails or neglects to obey any order made under the provisions of this section shall forfeit to the United States the sum of $12,000 for each offense. Every distinct violation shall be a separate offense, and in case of continuing violation each day shall be deemed a separate offense.

§ 206. Carriers' Liability for Damages

In case any common carrier shall do, or cause or permit to be done, any act, matter, or thing in this Act prohibited or declared to be unlawful, or shall omit to do any act, matter, or thing in this Act required to be done, such common carrier shall be liable to the person or persons injured thereby for the full amount of damages sustained in consequence of any such violation of the provisions of this Act, together with a reasonable counsel or attorney's fee, to be fixed by the court in every case of recovery, which attorney's fee shall be taxed and collected as part of the costs in the case.

§ 207. Recovery of Damages

Any person claiming to be damaged by any common carrier subject to the provisions of this Act may either make complaint to the Commission as hereinafter provided for, or may bring suit for the recovery of the damages for which such common carrier may be liable under the provisions of this Act, in any district court of the United States of competent jurisdiction; but such person shall not have the right to pursue both such remedies.

§ 208. Complaints to Commission; Investigations; Duration of Investigation; Appeal of Order Concluding Investigation

(a) Any person, any body politic or municipal organization, or State commission, complaining of anything done or omitted to be done by any common carrier subject to this Act, in contravention of the provisions thereof, may apply to said Commission by petition which shall briefly state the facts, whereupon a statement of the complaint thus made shall be forwarded by the Commission to such common carrier, who shall be called upon to satisfy the complaint or to answer the same in writing within a reasonable time to be specified by the Commission. If such common carrier within the time specified shall make reparation for the injury alleged to have been caused, the common carrier shall be relieved of liability to the complainant only for the particular violation of law thus complained of. If such carrier or carriers shall not satisfy the complaint within the time specified or there shall appear to be any reasonable ground for investigating said complaint, it shall be the duty of the Commission to investigate the matters complained of in such manner and by such means as it shall deem proper. No complaint shall at any time be dismissed because of the absence of direct damage to the complainant.

(b) (1) Except as provided in paragraph (2), the Commission shall, with respect to any investigation under this section of the lawfulness of a charge, classification, regulation, or practice, issue an order concluding such investigation within 5 months after the date on which the complaint was filed.

(2) The Commission shall, with respect to any such investigation initiated prior to the date of enactment of this subsection [enacted Nov. 3, 1988], issue an order concluding the investigation not later than 12 months after such date of enactment.

(3) Any order concluding an investigation under paragraph (1) or (2) shall be a final order and may be appealed under section 402(a).

§ 209. Orders for Payment of Money

If, after hearing on a complaint, the Commission shall determine that any party complainant is entitled to an award of damages under the provisions of this Act, the Commission shall make an order directing the carrier to pay to the complainant the sum to which he is entitled on or before a day named.

§ 211. Contracts of Carriers; Filing with Commission

(a) Every carrier subject to this Act shall file with the Commission copies of all contracts, agreements, or arrangements with other carriers, or with common carriers not subject to the provisions of this Act, in relation to any traffic affected by the provisions of this Act to which it may be a party.

(b) The Commission shall have authority to require the filing of any other contracts of any carrier, and shall also have authority to exempt any carrier from submitting copies of such minor contracts as the Commission may determine.

§ 214. Extension of Lines or Discontinuance of Service; Certificate of Public Convenience and Necessity

(a) Exceptions; temporary or emergency service or discontinuance of service; changes in plant, operation or equipment. No carrier shall undertake the construction of a new line or of an extension of any line, or shall acquire or operate any line, or extension thereof, or shall engage in transmission over or by means of such additional or extended line, unless and until there shall first have been obtained from the Commission a certificate that the present or future public convenience and necessity require or will require the construction, or operation, or construction and operation, of such additional or extended line: Provided, That no such certificate shall be required under this section for the construction, acquisition, or operation of (1) a line within a single State unless such line constitutes part of an interstate line, (2) local, branch, or terminal lines not exceeding ten miles in length, or (3) any line acquired under section 221 of this Act: Provided further, That the Commission may, upon appropriate request being made, authorize temporary or emergency service, or the supplementing of existing facilities, without regard to the provisions of this section. No carrier shall discontinue, reduce, or impair service to a community, or part of a community, unless and until there shall first have been obtained from the Commission a certificate that neither the present nor future public convenience and necessity will be adversely affected thereby; except that the Commission may, upon appropriate request being made, authorize temporary or emergency discontinuance, reduction, or impairment of service, or partial discontinuance, reduction, or impairment of service, without regard to the provisions of this section. As used in this section the term "line" means any channel of communication established by the use of appropriate equipment, other than a channel of communication established by the interconnection of two or more existing channels: Provided, however, That nothing in this section shall be construed to require a certificate or other authorization from the Commission for any installation, replacement, or other changes in plant, operation, or equipment, other than new construction, which will not impair the adequacy or quality of service provided.

(b) Notification of Secretary of Defense, Secretary of State and State Governor. Upon receipt of an application for any such certificate, the Commission shall cause notice thereof to be given to, and shall cause a copy of such application to be filed with, the Secretary of Defense, the Secretary of State (with respect to such applications involving service to foreign points), and the Governor of each State in which such line is proposed to be constructed, extended, acquired, or operated, or in which such discontinuance, reduction, or impairment of service is proposed, with the right to those notified to be heard; and the Commission may require such published notice as it shall determine.

(c) Approval or disapproval; injunction. The Commission shall have power to issue such certificate as applied for, or to refuse to issue it, or to issue it for a portion or portions of a line, or extension thereof, or discontinuance, reduction, or impairment of service, described in the application, or for the partial exercise only of such right or privilege, and may attach to the issuance of the certificate such terms and conditions as in its judgment the public convenience and necessity may require. After issuance of such certificate, and not before, the carrier may, without securing approval other than such certificate, comply with the terms and conditions contained in or attached to the issuance of such certificate and proceed with the construction, extension, acquisition, operation, or discontinuance, reduction, or impairment of service covered thereby. Any construction, extension, acquisition, operation, discontinuance, reduction, or impairment of service contrary to the provisions of this section may be enjoined by any court of competent jurisdiction at the suit of the United States, the Commission, the State commission, any State affected, or any party in interest.

(d) Order of Commission; hearing; penalty. The Commission may, after full opportunity for hearing, in a proceeding upon complaint or upon its own initiative without complaint, authorize or require by order any carrier, party to such proceeding, to provide itself with adequate facilities for the expeditious and efficient performance of its service as a common carrier and to extend its line or to establish a public office; but no such authorization or order shall be made unless the Commission finds, as to such provision of facilities, as to such establishment of public offices, or as to such extension, that it is reasonably required in the interest of public convenience and necessity, or as to such extension or facilities that the expense involved therein will not impair the ability of the carrier to perform its duty to the public. Any carrier which refuses or neglects to comply with any order of the Commission made in pursuance of this paragraph shall forfeit to the United States $1,200 for each day during which such refusal or neglect continues.

(e) Provision of universal service.

(1) Eligible telecommunications carriers. A common carrier designated as an eligible telecommunications carrier under paragraph (2), (3), or (6) shall be eligible to receive universal service support in accordance with section 254 and shall, throughout the service area for which the designation is received—

(A) offer the services that are supported by Federal universal service support mechanisms under section 254(c), either using its own facilities or a combination of its own facilities and resale of another carrier's services (including the services offered by another eligible telecommunications carrier); and

(B) advertise the availability of such services and the charges therefor using media of general distribution.

(2) Designation of eligible telecommunications carriers. A State commission shall upon its own motion or upon request designate a common carrier that meets the requirements of paragraph (1) as an eligible telecommunications carrier for a service area designated by the State commission. Upon request and consistent with the public interest, convenience, and necessity, the State commission may, in the case of an area served by a rural telephone company, and shall, in the case of all other areas, designate more than one common carrier as an eligible telecommunications carrier for a service area designated by the State commission, so long as each additional requesting carrier meets the requirements of paragraph (1). Before designating an additional eligible telecommunications carrier for an area served by a rural telephone company, the State commission shall find that the designation is in the public interest.

(3) Designation of eligible telecommunications carriers for unserved areas. If no common carrier will provide the services that are supported by Federal universal service support mechanisms under section 254(c) to an unserved community or any portion thereof that requests such service, the

Commission, with respect to interstate services or an area served by a common carrier to which paragraph (6) applies, or a State commission, with respect to intrastate services, shall determine which common carrier or carriers are best able to provide such service to the requesting unserved community or portion thereof and shall order such carrier or carriers to provide such service for that unserved community or portion thereof. Any carrier or carriers ordered to provide such service under this paragraph shall meet the requirements of paragraph (1) and shall be designated as an eligible telecommunications carrier for that community or portion thereof.

(4) Relinquishment of universal service. A State commission (or the Commission in the case of a common carrier designated under paragraph (6)) shall permit an eligible telecommunications carrier to relinquish its designation as such a carrier in any area served by more than one eligible telecommunications carrier. An eligible telecommunications carrier that seeks to relinquish its eligible telecommunications carrier designation for an area served by more than one eligible telecommunications carrier shall give advance notice to the State commission (or the Commission in the case of a common carrier designated under paragraph (6)) of such relinquishment. Prior to permitting a telecommunications carrier designated as an eligible telecommunications carrier to cease providing universal service in an area served by more than one eligible telecommunications carrier, the State commission (or the Commission in the case of a common carrier designated under paragraph (6)) shall require the remaining eligible telecommunications carrier or carriers to ensure that all customers served by the relinquishing carrier will continue to be served, and shall require sufficient notice to permit the purchase or construction of adequate facilities by any remaining eligible telecommunications carrier. The State commission (or the Commission in the case of a common carrier designated under paragraph (6)) shall establish a time, not to exceed one year after the State commission (or the Commission in the case of a common carrier designated under paragraph (6)) approves such relinquishment under this paragraph, within which such purchase or construction shall be completed.

(5) Service area defined. The term "service area" means a geographic area established by a State commission (or the Commission under paragraph (6)) for the purpose of determining universal service obligations and support mechanisms. In the case of an area served by a rural telephone company, "service area" means such company's "study area" unless and until the Commission and the States, after taking into account recommendations of a Federal-State Joint Board instituted under section 410(c), establish a different definition of service area for such company.

(6) Common carriers not subject to State commission jurisdiction. In the case of a common carrier providing telephone exchange service and exchange access that is not subject to the jurisdiction of a State commission, the Commission shall upon request designate such a common carrier that meets the requirements of paragraph (1) as an eligible telecommunications carrier for a service area designated by the Commission consistent with applicable Federal and State law. Upon request and consistent with the public interest, convenience and necessity, the Commission may, with respect to an area served by a rural telephone company, and shall, in the case of all other areas, designate more than one common carrier as an eligible telecommunications carrier for a service area designated under this paragraph, so long as each additional requesting carrier meets the requirements of paragraph (1). Before designating an additional eligible telecommunications carrier for an area served by a rural telephone company, the Commission shall find that the designation is in the public interest.

§ 222. Privacy of Customer Information

(a) In general. Every telecommunications carrier has a duty to protect the confidentiality of proprietary information of, and relating to, other telecommunication carriers, equipment manufacturers, and customers, including telecommunication carriers reselling telecommunications services provided by a telecommunications carrier.

(b) Confidentiality of carrier information. A telecommunications carrier that receives or obtains proprietary information from another carrier for purposes of providing any telecommunications service shall use such information only for such purpose, and shall not use such information for its own marketing efforts.

(c) Confidentiality of customer proprietary network information.
  (1) Privacy requirements for telecommunications carriers. Except as required by law or with the approval of the customer, a telecommunications carrier that receives or obtains customer proprietary network information by virtue of its provision of a telecommunications service shall only use, disclose, or permit access to individually identifiable customer proprietary network information in its provision of (A) the telecommunications service from which such information is derived, or (B) services necessary to, or used in, the provision of such telecommunications service, including the publishing of directories.

(2) Disclosure on request by customers. A telecommunications carrier shall disclose customer proprietary network information, upon affirmative written request by the customer, to any person designated by the customer.

(3) Aggregate customer information. A telecommunications carrier that receives or obtains customer proprietary network information by virtue of its provision of a telecommunications service may use, disclose, or permit access to aggregate customer information other than for the purposes described in paragraph (1). A local exchange carrier may use, disclose, or permit access to aggregate customer information other than for purposes described in paragraph (1) only if it provides such aggregate information to other carriers or persons on reasonable and nondiscriminatory terms and conditions upon reasonable request therefor.

(d) Exceptions. Nothing in this section prohibits a telecommunications carrier from using, disclosing, or permitting access to customer proprietary network information obtained from its customers, either directly or indirectly through its agents—

(1) to initiate, render, bill, and collect for telecommunications services;

(2) to protect the rights or property of the carrier, or to protect users of those services and other carriers from fraudulent, abusive, or unlawful use of, or subscription to, such services;

(3) to provide any inbound telemarketing, referral, or administrative services to the customer for the duration of the call, if such call was initiated by the customer and the customer approves of the use of such information to provide such service; and

(4) to provide call location information concerning the user of a commercial mobile service (as such term is defined in section 332(d))—

(A) to a public safety answering point, emergency medical service provider or emergency dispatch provider, public safety, fire service, or law enforcement official, or hospital emergency or trauma care facility, in order to respond to the user's call for emergency services;

(B) to inform the user's legal guardian or members of the user's immediate family of the user's location in an emergency situation that involves the risk of death or serious physical harm; or

(C) to providers of information or database management services solely for purposes of assisting in the delivery of emergency services in response to an emergency.

(e) Subscriber list information. Notwithstanding subsections (b), (c), and (d), a telecommunications carrier that provides telephone exchange service shall provide subscriber list information gathered in its capacity as a provider of such service on a timely and unbundled basis, under nondiscriminatory and reasonable rates, terms, and conditions, to any person upon request for the purpose of publishing directories in any format.

(f) Authority to use wireless location information. For purposes of subsection (c)(1), without the express prior authorization of the customer, a customer shall not be considered to have approved the use or disclosure of or access to—

(1) call location information concerning the user of a commercial mobile service (as such term is defined in section 332(d)), other than in accordance with subsection (d)(4); or

(2) automatic crash notification information to any person other than for use in the operation of an automatic crash notification system.

(g) Subscriber listed and unlisted information for emergency services. Notwithstanding subsections (b), (c), and (d), a telecommunications carrier that provides telephone exchange service shall provide information described in subsection (i)(3)(A) (including information pertaining to subscribers whose information is unlisted or unpublished) that is in its possession or control (including information pertaining to subscribers of other carriers) on a timely and unbundled basis, under nondiscriminatory and reasonable rates, terms, and conditions to providers of emergency services, and providers of emergency support services, solely for purposes of delivering or assisting in the delivery of emergency services.

(h) Definitions. As used in this section:

(1) Customer proprietary network information. The term "customer proprietary network information" means—

(A) information that relates to the quantity, technical configuration, type, destination, location, and amount of use of a telecommunications service subscribed to by any customer of a telecommunications carrier, and that is made available to the carrier by the customer solely by virtue of the carrier-customer relationship; and

(B) information contained in the bills pertaining to telephone exchange service or telephone toll service received by a customer of a carrier; except that such term does not include subscriber list information.

(2) Aggregate information. The term "aggregate customer information" means collective data that relates to a group or category of services or customers, from which individual customer identities and characteristics have been removed.

(3) Subscriber list information. The term "subscriber list information" means any information—

(A) identifying the listed names of subscribers of a carrier and such subscribers' telephone numbers, addresses, or primary advertising classifications (as such classifications are assigned at the time of the establishment of such service), or any combination of such listed names, numbers, addresses, or classifications; and

(B) that the carrier or an affiliate has published, caused to be published, or accepted for publication in any directory format.

(4) Public safety answering point. The term "public safety answering point" means a facility that has been designated to receive emergency calls and route them to emergency service personnel.

(5) Emergency services. The term "emergency services" means 9-1-1 emergency services and emergency notification services.

(6) Emergency notification services. The term "emergency notification services" means services that notify the public of an emergency.

(7) Emergency support services. The term "emergency support services" means information or data base management services used in support of emergency services.

§ 224. Pole Attachments

(a) Definitions. As used in this section:

(1) The term "utility" means any person who is a local exchange carrier or an electric, gas, water, steam, or other public utility, and who owns or controls poles, ducts, conduits, or rights-of-way used, in whole or in part, for any wire communications. Such term does not include any railroad, any person who is cooperatively organized, or any person owned by the Federal Government or any State.

(2) The term "Federal Government" means the Government of the United States or any agency or instrumentality thereof.

(3) The term "State" means any State, territory, or possession of the United States, the District of Columbia, or any political subdivision, agency, or instrumentality thereof.

(4) The term "pole attachment" means any attachment by a cable television system or provider of telecommunications service to a pole, duct, conduit, or right-of-way owned or controlled by a utility.

(5) For purposes of this section, the term "telecommunications carrier" (as defined in section 3 of this Act does not include any incumbent local exchange carrier as defined in section 251(h).

(b) Authority of Commission to regulate rates, terms, and conditions; enforcement powers; promulgation of regulations.

(1) Subject to the provisions of subsection (c) of this section, the Commission shall regulate the rates, terms, and conditions for pole attachments to provide that such rates, terms, and conditions are just and reasonable, and shall adopt procedures necessary and appropriate to hear and resolve complaints concerning such rates, terms, and conditions. For purposes of enforcing any determinations resulting from complaint procedures established pursuant to this subsection, the Commission shall take such action as it deems appropriate and necessary, including issuing cease and desist orders, as authorized by section 312(b) of title III of the Communications Act of 1934, as amended.

(2) The Commission shall prescribe by rule regulations to carry out the provisions of this section.

(c) State regulatory authority over rates, terms, and conditions; preemption; certification; circumstances constituting State regulation.

(1) Nothing in this section shall be construed to apply to, or to give the Commission jurisdiction with respect to rates, terms, and conditions, or access to poles, ducts, conduits, and rights-of-way as provided in subsection (f), for pole attachments in any case where such matters are regulated by a State.

(2) Each State which regulates the rates, terms, and conditions for pole attachments shall certify to the Commission that—

(A) it regulates such rates, terms, and conditions; and

(B) in so regulating such rates, terms, and conditions, the State has the authority to consider and does consider the interests of the subscribers of the services offered via such attachments, as well as the interests of the consumers of the utility services.

(3) For purposes of this subsection, a State shall not be considered to regulate the rates, terms, and conditions for pole attachments—

(A) unless the State has issued and made effective rules and regulations implementing the State's regulatory authority over pole attachments; and

(B) with respect to any individual matter, unless the State takes final action on a complaint regarding such matter—
(i) within 180 days after the complaint is filed with the State, or
(ii) within the applicable period prescribed for such final action in such rules and regulations of the State, if the prescribed period does not extend beyond 360 days after the filing of such complaint.

(d) Determination of just and reasonable rates; definition of "usable space."

(1) For purposes of subsection (b) of this section, a rate is just and reasonable if it assures a utility the recovery of not less than the additional costs of providing pole attachments, nor more than an amount determined by multiplying the percentage of the total usable space, or the percentage of the total duct or conduit capacity, which is occupied by the pole attachment by the sum of the operating expenses and actual capital costs of the utility attributable to the entire pole, duct, conduit, or right-of-way.

(2) As used in this subsection, the term "usable space" means the space above the minimum grade level which can be used for the attachment of wires, cables, and associated equipment.

(3) This subsection shall apply to the rate for any pole attachment used by a cable television system solely to provide cable service. Until the effective date of the regulations required under subsection (e), this subsection shall also apply to the rate for any pole attachment used by a cable system or any telecommunications carrier (to the extent such carrier is not a party to a pole attachment agreement) to provide any telecommunications service.

(e) (1) The Commission shall, no later than 2 years after the date of enactment of the Telecommunications Act of 1996, prescribe regulations in accordance with this subsection to govern the charges for pole attachments used by telecommunications carriers to provide telecommunications services, when the parties fail to resolve a dispute over such charges. Such regulations shall ensure that a utility charges just, reasonable, and nondiscriminatory rates for pole attachments.

(2) A utility shall apportion the cost of providing space on a pole, duct, conduit, or right-of-way other than the usable space among entities so that such apportionment equals two-thirds of the costs of providing space other than the usable space that would be allocated to such entity under an equal apportionment of such costs among all attaching entities.

(3) A utility shall apportion the cost of providing usable space among all entities according to the percentage of usable space required for each entity.

(4) The regulations required under paragraph (1) shall become effective 5 years after the date of enactment of the Telecommunications Act of 1996. Any increase in the rates for pole attachments that result from the adoption of the regulations required by this subsection shall be phased in equal annual increments over a period of 5 years beginning on the effective date of such regulations.

(f) (1) A utility shall provide a cable television system or any telecommunications carrier with nondiscriminatory access to any pole, duct, conduit, or right-of-way owned or controlled by it.

(2) Notwithstanding paragraph (1), a utility providing electric service may deny a cable television system or any telecommunications carrier access to its poles, ducts, conduits, or rights-of-way, on a non-discriminatory basis where there is insufficient capacity and for reasons of safety, reliability and generally applicable engineering purposes.

(g) A utility that engages in the provision of telecommunications services or cable services shall impute to its costs of providing such services (and charge any affiliate, subsidiary, or associate company engaged in the provision of such services) an equal amount to the pole attachment rate for which such company would be liable under this section.

(h) Whenever the owner of a pole, duct, conduit, or right-of-way intends to modify or alter such pole, duct, conduit, or right-of-way, the owner shall provide written notification of such action to any entity that has obtained an attachment to such conduit or right-of-way so that such entity may have a reasonable opportunity to add to or modify its existing attachment. Any entity that adds to or modifies its existing attachment after receiving such notification shall bear a proportionate share of the costs incurred by the owner in making such pole, duct, conduit, or right-of-way accessible.

(i) An entity that obtains an attachment to a pole, conduit, or right-of-way shall not be required to bear any of the costs of rearranging or replacing its attachment, if such rearrangement or replacement is required as a result of an additional attachment or the modification of an existing attachment sought by any other entity (including the owner of such pole, duct, conduit, or right-of-way).

§ 226. Telephone Operator Services

(a) Definitions. As used in this section—

(1) The term "access code" means a sequence of numbers that, when dialed, connect the caller to the provider of operator services associated with that sequence.

(2) The term "aggregator" means any person that, in the ordinary course of its operations, makes telephones available to the public or to transient users of its premises, for interstate telephone calls using a provider of operator services.

(3) The term "call splashing" means the transfer of a telephone call from one provider of operator services to another such provider in such a manner that the subsequent provider is unable or unwilling to determine the location of the origination of the call and, because of such inability or unwillingness, is prevented from billing the call on the basis of such location.

(4) The term "consumer" means a person initiating any interstate telephone call using operator services.

(5) The term "equal access" has the meaning given that term in Appendix B of the Modification of Final Judgment entered August 24, 1982, in United States v. Western Electric, Civil Action No. 82-0192 (United States District Court, District of Columbia), as amended by the Court in its orders issued prior to the enactment of this section.

(6) The term "equal access code" means an access code that allows the public to obtain an equal access connection to the carrier associated with that code.

(7) The term "operator services" means any interstate telecommunications service initiated from an aggregator location that includes, as a component, any automatic or live assistance to a consumer to arrange for billing or completion, or both, of an interstate telephone call through a method other than—

(A) automatic completion with billing to the telephone from which the call originated; or

(B) completion through an access code used by the consumer, with billing to an account previously established with the carrier by the consumer.

(8) The term "presubscribed provider of operator services" means the interstate provider of operator services to which the consumer is

connected when the consumer places a call using a provider of operator services without dialing an access code.

(9) The term "provider of operator services" means any common carrier that provides operator services or any other person determined by the Commission to be providing operator services.

(b) Requirements for providers of operator services.
    (1) In general. Beginning not later than 90 days after the date of enactment of this section, each provider of operator services shall, at a minimum—
        (A) identify itself, audibly and distinctly, to the consumer at the beginning of each telephone call and before the consumer incurs any charge for the call;
        (B) permit the consumer to terminate the telephone call at no charge before the call is connected;
        (C) disclose immediately to the consumer, upon request and at no charge to the consumer—
            (i) a quote of its rates or charges for the call;
            (ii) the methods by which such rates or charges will be collected; and
            (iii) the methods by which complaints concerning such rates, charges, or collection practices will be resolved;
        (D) ensure, by contract or tariff, that each aggregator for which such provider is the presubscribed provider of operator services is in compliance with the requirements of subsection (c) and, if applicable, subsection (e)(1);
        (E) withhold payment (on a location-by-location basis) of any compensation, including commissions, to aggregators if such provider reasonably believes that the aggregator (i) is blocking access by means of "950" or "800" numbers to interstate common carriers in violation of subsection (c)(1)(B) or (ii) is blocking access to equal access codes in violation of rules the Commission may prescribe under subsection (e)(1);
        (F) not bill for unanswered telephone calls in areas where equal access is available;
        (G) not knowingly bill for unanswered telephone calls where equal access is not available;
        (H) not engage in call splashing, unless the consumer requests to be transferred to another provider of operator services, the consumer is informed prior to incurring any charges that the rates for the call may not

reflect the rates from the actual originating location of the call, and the consumer then consents to be transferred; and

(I) except as provided in subparagraph (H), not bill for a call that does not reflect the location of the origination of the call.

(J) [Repealed]

(2) Additional requirements for first 3 years. In addition to meeting the requirements of paragraph (1), during the 3-year period beginning on the date that is 90 days after the date of enactment of this section, each presubscribed provider of operator services shall identify itself audibly and distinctly to the consumer, not only as required in paragraph (1)(A), but also for a second time before connecting the call and before the consumer incurs any charge.

(c) Requirements for aggregators.

(1) In general. Each aggregator, beginning not later than 90 days after the date of enactment of this section, shall—

(A) post on or near the telephone instrument, in plain view of consumers—

(i) the name, address, and toll-free telephone number of the provider of operator services;

(ii) a written disclosure that the rates for all operator-assisted calls are available on request, and that consumers have a right to obtain access to the interstate common carrier of their choice and may contact their preferred interstate common carriers for information on accessing that carrier's service using that telephone; and

(iii) the name and address of the enforcement division of the Common Carrier Bureau of the Commission, to which the consumer may direct complaints regarding operator services;

(B) ensure that each of its telephones presubscribed to a provider of operator services allows the consumer to use "800" and "950" access code numbers to obtain access to the provider of operator services desired by the consumer; and

(C) ensure that no charge by the aggregator to the consumer for using an "800" or "950" access code number, or any other access code number, is greater than the amount the aggregator charges for calls placed using the presubscribed provider of operator services.

(2) Effect of State law or regulation. The requirements of paragraph (1)(A) shall not apply to an aggregator in any case in which State law or State regulation requires the aggregator to take actions that are substantially the same as those required in paragraph (1)(A).

(d) General rulemaking required.

(1) Rulemaking proceeding. The Commission shall conduct a rulemaking proceeding pursuant to this title to prescribe regulations to—

(A) protect consumers from unfair and deceptive practices relating to their use of operator services to place interstate telephone calls; and

(B) ensure that consumers have the opportunity to make informed choices in making such calls.

(2) Contents of regulations. The regulations prescribed under this section shall—

(A) contain provisions to implement each of the requirements of this section, other than the requirements established by the rulemaking under subsection (e) on access and compensation; and

(B) contain such other provisions as the Commission determines necessary to carry out this section and the purposes and policies of this section.

(3) Additional requirements to be implemented by regulations. The regulations prescribed under this section shall, at a minimum—

(A) establish minimum standards for providers of operator services and aggregators to use in the routing and handling of emergency telephone calls; and

(B) establish a policy for requiring providers of operator services to make public information about recent changes in operator services and choices available to consumers in that market.

(4) [Redesignated]

(e) Separate rulemaking on access and compensation.

(1) Access. The Commission[,] shall require—

(A) that each aggregator ensure within a reasonable time that each of its telephones presubscribed to a provider of operator services allows the consumer to obtain access to the provider of operator services desired by the consumer through the use of an equal access code; or

(B) that all providers of operator services, within a reasonable time, make available to their customers a "950" or "800" access code number for use in making operator services calls from anywhere in the United States; or

(C) that the requirements described under both subparagraphs (A) and (B) apply.

(2) Compensation. The Commission shall consider the need to prescribe compensation (other than advance payment by consumers) for owners of competitive public pay telephones for calls routed to providers of operator services that are other than the presubscribed provider of operator services for

such telephones. Within 9 months after the date of enactment of this section, the Commission shall reach a final decision on whether to prescribe such compensation.

(f) Technological capability of equipment. Any equipment and software manufactured or imported more than 18 months after the date of enactment of this section and installed by any aggregator shall be technologically capable of providing consumers with access to interstate providers of operator services through the use of equal access codes.

(g) Fraud. In any proceeding to carry out the provisions of this section, the Commission shall require such actions or measures as are necessary to ensure that aggregators are not exposed to undue risk of fraud.

(h) Determinations of rate compliance.
    (1) Filing of informational tariff.
        (A) In general. Each provider of operator services shall file, within 90 days after the date of enactment of this section, and shall maintain, update regularly, and keep open for public inspection, an informational tariff specifying rates, terms, and conditions, and including commissions, surcharges, any fees which are collected from consumers, and reasonable estimates of the amount of traffic priced at each rate, with respect to calls for which operator services are provided. Any changes in such rates, terms, or conditions shall be filed no later than the first day on which the changed rates, terms, or conditions are in effect.
        (B) Waiver authority. The Commission may, after 4 years following the date of enactment of this section, waive the requirements of this paragraph only if—
            (i) the findings and conclusions of the Commission in the final report issued under paragraph (3)(B)(iii) state that the regulatory objectives specified in subsection (d)(1)(A) and (B) have been achieved; and
            (ii) the Commission determines that such waiver will not adversely affect the continued achievement of such regulatory objectives.
    (2) Review of informational tariffs. If the rates and charges filed by any provider of operator services under paragraph (1) appear upon review by the Commission to be unjust or unreasonable, the Commission may require such provider of operator services to do either or both of the following:
        (A) demonstrate that its rates and charges are just and reasonable, and

(B) announce that its rates are available on request at the beginning of each call.

(3) Proceeding required.

(A) In general. Within 60 days after the date of enactment of this section, the Commission shall initiate a proceeding to determine whether the regulatory objectives specified in subsection (d)(1) (A) and (B) are being achieved. The proceeding shall—

(i) monitor operator service rates;

(ii) determine the extent to which offerings made by providers of operator services are improvements, in terms of service quality, price, innovation, and other factors, over those available before the entry of new providers of operator services into the market;

(iii) report on (in the aggregate and by individual provider) operator service rates, incidence of service complaints, and service offerings;

(iv) consider the effect that commissions and surcharges, billing and validation costs, and other costs of doing business have on the overall rates charged to consumers; and

(v) monitor compliance with the provisions of this section, including the periodic placement of telephone calls from aggregator locations.

(B) Reports.

(i) The Commission shall, during the pendency of such proceeding and not later than 5 months after its commencement, provide the Congress with an interim report on the Commission's activities and progress to date.

(ii) Not later than 11 months after the commencement of such proceeding, the Commission shall report to the Congress on its interim findings as a result of the proceeding.

(iii) Not later than 23 months after the commencement of such proceeding, the Commission shall submit a final report to the Congress on its findings and conclusions.

(4) Implementing regulations.

(A) In general. Unless the Commission makes the determination described in subparagraph (B), the Commission shall, within 180 days after submission of the report required under paragraph (3)(B)(iii), complete a rulemaking proceeding pursuant to this title to establish regulations for implementing the requirements of this title (and paragraphs (1) and (2) of this subsection) that rates and charges for operator services be just and reasonable. Such regulations shall include limitations on the amount of commissions or any other compensation given to aggregators by providers of operator service.

(B) Limitation. The requirement of subparagraph (A) shall not apply if, on the basis of the proceeding under paragraph (3)(A), the Commission makes (and includes in the report required by paragraph (3)(B)(iii)) a factual determination that market forces are securing rates and charges that are just and reasonable, as evidenced by rate levels, costs, complaints, service quality, and other relevant factors.

(i) Statutory construction. Nothing in this section shall be construed to alter the obligations, powers, or duties of common carriers or the Commission under the other sections of this Act.

§ 230. Protection for Private Blocking and Screening of Offensive Material

(a) Findings. The Congress finds the following:
   (1) The rapidly developing array of Internet and other interactive computer services available to individual Americans represent an extraordinary advance in the availability of educational and informational resources to our citizens.
   (2) These services offer users a great degree of control over the information that they receive, as well as the potential for even greater control in the future as technology develops.
   (3) The Internet and other interactive computer services offer a forum for a true diversity of political discourse, unique opportunities for cultural development, and myriad avenues for intellectual activity.
   (4) The Internet and other interactive computer services have flourished, to the benefit of all Americans, with a minimum of government regulation.
   (5) Increasingly Americans are relying on interactive media for a variety of political, educational, cultural, and entertainment services.

(b) Policy. It is the policy of the United States—
   (1) to promote the continued development of the Internet and other interactive computer services and other interactive media;
   (2) to preserve the vibrant and competitive free market that presently exists for the Internet and other interactive computer services, unfettered by Federal or State regulation;
   (3) to encourage the development of technologies which maximize user control over what information is received by individuals, families, and schools who use the Internet and other interactive computer services;

(4) to remove disincentives for the development and utilization of blocking and filtering technologies that empower parents to restrict their children's access to objectionable or inappropriate online material; and

(5) to ensure vigorous enforcement of Federal criminal laws to deter and punish trafficking in obscenity, stalking, and harassment by means of computer.

(c) Protection for "good samaritan" blocking and screening of offensive material.

(1) Treatment of publisher or speaker. No provider or user of an interactive computer service shall be treated as the publisher or speaker of any information provided by another information content provider.

(2) Civil liability. No provider or user of an interactive computer service shall be held liable on account of—

(A) any action voluntarily taken in good faith to restrict access to or availability of material that the provider or user considers to be obscene, lewd, lascivious, filthy, excessively violent, harassing, or otherwise objectionable, whether or not such material is constitutionally protected; or

(B) any action taken to enable or make available to information content providers or others the technical means to restrict access to material described in paragraph (1).

(d) Obligations of interactive computer service. A provider of interactive computer service shall, at the time of entering an agreement with a customer for the provision of interactive computer service and in a manner deemed appropriate by the provider, notify such customer that parental control protections (such as computer hardware, software, or filtering services) are commercially available that may assist the customer in limiting access to material that is harmful to minors. Such notice shall identify, or provide the customer with access to information identifying, current providers of such protections.

(e) Effect on other laws.

(1) No effect on criminal law. Nothing in this section shall be construed to impair the enforcement of section 223 or 231 of this Act, chapter 71 (relating to obscenity) or 110 (relating to sexual exploitation of children) of title 18, United States Code, or any other Federal criminal statute.

(2) No effect on intellectual property law. Nothing in this section shall be construed to limit or expand any law pertaining to intellectual property.

(3) State law. Nothing in this section shall be construed to prevent any State from enforcing any State law that is consistent with this section. No cause of action may be brought and no liability may be imposed under any State or local law that is inconsistent with this section.

(4) No effect on communications privacy law. Nothing in this section shall be construed to limit the application of the Electronic Communications Privacy Act of 1986 or any of the amendments made by such Act, or any similar State law.

(f) Definitions. As used in this section:

(1) Internet. The term "Internet" means the international computer network of both Federal and non-Federal interoperable packet switched data networks.

(2) Interactive computer service. The term "interactive computer service" means any information service, system, or access software provider that provides or enables computer access by multiple users to a computer server, including specifically a service or system that provides access to the Internet and such systems operated or services offered by libraries or educational institutions.

(3) Information content provider. The term "information content provider" means any person or entity that is responsible, in whole or in part, for the creation or development of information provided through the Internet or any other interactive computer service.

(4) Access software provider. The term "access software provider" means a provider of software (including client or server software), or enabling tools that do any one or more of the following:

(A) filter, screen, allow, or disallow content;

(B) pick, choose, analyze, or digest content; or

(C) transmit, receive, display, forward, cache, search, subset, organize, reorganize, or translate content.

§ 251. Interconnection

(a) General duty of telecommunications carriers. Each telecommunications carrier has the duty—

(1) to interconnect directly or indirectly with the facilities and equipment of other telecommunications carriers; and

(2) not to install network features, functions, or capabilities that do not comply with the guidelines and standards established pursuant to section 255 or 256.

(b) Obligations of all local exchange carriers. Each local exchange carrier has the following duties:

(1) Resale. The duty not to prohibit, and not to impose unreasonable or discriminatory conditions or limitations on, the resale of its telecommunications services.

(2) Number portability. The duty to provide, to the extent technically feasible, number portability in accordance with requirements prescribed by the Commission.

(3) Dialing parity. The duty to provide dialing parity to competing providers of telephone exchange service and telephone toll service, and the duty to permit all such providers to have nondiscriminatory access to telephone numbers, operator services, directory assistance, and directory listing, with no unreasonable dialing delays.

(4) Access to rights-of-way. The duty to afford access to the poles, ducts, conduits, and rights-of-way of such carrier to competing providers of telecommunications services on rates, terms, and conditions that are consistent with section 224.

(5) Reciprocal compensation. The duty to establish reciprocal compensation arrangements for the transport and termination of telecommunications.

(c) Additional obligations of incumbent local exchange carriers. In addition to the duties contained in subsection (b), each incumbent local exchange carrier has the following duties:

(1) Duty to negotiate. The duty to negotiate in good faith in accordance with section 252 the particular terms and conditions of agreements to fulfill the duties described in paragraphs (1) through (5) of subsection (b) and this subsection. The requesting telecommunications carrier also has the duty to negotiate in good faith the terms and conditions of such agreements.

(2) Interconnection. The duty to provide, for the facilities and equipment of any requesting telecommunications carrier, interconnection with the local exchange carrier's network—

(A) for the transmission and routing of telephone exchange service and exchange access;

(B) at any technically feasible point within the carrier's network;

(C) that is at least equal in quality to that provided by the local exchange carrier to itself or to any subsidiary, affiliate, or any other party to which the carrier provides interconnection; and

(D) on rates, terms, and conditions that are just, reasonable, and nondiscriminatory, in accordance with the terms and conditions of the agreement and the requirements of this section and section 252.

(3) Unbundled access. The duty to provide, to any requesting telecommunications carrier for the provision of a telecommunications service, nondiscriminatory access to network elements on an unbundled basis at any technically feasible point on rates, terms, and conditions that are just, reasonable, and nondiscriminatory in accordance with the terms and conditions of the agreement and the requirements of this section and section 252. An incumbent local exchange carrier shall provide such unbundled network elements in a manner that allows requesting carriers to combine such elements in order to provide such telecommunications service.

(4) Resale. The duty—

(A) to offer for resale at wholesale rates any telecommunications service that the carrier provides at retail to subscribers who are not telecommunications carriers; and

(B) not to prohibit, and not to impose unreasonable or discriminatory conditions or limitations on, the resale of such telecommunications service, except that a State commission may, consistent with regulations prescribed by the Commission under this section, prohibit a reseller that obtains at wholesale rates a telecommunications service that is available at retail only to a category of subscribers from offering such service to a different category of subscribers.

(5) Notice of changes. The duty to provide reasonable public notice of changes in the information necessary for the transmission and routing of services using that local exchange carrier's facilities or networks, as well as of any other changes that would affect the interoperability of those facilities and networks.

(6) Collocation. The duty to provide, on rates, terms, and conditions that are just, reasonable, and nondiscriminatory, for physical collocation of equipment necessary for interconnection or access to unbundled network elements at the premises of the local exchange carrier, except that the carrier may provide for virtual collocation if the local exchange carrier demonstrates to the State commission that physical collocation is not practical for technical reasons or because of space limitations.

(d) Implementation.

(1) In general. Within 6 months after the date of enactment of the Telecommunications Act of 1996, the Commission shall complete all actions

necessary to establish regulations to implement the requirements of this section.

(2) Access standards. In determining what network elements should be made available for purposes of subsection (c)(3), the Commission shall consider, at a minimum, whether—

(A) access to such network elements as are proprietary in nature is necessary; and

(B) the failure to provide access to such network elements would impair the ability of the telecommunications carrier seeking access to provide the services that it seeks to offer.

(3) Preservation of State access regulations. In prescribing and enforcing regulations to implement the requirements of this section, the Commission shall not preclude the enforcement of any regulation, order, or policy of a State commission that—

(A) establishes access and interconnection obligations of local exchange carriers;

(B) is consistent with the requirements of this section; and

(C) does not substantially prevent implementation of the requirements of this section and the purposes of this part.

(e) Numbering administration.

(1) Commission authority and jurisdiction. The Commission shall create or designate one or more impartial entities to administer telecommunications numbering and to make such numbers available on an equitable basis. The Commission shall have exclusive jurisdiction over those portions of the North American Numbering Plan that pertain to the United States. Nothing in this paragraph shall preclude the Commission from delegating to State commissions or other entities all or any portion of such jurisdiction.

(2) Costs. The cost of establishing telecommunications numbering administration arrangements and number portability shall be borne by all telecommunications carriers on a competitively neutral basis as determined by the Commission.

(3) Universal emergency telephone number. The Commission and any agency or entity to which the Commission has delegated authority under this subsection shall designate 9-1-1 as the universal emergency telephone number within the United States for reporting an emergency to appropriate authorities and requesting assistance. The designation shall apply to both wireline and wireless telephone service. In making the designation, the Commission (and any such agency or entity) shall provide appropriate transition periods for areas in which 9-1-1 is not in use as an emergency telephone

number on the date of enactment of the Wireless Communications and Public Safety Act of 1999.

(f) Exemptions, suspensions, and modifications.

(1) Exemption for certain rural telephone companies.

(A) Exemption. Subsection (c) of this section shall not apply to a rural telephone company until (i) such company has received a bona fide request for interconnection, services, or network elements, and (ii) the State commission determines (under subparagraph (B)) that such request is not unduly economically burdensome, is technically feasible, and is consistent with section 254 (other than subsections (b)(7) and (c)(1)(D) thereof).

(B) State termination of exemption and implementation schedule. The party making a bona fide request of a rural telephone company for interconnection, services, or network elements shall submit a notice of its request to the State commission. The State commission shall conduct an inquiry for the purpose of determining whether to terminate the exemption under subparagraph (A). Within 120 days after the State commission receives notice of the request, the State commission shall terminate the exemption if the request is not unduly economically burdensome, is technically feasible, and is consistent with section 254 (other than subsections (b)(7) and (c)(1)(D) thereof). Upon termination of the exemption, a State commission shall establish an implementation schedule for compliance with the request that is consistent in time and manner with Commission regulations.

(C) Limitation on exemption. The exemption provided by this paragraph shall not apply with respect to a request under subsection (c) from a cable operator providing video programming, and seeking to provide any telecommunications service, in the area in which the rural telephone company provides video programming. The limitation contained in this subparagraph shall not apply to a rural telephone company that is providing video programming on the date of enactment of the Telecommunications Act of 1996.

(2) Suspensions and modifications for rural carriers. A local exchange carrier with fewer than 2 percent of the Nation's subscriber lines installed in the aggregate nationwide may petition a State commission for a suspension or modification of the application of a requirement or requirements of subsection (b) or (c) to telephone exchange service facilities specified in such petition. The State commission shall grant such petition to the extent that, and for such duration as, the State commission determines that such suspension or modification—

(A) is necessary—

(i) to avoid a significant adverse economic impact on users of telecommunications services generally;

(ii) to avoid imposing a requirement that is unduly economically burdensome; or

(iii) to avoid imposing a requirement that is technically infeasible; and

(B) is consistent with the public interest, convenience, and necessity.

The State commission shall act upon any petition filed under this paragraph within 180 days after receiving such petition. Pending such action, the State commission may suspend enforcement of the requirement or requirements to which the petition applies with respect to the petitioning carrier or carriers.

(g) Continued enforcement of exchange access and interconnection requirements. On and after the date of enactment of the Telecommunications Act of 1996, each local exchange carrier, to the extent that it provides wireline services, shall provide exchange access, information access, and exchange services for such access to interexchange carriers and information service providers in accordance with the same equal access and nondiscriminatory interconnection restrictions and obligations (including receipt of compensation) that apply to such carrier on the date immediately preceding the date of enactment of the Telecommunications Act of 1996 under any court order, consent decree, or regulation, order, or policy of the Commission, until such restrictions and obligations are explicitly superseded by regulations prescribed by the Commission after such date of enactment. During the period beginning on such date of enactment and until such restrictions and obligations are so superseded, such restrictions and obligations shall be enforceable in the same manner as regulations of the Commission.

(h) Definition of incumbent local exchange carrier.

(1) Definition. For purposes of this section, the term "incumbent local exchange carrier" means, with respect to an area, the local exchange carrier that—

(A) on the date of enactment of the Telecommunications Act of 1996, provided telephone exchange service in such area; and

(B) (i) on such date of enactment, was deemed to be a member of the exchange carrier association pursuant to section 69.601(b) of the Commission's regulations (47 C.F.R. 69.601(b)); or

(ii) is a person or entity that, on or after such date of enactment, became a successor or assign of a member described in clause (i).

(2) Treatment of comparable carriers as incumbents. The Commission may, by rule, provide for the treatment of a local exchange carrier (or class or category thereof) as an incumbent local exchange carrier for purposes of this section if—

(A) such carrier occupies a position in the market for telephone exchange service within an area that is comparable to the position occupied by a carrier described in paragraph (1);

(B) such carrier has substantially replaced an incumbent local exchange carrier described in paragraph (1); and

(C) such treatment is consistent with the public interest, convenience, and necessity and the purposes of this section.

(i) Savings provision. Nothing in this section shall be construed to limit or otherwise affect the Commission's authority under section 201.

§ 252. Procedures for Negotiation, Arbitration, and Approval of Agreements

(a) Agreements arrived at through negotiation.

(1) Voluntary negotiations. Upon receiving a request for interconnection, services, or network elements pursuant to section 251, an incumbent local exchange carrier may negotiate and enter into a binding agreement with the requesting telecommunications carrier or carriers without regard to the standards set forth in subsections (b) and (c) of section 251. The agreement shall include a detailed schedule of itemized charges for interconnection and each service or network element included in the agreement. The agreement, including any interconnection agreement negotiated before the date of enactment of the Telecommunications Act of 1996, shall be submitted to the State commission under subsection (e) of this section.

(2) Mediation. Any party negotiating an agreement under this section may, at any point in the negotiation, ask a State commission to participate in the negotiation and to mediate any differences arising in the course of the negotiation.

(b) Agreements arrived at through compulsory arbitration.

(1) Arbitration. During the period from the 135th to the 160th day (inclusive) after the date on which an incumbent local exchange carrier receives a request for negotiation under this section, the carrier or any other

party to the negotiation may petition a State commission to arbitrate any open issues.

(2) Duty of petitioner.

(A) A party that petitions a State commission under paragraph (1) shall, at the same time as it submits the petition, provide the State commission all relevant documentation concerning—

(i) the unresolved issues;

(ii) the position of each of the parties with respect to those issues; and

(iii) any other issue discussed and resolved by the parties.

(B) A party petitioning a State commission under paragraph (1) shall provide a copy of the petition and any documentation to the other party or parties not later than the day on which the State commission receives the petition.

(3) Opportunity to respond. A non-petitioning party to a negotiation under this section may respond to the other party's petition and provide such additional information as it wishes within 25 days after the State commission receives the petition.

(4) Action by State commission.

(A) The State commission shall limit its consideration of any petition under paragraph (1) (and any response thereto) to the issues set forth in the petition and in the response, if any, filed under paragraph (3).

(B) The State commission may require the petitioning party and the responding party to provide such information as may be necessary for the State commission to reach a decision on the unresolved issues. If any party refuses or fails unreasonably to respond on a timely basis to any reasonable request from the State commission, then the State commission may proceed on the basis of the best information available to it from whatever source derived.

(C) The State commission shall resolve each issue set forth in the petition and the response, if any, by imposing appropriate conditions as required to implement subsection (c) upon the parties to the agreement, and shall conclude the resolution of any unresolved issues not later than 9 months after the date on which the local exchange carrier received the request under this section.

(5) Refusal to negotiate. The refusal of any other party to the negotiation to participate further in the negotiations, to cooperate with the State commission in carrying out its function as an arbitrator, or to continue to negotiate in good faith in the presence, or with the assistance, of the State commission shall be considered a failure to negotiate in good faith.

(c) Standards for arbitration. In resolving by arbitration under subsection (b) any open issues and imposing conditions upon the parties to the agreement, a State commission shall—

(1) ensure that such resolution and conditions meet the requirements of section 251, including the regulations prescribed by the Commission pursuant to section 251;

(2) establish any rates for interconnection, services, or network elements according to subsection (d); and

(3) provide a schedule for implementation of the terms and conditions by the parties to the agreement.

(d) Pricing standards.

(1) Interconnection and network element charges. Determinations by a State commission of the just and reasonable rate for the interconnection of facilities and equipment for purposes of subsection (c)(2) of section 251, and the just and reasonable rate for network elements for purposes of subsection (c)(3) of such section —

(A) shall be—

(i) based on the cost (determined without reference to a rate-of-return or other rate-based proceeding) of providing the interconnection or network element (whichever is applicable), and

(ii) nondiscriminatory, and

(B) may include a reasonable profit.

(2) Charges for transport and termination of traffic.

(A) In general. For the purposes of compliance by an incumbent local exchange carrier with section 251(b)(5), a State commission shall not consider the terms and conditions for reciprocal compensation to be just and reasonable unless—

(i) such terms and conditions provide for the mutual and reciprocal recovery by each carrier of costs associated with the transport and termination on each carrier's network facilities of calls that originate on the network facilities of the other carrier; and

(ii) such terms and conditions determine such costs on the basis of a reasonable approximation of the additional costs of terminating such calls.

(B) Rules of construction. This paragraph shall not be construed—

(i) to preclude arrangements that afford the mutual recovery of costs through the offsetting of reciprocal obligations, including arrangements that waive mutual recovery (such as bill-and-keep arrangements); or

(ii) to authorize the Commission or any State commission to engage in any rate regulation proceeding to establish with particularity the additional costs of transporting or terminating calls, or to require carriers to maintain records with respect to the additional costs of such calls.

(3) Wholesale prices for telecommunications services. For the purposes of section 251(c)(4), a State commission shall determine wholesale rates on the basis of retail rates charged to subscribers for the telecommunications service requested, excluding the portion thereof attributable to any marketing, billing, collection, and other costs that will be avoided by the local exchange carrier.

(e) Approval by State commission.

(1) Approval required. Any interconnection agreement adopted by negotiation or arbitration shall be submitted for approval to the State commission. A State commission to which an agreement is submitted shall approve or reject the agreement, with written findings as to any deficiencies.

(2) Grounds for rejection. The State commission may only reject—

(A) an agreement (or any portion thereof) adopted by negotiation under subsection (a) if it finds that—

(i) the agreement (or portion thereof) discriminates against a telecommunications carrier not a party to the agreement; or

(ii) the implementation of such agreement or portion is not consistent with the public interest, convenience, and necessity; or

(B) an agreement (or any portion thereof) adopted by arbitration under subsection (b) if it finds that the agreement does not meet the requirements of section 251, including the regulations prescribed by the Commission pursuant to section 251, or the standards set forth in subsection (d) of this section.

(3) Preservation of authority. Notwithstanding paragraph (2), but subject to section 253, nothing in this section shall prohibit a State commission from establishing or enforcing other requirements of State law in its review of an agreement, including requiring compliance with intrastate telecommunications service quality standards or requirements.

(4) Schedule for decision. If the State commission does not act to approve or reject the agreement within 90 days after submission by the parties of an agreement adopted by negotiation under subsection (a), or within 30 days after submission by the parties of an agreement adopted by arbitration under subsection (b), the agreement shall be deemed approved. No State court shall have jurisdiction to review the action of a State commission in approving or rejecting an agreement under this section.

(5) Commission to act if State will not act. If a State commission fails to act to carry out its responsibility under this section in any proceeding

or other matter under this section, then the Commission shall issue an order preempting the State commission's jurisdiction of that proceeding or matter within 90 days after being notified (or taking notice) of such failure, and shall assume the responsibility of the State commission under this section with respect to the proceeding or matter and act for the State commission.

(6) Review of State commission actions. In a case in which a State fails to act as described in paragraph (5), the proceeding by the Commission under such paragraph and any judicial review of the Commission's actions shall be the exclusive remedies for a State commission's failure to act. In any case in which a State commission makes a determination under this section, any party aggrieved by such determination may bring an action in an appropriate Federal district court to determine whether the agreement or statement meets the requirements of section 251 and this section.

(f) Statements of generally available terms.
(1) In general. A Bell operating company may prepare and file with a State commission a statement of the terms and conditions that such company generally offers within that State to comply with the requirements of section 251 and the regulations thereunder and the standards applicable under this section.

(2) State commission review. A State commission may not approve such statement unless such statement complies with subsection (d) of this section and section 251 and the regulations thereunder. Except as provided in section 253, nothing in this section shall prohibit a State commission from establishing or enforcing other requirements of State law in its review of such statement, including requiring compliance with intrastate telecommunications service quality standards or requirements.

(3) Schedule for review. The State commission to which a statement is submitted shall, not later than 60 days after the date of such submission—
(A) complete the review of such statement under paragraph (2) (including any reconsideration thereof), unless the submitting carrier agrees to an extension of the period for such review; or
(B) permit such statement to take effect.

(4) Authority to continue review. Paragraph (3) shall not preclude the State commission from continuing to review a statement that has been permitted to take effect under subparagraph (B) of such paragraph or from approving or disapproving such statement under paragraph (2).

(5) Duty to negotiate not affected. The submission or approval of a statement under this subsection shall not relieve a Bell operating company

of its duty to negotiate the terms and conditions of an agreement under section 251.

(g) Consolidation of State proceedings. Where not inconsistent with the requirements of this Act, a State commission may, to the extent practical, consolidate proceedings under sections 214(e), 251(f), 253, and this section in order to reduce administrative burdens on telecommunications carriers, other parties to the proceedings, and the State commission in carrying out its responsibilities under this Act.

(h) Filing required. A State commission shall make a copy of each agreement approved under subsection (e) and each statement approved under subsection (f) available for public inspection and copying within 10 days after the agreement or statement is approved. The State commission may charge a reasonable and nondiscriminatory fee to the parties to the agreement or to the party filing the statement to cover the costs of approving and filing such agreement or statement.

(i) Availability to other telecommunications carriers. A local exchange carrier shall make available any interconnection, service, or network element provided under an agreement approved under this section to which it is a party to any other requesting telecommunications carrier upon the same terms and conditions as those provided in the agreement.

(j) Definition of incumbent local exchange carrier. For purposes of this section, the term "incumbent local exchange carrier" has the meaning provided in section 251.

§ 253. Removal of Barriers to Entry

(a) In general. No State or local statute or regulation, or other State or local legal requirement, may prohibit or have the effect of prohibiting the ability of any entity to provide any interstate or intrastate telecommunications service.

(b) State regulatory authority. Nothing in this section shall affect the ability of a State to impose, on a competitively neutral basis and consistent with section 254, requirements necessary to preserve and advance universal service, protect the public safety and welfare, ensure the continued quality of telecommunications services, and safeguard the rights of consumers.

(c) State and local government authority. Nothing in this section affects the authority of a State or local government to manage the public rights-of-way or to require fair and reasonable compensation from telecommunications providers, on a competitively neutral and nondiscriminatory basis, for use of public rights-of-way on a nondiscriminatory basis, if the compensation required is publicly disclosed by such government.

(d) Preemption. If, after notice and an opportunity for public comment, the Commission determines that a State or local government has permitted or imposed any statute, regulation, or legal requirement that violates subsection (a) or (b), the Commission shall preempt the enforcement of such statute, regulation, or legal requirement to the extent necessary to correct such violation or inconsistency.

(e) Commercial mobile service providers. Nothing in this section shall affect the application of section 332(c)(3) to commercial mobile service providers.

(f) Rural Markets. It shall not be a violation of this section for a State to require a telecommunications carrier that seeks to provide telephone exchange service or exchange access in a service area served by a rural telephone company to meet the requirements in section 214(e)(1) for designation as an eligible telecommunications carrier for that area before being permitted to provide such service. This subsection shall not apply—
    (1) to a service area served by a rural telephone company that has obtained an exemption, suspension, or modification of section 251(c)(4) that effectively prevents a competitor from meeting the requirements of section 214(e)(1); and
    (2) to a provider of commercial mobile services.

§ 254. Universal Service

(a) Procedures to review universal service requirements.
    (1) Federal-State Joint Board on Universal Service. Within one month after the date of enactment of the Telecommunications Act of 1996, the Commission shall institute and refer to a Federal-State Joint Board under section 410(c) a proceeding to recommend changes to any of its regulations in order to implement sections 214(e) and this section, including the definition of the services that are supported by Federal universal service support mechanisms and a specific timetable for completion of such recommendations. In addition to the members of the Joint Board required

under section 410(c), one member of such Joint Board shall be a State-appointed utility consumer advocate nominated by a national organization of State utility consumer advocates. The Joint Board shall, after notice and opportunity for public comment, make its recommendations to the Commission 9 months after the date of enactment of the Telecommunications Act of 1996.

(2) Commission action. The Commission shall initiate a single proceeding to implement the recommendations from the Joint Board required by paragraph (1) and shall complete such proceeding within 15 months after the date of enactment of the Telecommunications Act of 1996. The rules established by such proceeding shall include a definition of the services that are supported by Federal universal service support mechanisms and a specific timetable for implementation. Thereafter, the Commission shall complete any proceeding to implement subsequent recommendations from any Joint Board on universal service within one year after receiving such recommendations.

(b) Universal service principles. The Joint Board and the Commission shall base policies for the preservation and advancement of universal service on the following principles:

(1) Quality and rates. Quality services should be available at just, reasonable, and affordable rates.

(2) Access to advanced services. Access to advanced telecommunications and information services should be provided in all regions of the Nation.

(3) Access in rural and high cost areas. Consumers in all regions of the Nation, including low-income consumers and those in rural, insular, and high cost areas, should have access to telecommunications and information services, including interexchange services and advanced telecommunications and information services, that are reasonably comparable to those services provided in urban areas and that are available at rates that are reasonably comparable to rates charged for similar services in urban areas.

(4) Equitable and nondiscriminatory contributions. All providers of telecommunications services should make an equitable and nondiscriminatory contribution to the preservation and advancement of universal service.

(5) Specific and predictable support mechanisms. There should be specific, predictable and sufficient Federal and State mechanisms to preserve and advance universal service.

(6) Access to advanced telecommunications services for schools, health care, and libraries. Elementary and secondary schools and classrooms,

health care providers, and libraries should have access to advanced telecommunications services as described in subsection (h).

(7) Additional principles. Such other principles as the Joint Board and the Commission determine are necessary and appropriate for the protection of the public interest, convenience, and necessity and are consistent with this Act.

(c) Definition.

(1) In general. Universal service is an evolving level of telecommunications services that the Commission shall establish periodically under this section, taking into account advances in telecommunications and information technologies and services. The Joint Board in recommending, and the Commission in establishing, the definition of the services that are supported by Federal universal service support mechanisms shall consider the extent to which such telecommunications services—

(A) are essential to education, public health, or public safety;

(B) have, through the operation of market choices by customers, been subscribed to by a substantial majority of residential customers;

(C) are being deployed in public telecommunications networks by telecommunications carriers; and

(D) are consistent with the public interest, convenience, and necessity.

(2) Alterations and modifications. The Joint Board may, from time to time, recommend to the Commission modifications in the definition of the services that are supported by Federal universal service support mechanisms.

(3) Special services. In addition to the services included in the definition of universal service under paragraph (1), the Commission may designate additional services for such support mechanisms for schools, libraries, and health care providers for the purposes of subsection (h).

(d) Telecommunications carrier contribution. Every telecommunications carrier that provides interstate telecommunications services shall contribute, on an equitable and nondiscriminatory basis, to the specific, predictable, and sufficient mechanisms established by the Commission to preserve and advance universal service. The Commission may exempt a carrier or class of carriers from this requirement if the carrier's telecommunications activities are limited to such an extent that the level of such carrier's contribution to the preservation and advancement of universal service would be de minimis. Any other provider of interstate telecommunications may be required to contribute to the preservation and advancement of universal service if the public interest so requires.

(e) Universal service support. After the date on which Commission regulations implementing this section take effect, only an eligible telecommunications carrier designated under section 214(e) [*47 USC § 214(e)*] shall be eligible to receive specific Federal universal service support. A carrier that receives such support shall use that support only for the provision, maintenance, and upgrading of facilities and services for which the support is intended. Any such support should be explicit and sufficient to achieve the purposes of this section.

(f) State authority. A State may adopt regulations not inconsistent with the Commission's rules to preserve and advance universal service. Every telecommunications carrier that provides intrastate telecommunications services shall contribute, on an equitable and nondiscriminatory basis, in a manner determined by the State to the preservation and advancement of universal service in that State. A State may adopt regulations to provide for additional definitions and standards to preserve and advance universal service within that State only to the extent that such regulations adopt additional specific, predictable, and sufficient mechanisms to support such definitions or standards that do not rely on or burden Federal universal service support mechanisms.

(g) Interexchange and interstate services. Within 6 months after the date of enactment of the Telecommunications Act of 1996, the Commission shall adopt rules to require that the rates charged by providers of interexchange telecommunications services to subscribers in rural and high cost areas shall be no higher than the rates charged by each such provider to its subscribers in urban areas. Such rules shall also require that a provider of interstate interexchange telecommunications services shall provide such services to its subscribers in each State at rates no higher than the rates charged to its subscribers in any other State.

(h) Telecommunications services for certain providers.
    (1) In general.
        (A) Health care providers for rural areas. A telecommunications carrier shall, upon receiving a bona fide request, provide telecommunications services which are necessary for the provision of health care services in a State, including instruction relating to such services, to any public or

nonprofit health care provider that serves persons who reside in rural areas in that State at rates that are reasonably comparable to rates charged for similar services in urban areas in that State. A telecommunications carrier providing service under this paragraph shall be entitled to have an amount equal to the difference, if any, between the rates for services provided to health care providers for rural areas in a State and the rates for similar services provided to other customers in comparable rural areas in that State treated as a service obligation as a part of its obligation to participate in the mechanisms to preserve and advance universal service.

(B) Educational providers and libraries. All telecommunications carriers serving a geographic area shall, upon a bona fide request for any of its services that are within the definition of universal service under subsection (c)(3), provide such services to elementary schools, secondary schools, and libraries for educational purposes at rates less than the amounts charged for similar services to other parties. The discount shall be an amount that the Commission, with respect to interstate services, and the States, with respect to intrastate services, determine is appropriate and necessary to ensure affordable access to and use of such services by such entities. A telecommunications carrier providing service under this paragraph shall—

(i) have an amount equal to the amount of the discount treated as an offset to its obligation to contribute to the mechanisms to preserve and advance universal service, or

(ii) notwithstanding the provisions of subsection (e) of this section, receive reimbursement utilizing the support mechanisms to preserve and advance universal service.

(2) Advanced services. The Commission shall establish competitively neutral rules—

(A) to enhance, to the extent technically feasible and economically reasonable, access to advanced telecommunications and information services for all public and nonprofit elementary and secondary school classrooms, health care providers, and libraries; and

(B) to define the circumstances under which a telecommunications carrier may be required to connect its network to such public institutional telecommunications users.

(3) Terms and conditions. Telecommunications services and network capacity provided to a public institutional telecommunications user under this subsection may not be sold, resold, or otherwise transferred by such user in consideration for money or any other thing of value.

(4) Eligibility of users. No entity listed in this subsection shall be entitled to preferential rates or treatment as required by this subsection, if

such entity operates as a for-profit business, is a school described in paragraph (5)(A) with an endowment of more than $50,000,000, or is a library or library consortium not eligible for assistance from a State library administrative agency under the Library Services and Technology Act.

(5) Definitions. For purposes of this subsection:

(A) Elementary and secondary schools. The term "elementary and secondary schools" means elementary schools and secondary schools, as defined in paragraphs (14) and (25), respectively, of section 14101 of the Elementary and Secondary Education Act of 1965.

(B) Health care provider. The term "health care provider" means—

(i) post-secondary educational institutions offering health care instruction, teaching hospitals, and medical schools;

(ii) community health centers or health centers providing health care to migrants;

(iii) local health departments or agencies;

(iv) community mental health centers;

(v) not-for-profit hospitals;

(vi) rural health clinics; and

(vii) consortia of health care providers consisting of one or more entities described in clauses (i) through (vi).

(C) Public institutional telecommunications user. The term "public institutional telecommunications user" means an elementary or secondary school, a library, or a health care provider as those terms are defined in this paragraph.

(i) Consumer protection. The Commission and the States should ensure that universal service is available at rates that are just, reasonable, and affordable.

(j) Lifeline assistance. Nothing in this section shall affect the collection, distribution, or administration of the Lifeline Assistance Program provided for by the Commission under regulations set forth in section 69.117 of title 47, Code of Federal Regulations, and other related sections of such title.

(k) Subsidy of competitive services prohibited. A telecommunications carrier may not use services that are not competitive to subsidize services that are subject to competition. The Commission, with respect to interstate services, and the States, with respect to intrastate services, shall establish any necessary cost allocation rules, accounting safeguards, and guidelines to ensure that services included in the definition of universal service bear no more than a reasonable share of the joint and common costs of facilities used to provide those services.

§ 271. Bell Operating Company Entry into Interlata Services

(a) General limitation. Neither a Bell operating company, nor any affiliate of a Bell operating company, may provide interLATA services except as provided in this section.

(b) InterLATA services to which this section applies.

(1) In-region services. A Bell operating company, or any affiliate of that Bell operating company, may provide interLATA services originating in any of its in-region States (as defined in subsection (i)) if the Commission approves the application of such company for such State under subsection (d)(3).

(2) Out-of-region services. A Bell operating company, or any affiliate of that Bell operating company, may provide interLATA services originating outside its in-region States after the date of enactment of the Telecommunications Act of 1996, subject to subsection (j).

(3) Incidental interLATA services. A Bell operating company, or any affiliate of a Bell operating company, may provide incidental interLATA services (as defined in subsection (g)) originating in any State after the date of enactment of the Telecommunications Act of 1996.

(4) Termination. Nothing in this section prohibits a Bell operating company or any of its affiliates from providing termination for interLATA services, subject to subsection (j).

(c) Requirements for providing certain in-region interLATA services.

(1) Agreement or statement. A Bell operating company meets the requirements of this paragraph if it meets the requirements of subparagraph (A) or subparagraph (B) of this paragraph for each State for which the authorization is sought.

(A) Presence of a facilities-based competitor. A Bell operating company meets the requirements of this subparagraph if it has entered into one or more binding agreements that have been approved under section 252 specifying the terms and conditions under which the Bell operating company is providing access and interconnection to its network facilities for the network facilities of one or more unaffiliated competing providers of telephone exchange service (as defined in section 3(47)(A), but excluding exchange access) to residential and business subscribers. For the purpose of this subparagraph, such telephone exchange service may be offered by such competing providers either exclusively over their own telephone exchange service facilities or predominantly over their own telephone exchange service facilities in combination with the resale of the telecommunications services of another carrier. For the purpose of this subparagraph, services provided pursuant to subpart K of part 22 of the Commission's regulations (47 C.F.R. 22.901 et seq.) shall not be considered to be telephone exchange services.

(B) Failure to request access. A Bell operating company meets the requirements of this subparagraph if, after 10 months after the date of enactment of the Telecommunications Act of 1996, no such provider has requested the access and interconnection described in subparagraph (A) before the date which is 3 months before the date the company makes its application under subsection (d)(1), and a statement of the terms and conditions that the company generally offers to provide such access and interconnection has been approved or permitted to take effect by the State commission under section 252(f). For purposes of this subparagraph, a Bell operating company shall be considered not to have received any request for access and interconnection if the State commission of such State certifies that the only provider or providers making such a request have (i) failed to negotiate in good faith as required by section 252, or (ii) violated the terms of an agreement approved under section 252 by the provider's failure to comply, within a reasonable period of time, with the implementation schedule contained in such agreement.

(2) Specific interconnection requirements.

(A) Agreement required. A Bell operating company meets the requirements of this paragraph if, within the State for which the authorization is sought—

(i) (I) such company is providing access and interconnection pursuant to one or more agreements described in paragraph (1)(A), or

(II) such company is generally offering access and interconnection pursuant to a statement described in paragraph (1)(B), and

(ii) such access and interconnection meets the requirements of subparagraph (B) of this paragraph.

(B) Competitive checklist. Access or interconnection provided or generally offered by a Bell operating company to other telecommunications carriers meets the requirements of this subparagraph if such access and interconnection includes each of the following:

(i) Interconnection in accordance with the requirements of sections 251(c)(2) and 252(d)(1).

(ii) Nondiscriminatory access to network elements in accordance with the requirements of sections 251(c)(3) and 252(d)(1).

(iii) Nondiscriminatory access to the poles, ducts, conduits, and rights-of-way owned or controlled by the Bell operating company at just and reasonable rates in accordance with the requirements of section 224.

(iv) Local loop transmission from the central office to the customer's premises, unbundled from local switching or other services.

(v) Local transport from the trunk side of a wireline local exchange carrier switch unbundled from switching or other services.

(vi) Local switching unbundled from transport, local loop transmission, or other services.

(vii) Nondiscriminatory access to—

(I) 911 and E911 services;

(II) directory assistance services to allow the other carrier's customers to obtain telephone numbers; and

(III) operator call completion services.

(viii) White pages directory listings for customers of the other carrier's telephone exchange service.

(ix) Until the date by which telecommunications numbering administration guidelines, plan, or rules are established, nondiscriminatory access to telephone numbers for assignment to the other carrier's telephone exchange service customers. After that date, compliance with such guidelines, plan, or rules.

(x) Nondiscriminatory access to databases and associated signaling necessary for call routing and completion.

(xi) Until the date by which the Commission issues regulations pursuant to section 251 to require number portability, interim telecommunications number portability through remote call forwarding, direct inward dialing trunks, or other comparable arrangements, with as little impairment of functioning, quality, reliability, and convenience as possible. After that date, full compliance with such regulations.

(xii) Nondiscriminatory access to such services or information as are necessary to allow the requesting carrier to implement local dialing parity in accordance with the requirements of section 251(b)(3).

(xiii) Reciprocal compensation arrangements in accordance with the requirements of section 252(d)(2).

(xiv) Telecommunications services are available for resale in accordance with the requirements of sections 251(c)(4) and 252(d)(3).

(d) Administrative provisions.

(1) Application to Commission. On and after the date of enactment of the Telecommunications Act of 1996, a Bell operating company or its affiliate may apply to the Commission for authorization to provide interLATA services originating in any in-region State. The application shall identify each State for which the authorization is sought.

(2) Consultation.

(A) Consultation with the Attorney General. The Commission shall notify the Attorney General promptly of any application under paragraph (1). Before making any determination under this subsection, the Commission shall consult with the Attorney General, and if the Attorney General submits any comments in writing, such comments shall be included in the record of the Commission's decision. In consulting with and submitting comments to the Commission under this paragraph, the Attorney General shall provide to the Commission an evaluation of the application using any standard the Attorney General considers appropriate. The Commission shall give substantial weight to the Attorney General's evaluation, but such evaluation shall not have any preclusive effect on any Commission decision under paragraph (3).

(B) Consultation with State commissions. Before making any determination under this subsection, the Commission shall consult with the State commission of any State that is the subject of the application in order to verify the compliance of the Bell operating company with the requirements of subsection (c).

(3) Determination. Not later than 90 days after receiving an application under paragraph (1), the Commission shall issue a written determination approving or denying the authorization requested in the application for each State. The Commission shall not approve the authorization requested in an application submitted under paragraph (1) unless it finds that—

(A) the petitioning Bell operating company has met the requirements of subsection (c)(1) and—

(i) with respect to access and interconnection provided pursuant to subsection (c)(1)(A), has fully implemented the competitive checklist in subsection (c)(2)(B); or

(ii) with respect to access and interconnection generally offered pursuant to a statement under subsection (c)(1)(B), such statement offers all of the items included in the competitive checklist in subsection (c)(2)(B);

(B) the requested authorization will be carried out in accordance with the requirements of section 272; and

(C) the requested authorization is consistent with the public interest, convenience, and necessity.

The Commission shall state the basis for its approval or denial of the application.

(4) Limitation on commission. The Commission may not, by rule or otherwise, limit or extend the terms used in the competitive checklist set forth in subsection (c)(2)(B).

(5) Publication. Not later than 10 days after issuing a determination under paragraph (3), the Commission shall publish in the Federal Register a brief description of the determination.

(6) Enforcement of conditions.

(A) Commission authority. If at any time after the approval of an application under paragraph (3), the Commission determines that a Bell operating company has ceased to meet any of the conditions required for such approval, the Commission may, after notice and opportunity for a hearing—

(i) issue an order to such company to correct the deficiency;

(ii) impose a penalty on such company pursuant to title V; or

(iii) suspend or revoke such approval.

(B) Receipt and review of complaints. The Commission shall establish procedures for the review of complaints concerning failures by Bell operating companies to meet conditions required for approval under paragraph (3). Unless the parties otherwise agree, the Commission shall act on such complaint within 90 days.

(e) Limitations.

(1) Joint marketing of local and long distance services. Until a Bell operating company is authorized pursuant to subsection (d) to provide interLATA services in an in-region State, or until 36 months have passed since the date of enactment of the Telecommunications Act of 1996, whichever is earlier, a telecommunications carrier that serves greater than 5 percent of the Nation's presubscribed access lines may not jointly market in such State telephone exchange service obtained from such company pursuant to section 251(c)(4) with interLATA services offered by that telecommunications carrier.

(2) IntraLATA toll dialing parity.

(A) Provision required. A Bell operating company granted authority to provide interLATA services under subsection (d) shall provide intraLATA toll dialing parity throughout that State coincident with its exercise of that authority.

(B) Limitation. Except for single-LATA States and States that have issued an order by December 19, 1995, requiring a Bell operating company to implement intraLATA toll dialing parity, a State may not require a Bell operating company to implement intraLATA toll dialing parity in that State before a Bell operating company has been granted authority under this section to provide interLATA services originating in that State or before 3 years after the date of enactment of the Telecommunications Act of 1996, whichever is earlier. Nothing in this subparagraph precludes a State from issuing an order requiring intraLATA toll dialing parity in that State prior to either such date so long as such order does not take effect until after the earlier of either such dates.

(f) Exception for previously authorized activities. Neither subsection (a) nor section 273 shall prohibit a Bell operating company or affiliate from engaging, at any time after the date of enactment of the Telecommunications Act of 1996, in any activity to the extent authorized by, and subject to the terms and conditions contained in, an order entered by the United States District Court for the District of Columbia pursuant to section VII or VIII(C) of the AT&T Consent Decree if such order was entered on or before such date of enactment, to the extent such order is not reversed or vacated on appeal. Nothing in this subsection shall be construed to limit, or to impose terms or conditions on, an activity in which a Bell operating company is otherwise authorized to engage under any other provision of this section.

(g) Definition of incidental interLATA services. For purposes of this section, the term "incidental interLATA services" means the interLATA provision by a Bell operating company or its affiliate—

(1) (A) of audio programming, video programming, or other programming services to subscribers to such services of such company or affiliate;

(B) of the capability for interaction by such subscribers to select or respond to such audio programming, video programming, or other programming services;

(C) to distributors of audio programming or video programming that such company or affiliate owns or controls, or is licensed by the copyright owner of such programming (or by an assignee of such owner) to distribute; or

(D) of alarm monitoring services;

(2) of two-way interactive video services or Internet services over dedicated facilities to or for elementary and secondary schools as defined in section 254(h)(5);

(3) of commercial mobile services in accordance with section 332(c) of this Act and with the regulations prescribed by the Commission pursuant to paragraph (8) of such section;

(4) of a service that permits a customer that is located in one LATA to retrieve stored information from, or file information for storage in, information storage facilities of such company that are located in another LATA;

(5) of signaling information used in connection with the provision of telephone exchange services or exchange access by a local exchange carrier; or

(6) of network control signaling information to, and receipt of such signaling information from, common carriers offering interLATA services at any location within the area in which such Bell operating company provides telephone exchange services or exchange access.

(h) Limitations. The provisions of subsection (g) are intended to be narrowly construed. The interLATA services provided under subparagraph (A), (B), or (C) of subsection (g)(1) are limited to those interLATA transmissions incidental to the provision by a Bell operating company or its affiliate of video, audio, and other programming services that the company or its affiliate is engaged in providing to the public. The Commission shall ensure that the provision of services authorized under subsection (g) by a Bell operating company or its affiliate will not adversely affect telephone exchange service ratepayers or competition in any telecommunications market.

(i) Additional definitions. As used in this section—

(1) In-region State. The term "in-region State" means a State in which a Bell operating company or any of its affiliates was authorized to provide wireline telephone exchange service pursuant to the reorganization plan approved under the AT&T Consent Decree, as in effect on the day before the date of enactment of the Telecommunications Act of 1996.

(2) Audio programming services. The term "audio programming services" means programming provided by, or generally considered to be comparable to programming provided by, a radio broadcast station.

(3) Video programming services; other programming services. The terms "video programming service" and "other programming services" have the same meanings as such terms have under section 602 of this Act.

(j) Certain service applications treated as in-region service applications. For purposes of this section, a Bell operating company application to provide 800 service, private line service, or their equivalents that—

(1) terminate in an in-region State of that Bell operating company, and

(2) allow the called party to determine the interLATA carrier, shall be considered an in-region service subject to the requirements of subsection (b)(1).

§ 272. Separate Affiliate; Safeguards

(a) Separate affiliate required for competitive activities.

(1) In general. A Bell operating company (including any affiliate) which is a local exchange carrier that is subject to the requirements of section 251(c) may not provide any service described in paragraph (2) unless it provides that service through one or more affiliates that—

(A) are separate from any operating company entity that is subject to the requirements of section 251(c); and

(B) meet the requirements of subsection (b).

(2) Services for which a separate affiliate is required. The services for which a separate affiliate is required by paragraph (1) are:

(A) Manufacturing activities (as defined in section 273(h)).

(B) Origination of interLATA telecommunications services, other than—

(i) incidental interLATA services described in paragraphs (1), (2), (3), (5), and (6) of section 271(g);

(ii) out-of-region services described in section 271(b)(2); or

(iii) previously authorized activities described in section 271(f).

(C) InterLATA information services, other than electronic publishing (as defined in section 274(h) [*47 USC § 274*(h)]) and alarm monitoring services (as defined in section 275(e)).

(b) Structural and transactional requirements. The separate affiliate required by this section—

(1) shall operate independently from the Bell operating company;

(2) shall maintain books, records, and accounts in the manner prescribed by the Commission which shall be separate from the books, records, and accounts maintained by the Bell operating company of which it is an affiliate;

(3) shall have separate officers, directors, and employees from the Bell operating company of which it is an affiliate;

(4) may not obtain credit under any arrangement that would permit a creditor, upon default, to have recourse to the assets of the Bell operating company; and

(5) shall conduct all transactions with the Bell operating company of which it is an affiliate on an arm's length basis with any such transactions reduced to writing and available for public inspection.

(c) Nondiscrimination safeguards. In its dealings with its affiliate described in subsection (a), a Bell operating company—

(1) may not discriminate between that company or affiliate and any other entity in the provision or procurement of goods, services, facilities, and information, or in the establishment of standards; and

(2) shall account for all transactions with an affiliate described in subsection (a) in accordance with accounting principles designated or approved by the Commission.

(d) Biennial audit.

(1) General requirement. A company required to operate a separate affiliate under this section shall obtain and pay for a joint Federal/State audit every 2 years conducted by an independent auditor to determine whether such company has complied with this section and the regulations promulgated under this section, and particularly whether such company has complied with the separate accounting requirements under subsection (b).

(2) Results submitted to Commission; State commissions. The auditor described in paragraph (1) shall submit the results of the audit to the Commission and to the State commission of each State in which the company audited provides service, which shall make such results available for public inspection. Any party may submit comments on the final audit report.

(3) Access to documents. For purposes of conducting audits and reviews under this subsection—

(A) the independent auditor, the Commission, and the State commission shall have access to the financial accounts and records of each company and of its affiliates necessary to verify transactions conducted with that company that are relevant to the specific activities permitted under this section and that are necessary for the regulation of rates;

(B) the Commission and the State commission shall have access to the working papers and supporting materials of any auditor who performs an audit under this section; and

(C) the State commission shall implement appropriate procedures to ensure the protection of any proprietary information submitted to it under this section.

(e) Fulfillment of certain requests. A Bell operating company and an affiliate that is subject to the requirements of section 251(c)—

(1) shall fulfill any requests from an unaffiliated entity for telephone exchange service and exchange access within a period no longer than the period in which it provides such telephone exchange service and exchange access to itself or to its affiliates;

(2) shall not provide any facilities, services, or information concerning its provision of exchange access to the affiliate described in subsection (a) unless such facilities, services, or information are made available to other providers of interLATA services in that market on the same terms and conditions;

(3) shall charge the affiliate described in subsection (a), or impute to itself (if using the access for its provision of its own services), an amount for access to its telephone exchange service and exchange access that is no less than the amount charged to any unaffiliated interexchange carriers for such service; and

(4) may provide any interLATA or intraLATA facilities or services to its interLATA affiliate if such services or facilities are made available to all carriers at the same rates and on the same terms and conditions, and so long as the costs are appropriately allocated.

(f) Sunset.

(1) Manufacturing and long distance. The provisions of this section (other than subsection (e)) shall cease to apply with respect to the manufacturing activities or the interLATA telecommunications services of a Bell operating company 3 years after the date such Bell operating company or any Bell operating company affiliate is authorized to provide interLATA telecommunications services under section 271(d), unless the Commission extends such 3-year period by rule or order.

(2) InterLATA information services. The provisions of this section (other than subsection (e)) shall cease to apply with respect to the interLATA information services of a Bell operating company 4 years after the date of enactment of the Telecommunications Act of 1996, unless the Commission extends such 4-year period by rule or order.

(3) Preservation of existing authority. Nothing in this subsection shall be construed to limit the authority of the Commission under any other

section of this Act to prescribe safeguards consistent with the public interest, convenience, and necessity.

(g) Joint marketing.

(1) Affiliate sales of telephone exchange services. A Bell operating company affiliate required by this section may not market or sell telephone exchange services provided by the Bell operating company unless that company permits other entities offering the same or similar service to market and sell its telephone exchange services.

(2) Bell operating company sales of affiliate services. A Bell operating company may not market or sell interLATA service provided by an affiliate required by this section within any of its in-region States until such company is authorized to provide interLATA services in such State under section 271(d).

(3) Rule of construction. The joint marketing and sale of services permitted under this subsection shall not be considered to violate the nondiscrimination provisions of subsection (c).

(h) Transition. With respect to any activity in which a Bell operating company is engaged on the date of enactment of the Telecommunications Act of 1996, such company shall have one year from such date of enactment to comply with the requirements of this section.

§ 273. Manufacturing by Bell Operating Companies

(a) Authorization. A Bell operating company may manufacture and provide telecommunications equipment, and manufacture customer premises equipment, if the Commission authorizes that Bell operating company or any Bell operating company affiliate to provide interLATA services under section 271(d), subject to the requirements of this section and the regulations prescribed thereunder, except that neither a Bell operating company nor any of its affiliates may engage in such manufacturing in conjunction with a Bell operating company not so affiliated or any of its affiliates.

(b) Collaboration; research and royalty agreements.

(1) Collaboration. Subsection (a) shall not prohibit a Bell operating company from engaging in close collaboration with any manufacturer of customer premises equipment or telecommunications equipment during the design and development of hardware, software, or combinations thereof related to such equipment.

(2) Certain research arrangements; royalty agreements. Subsection (a) shall not prohibit a Bell operating company from—
(A) engaging in research activities related to manufacturing, and
(B) entering into royalty agreements with manufacturers of telecommunications equipment.

(c) Information requirements.
(1) Information on protocols and technical requirements. Each Bell operating company shall, in accordance with regulations prescribed by the Commission, maintain and file with the Commission full and complete information with respect to the protocols and technical requirements for connection with and use of its telephone exchange service facilities. Each such company shall report promptly to the Commission any material changes or planned changes to such protocols and requirements, and the schedule for implementation of such changes or planned changes.
(2) Disclosure of information. A Bell operating company shall not disclose any information required to be filed under paragraph (1) unless that information has been filed promptly, as required by regulation by the Commission.
(3) Access by competitors to information. The Commission may prescribe such additional regulations under this subsection as may be necessary to ensure that manufacturers have access to the information with respect to the protocols and technical requirements for connection with and use of telephone exchange service facilities that a Bell operating company makes available to any manufacturing affiliate or any unaffiliated manufacturer.
(4) Planning information. Each Bell operating company shall provide, to interconnecting carriers providing telephone exchange service, timely information on the planned deployment of telecommunications equipment.

(d) Manufacturing limitations for standard-setting organizations.
(1) Application to Bell communications research or manufacturers. Bell Communications Research, Inc., or any successor entity or affiliate—
(A) shall not be considered a Bell operating company or a successor or assign of a Bell operating company at such time as it is no longer an affiliate of any Bell operating company; and
(B) notwithstanding paragraph (3), shall not engage in manufacturing telecommunications equipment or customer premises equipment as

long as it is an affiliate of more than 1 otherwise unaffiliated Bell operating company or successor or assign of any such company.

Nothing in this subsection prohibits Bell Communications Research, Inc., or any successor entity, from engaging in any activity in which it is lawfully engaged on the date of enactment of the Telecommunications Act of 1996. Nothing provided in this subsection shall render Bell Communications Research, Inc., or any successor entity, a common carrier under title II of this Act. Nothing in this subsection restricts any manufacturer from engaging in any activity in which it is lawfully engaged on the date of enactment of the Telecommunications Act of 1996.

(2) Proprietary information. Any entity which establishes standards for telecommunications equipment or customer premises equipment, or generic network requirements for such equipment, or certifies telecommunications equipment or customer premises equipment, shall be prohibited from releasing or otherwise using any proprietary information, designated as such by its owner, in its possession as a result of such activity, for any purpose other than purposes authorized in writing by the owner of such information, even after such entity ceases to be so engaged.

(3) Manufacturing safeguards.

(A) Except as prohibited in paragraph (1), and subject to paragraph (6), any entity which certifies telecommunications equipment or customer premises equipment manufactured by an unaffiliated entity shall only manufacture a particular class of telecommunications equipment or customer premises equipment for which it is undertaking or has undertaken, during the previous 18 months, certification activity for such class of equipment through a separate affiliate.

(B) Such separate affiliate shall—

(i) maintain books, records, and accounts separate from those of the entity that certifies such equipment, consistent with generally acceptable accounting principles;

(ii) not engage in any joint manufacturing activities with such entity; and

(iii) have segregated facilities and separate employees with such entity.

(C) Such entity that certifies such equipment shall—

(i) not discriminate in favor of its manufacturing affiliate in the establishment of standards, generic requirements, or product certification;

(ii) not disclose to the manufacturing affiliate any proprietary information that has been received at any time from an unaffiliated manufacturer, unless authorized in writing by the owner of the information; and

(iii) not permit any employee engaged in product certification for telecommunications equipment or customer premises equipment to engage jointly in sales or marketing of any such equipment with the affiliated manufacturer.

(4) Standard-setting entities. Any entity that is not an accredited standards development organization and that establishes industry-wide standards for telecommunications equipment or customer premises equipment, or industry-wide generic network requirements for such equipment, or that certifies telecommunications equipment or customer premises equipment manufactured by an unaffiliated entity, shall—

(A) establish and publish any industry-wide standard for, industry-wide generic requirement for, or any substantial modification of an existing industry-wide standard or industry-wide generic requirement for, telecommunications equipment or customer premises equipment only in compliance with the following procedure—

(i) such entity shall issue a public notice of its consideration of a proposed industry-wide standard or industry-wide generic requirement;

(ii) such entity shall issue a public invitation to interested industry parties to fund and participate in such efforts on a reasonable and nondiscriminatory basis, administered in such a manner as not to unreasonably exclude any interested industry party;

(iii) such entity shall publish a text for comment by such parties as have agreed to participate in the process pursuant to clause (ii), provide such parties a full opportunity to submit comments, and respond to comments from such parties;

(iv) such entity shall publish a final text of the industry-wide standard or industry-wide generic requirement, including the comments in their entirety, of any funding party which requests to have its comments so published; and

(v) such entity shall attempt, prior to publishing a text for comment, to agree with the funding parties as a group on a mutually satisfactory dispute resolution process which such parties shall utilize as their sole recourse in the event of a dispute on technical issues as to which there is disagreement between any funding party and the entity conducting such activities, except that if no dispute resolution process is agreed to by all the parties, a funding party may utilize the dispute resolution procedures established pursuant to paragraph (5) of this subsection;

(B) engage in product certification for telecommunications equipment or customer premises equipment manufactured by unaffiliated entities only if—

(i) such activity is performed pursuant to published criteria;

(ii) such activity is performed pursuant to auditable criteria; and

(iii) such activity is performed pursuant to available industry-accepted testing methods and standards, where applicable, unless otherwise agreed upon by the parties funding and performing such activity;

(C) not undertake any actions to monopolize or attempt to monopolize the market for such services; and

(D) not preferentially treat its own telecommunications equipment or customer premises equipment, or that of its affiliate, over that of any other entity in establishing and publishing industry-wide standards or industry-wide generic requirements for, and in certification of, telecommunications equipment and customer premises equipment.

(5) Alternate dispute resolution. Within 90 days after the date of enactment of the Telecommunications Act of 1996, the Commission shall prescribe a dispute resolution process to be utilized in the event that a dispute resolution process is not agreed upon by all the parties when establishing and publishing any industry-wide standard or industry-wide generic requirement for telecommunications equipment or customer premises equipment, pursuant to paragraph (4)(A)(v). The Commission shall not establish itself as a party to the dispute resolution process. Such dispute resolution process shall permit any funding party to resolve a dispute with the entity conducting the activity that significantly affects such funding party's interests, in an open, nondiscriminatory, and unbiased fashion, within 30 days after the filing of such dispute. Such disputes may be filed within 15 days after the date the funding party receives a response to its comments from the entity conducting the activity. The Commission shall establish penalties to be assessed for delays caused by referral of frivolous disputes to the dispute resolution process.

(6) Sunset. The requirements of paragraphs (3) and (4) shall terminate for the particular relevant activity when the Commission determines that there are alternative sources of industry-wide standards, industry-wide generic requirements, or product certification for a particular class of telecommunications equipment or customer premises equipment available in the United States. Alternative sources shall be deemed to exist when such sources provide commercially viable alternatives that are providing such services to customers. The Commission shall act on any application for such a determination within 90 days after receipt of such application, and shall receive public comment on such application.

(7) Administration and enforcement authority. For the purposes of administering this subsection and the regulations prescribed thereunder, the Commission shall have the same remedial authority as the Commission has

in administering and enforcing the provisions of this title with respect to any common carrier subject to this Act.

(8) Definitions. For purposes of this subsection:

(A) The term "affiliate" shall have the same meaning as in section 3 of this Act, except that, for purposes of paragraph (1)(B)—

(i) an aggregate voting equity interest in Bell Communications Research, Inc., of at least 5 percent of its total voting equity, owned directly or indirectly by more than 1 otherwise unaffiliated Bell operating company, shall constitute an affiliate relationship; and

(ii) a voting equity interest in Bell Communications Research, Inc., by any otherwise unaffiliated Bell operating company of less than 1 percent of Bell Communications Research's total voting equity shall not be considered to be an equity interest under this paragraph.

(B) The term "generic requirement" means a description of acceptable product attributes for use by local exchange carriers in establishing product specifications for the purchase of telecommunications equipment, customer premises equipment, and software integral thereto.

(C) The term "industry-wide" means activities funded by or performed on behalf of local exchange carriers for use in providing wireline telephone exchange service whose combined total of deployed access lines in the United States constitutes at least 30 percent of all access lines deployed by telecommunications carriers in the United States as of the date of enactment of the Telecommunications Act of 1996.

(D) The term "certification" means any technical process whereby a party determines whether a product, for use by more than one local exchange carrier, conforms with the specified requirements pertaining to such product.

(E) The term "accredited standards development organization" means an entity composed of industry members which has been accredited by an institution vested with the responsibility for standards accreditation by the industry.

(e) Bell operating company equipment procurement and sales.

(1) Nondiscrimination standards for manufacturing. In the procurement or awarding of supply contracts for telecommunications equipment, a Bell operating company, or any entity acting on its behalf, for the duration of the requirement for a separate subsidiary including manufacturing under this Act—

(A) shall consider such equipment, produced or supplied by unrelated persons; and

(B) may not discriminate in favor of equipment produced or supplied by an affiliate or related person.

(2) Procurement standards. Each Bell operating company or any entity acting on its behalf shall make procurement decisions and award all supply contracts for equipment, services, and software on the basis of an objective assessment of price, quality, delivery, and other commercial factors.

(3) Network planning and design. A Bell operating company shall, to the extent consistent with the antitrust laws, engage in joint network planning and design with local exchange carriers operating in the same area of interest. No participant in such planning shall be allowed to delay the introduction of new technology or the deployment of facilities to provide telecommunications services, and agreement with such other carriers shall not be required as a prerequisite for such introduction or deployment.

(4) Sales restrictions. Neither a Bell operating company engaged in manufacturing nor a manufacturing affiliate of such a company shall restrict sales to any local exchange carrier of telecommunications equipment, including software integral to the operation of such equipment and related upgrades.

(5) Protection of proprietary information. A Bell operating company and any entity it owns or otherwise controls shall protect the proprietary information submitted for procurement decisions from release not specifically authorized by the owner of such information.

(f) Administration and enforcement authority. For the purposes of administering and enforcing the provisions of this section and the regulations prescribed thereunder, the Commission shall have the same authority, power, and functions with respect to any Bell operating company or any affiliate thereof as the Commission has in administering and enforcing the provisions of this title with respect to any common carrier subject to this Act.

(g) Additional rules and regulations. The Commission may prescribe such additional rules and regulations as the Commission determines are necessary to carry out the provisions of this section, and otherwise to prevent discrimination and cross-subsidization in a Bell operating company's dealings with its affiliate and with third parties.

(h) Definition. As used in this section, the term "manufacturing" has the same meaning as such term has under the AT&T Consent Decree.

§ 274. Electronic Publishing by Bell Operating Companies

(a) Limitations. No Bell operating company or any affiliate may engage in the provision of electronic publishing that is disseminated by means of such Bell operating company's or any of its affiliates' basic telephone service, except that nothing in this section shall prohibit a separated affiliate or electronic publishing joint venture operated in accordance with this section from engaging in the provision of electronic publishing.

(b) Separated affiliate or electronic publishing joint venture requirements. A separated affiliate or electronic publishing joint venture shall be operated independently from the Bell operating company. Such separated affiliate or joint venture and the Bell operating company with which it is affiliated shall—
    (1) maintain separate books, records, and accounts and prepare separate financial statements;
    (2) not incur debt in a manner that would permit a creditor of the separated affiliate or joint venture upon default to have recourse to the assets of the Bell operating company;
    (3) carry out transactions (A) in a manner consistent with such independence, (B) pursuant to written contracts or tariffs that are filed with the Commission and made publicly available, and (C) in a manner that is auditable in accordance with generally accepted auditing standards;
    (4) value any assets that are transferred directly or indirectly from the Bell operating company to a separated affiliate or joint venture, and record any transactions by which such assets are transferred, in accordance with such regulations as may be prescribed by the Commission or a State commission to prevent improper cross subsidies;
    (5) between a separated affiliate and a Bell operating company—
        (A) have no officers, directors, and employees in common after the effective date of this section; and
        (B) own no property in common;
    (6) not use for the marketing of any product or service of the separated affiliate or joint venture, the name, trademarks, or service marks of an existing Bell operating company except for names, trademarks, or service marks that are owned by the entity that owns or controls the Bell operating company;
    (7) not permit the Bell operating company—
        (A) to perform hiring or training of personnel on behalf of a separated affiliate;

(B) to perform the purchasing, installation, or maintenance of equipment on behalf of a separated affiliate, except for telephone service that it provides under tariff or contract subject to the provisions of this section; or

(C) to perform research and development on behalf of a separated affiliate;

(8) each have performed annually a compliance review—

(A) that is conducted by an independent entity for the purpose of determining compliance during the preceding calendar year with any provision of this section; and

(B) the results of which are maintained by the separated affiliate or joint venture and the Bell operating company for a period of 5 years subject to review by any lawful authority; and

(9) within 90 days of receiving a review described in paragraph (8), file a report of any exceptions and corrective action with the Commission and allow any person to inspect and copy such report subject to reasonable safeguards to protect any proprietary information contained in such report from being used for purposes other than to enforce or pursue remedies under this section.

(c) Joint marketing.

(1) In general. Except as provided in paragraph (2)—

(A) a Bell operating company shall not carry out any promotion, marketing, sales, or advertising for or in conjunction with a separated affiliate; and

(B) a Bell operating company shall not carry out any promotion, marketing, sales, or advertising for or in conjunction with an affiliate that is related to the provision of electronic publishing.

(2) Permissible joint activities.

(A) Joint telemarketing. A Bell operating company may provide inbound telemarketing or referral services related to the provision of electronic publishing for a separated affiliate, electronic publishing joint venture, affiliate, or unaffiliated electronic publisher: Provided, That if such services are provided to a separated affiliate, electronic publishing joint venture, or affiliate, such services shall be made available to all electronic publishers on request, on nondiscriminatory terms.

(B) Teaming arrangements. A Bell operating company may engage in nondiscriminatory teaming or business arrangements to engage in electronic publishing with any separated affiliate or with any other electronic publisher if (i) the Bell operating company only provides facilities, services, and basic telephone service information as authorized by this section, and (ii) the Bell operating company does not own such teaming or business arrangement.

(C) Electronic publishing joint ventures. A Bell operating company or affiliate may participate on a nonexclusive basis in electronic publishing joint ventures with entities that are not a Bell operating company, affiliate, or separated affiliate to provide electronic publishing services, if the Bell operating company or affiliate has not more than a 50 percent direct or indirect equity interest (or the equivalent thereof) or the right to more than 50 percent of the gross revenues under a revenue sharing or royalty agreement in any electronic publishing joint venture. Officers and employees of a Bell operating company or affiliate participating in an electronic publishing joint venture may not have more than 50 percent of the voting control over the electronic publishing joint venture. In the case of joint ventures with small, local electronic publishers, the Commission for good cause shown may authorize the Bell operating company or affiliate to have a larger equity interest, revenue share, or voting control but not to exceed 80 percent. A Bell operating company participating in an electronic publishing joint venture may provide promotion, marketing, sales, or advertising personnel and services to such joint venture.

(d) Bell operating company requirement. A Bell operating company under common ownership or control with a separated affiliate or electronic publishing joint venture shall provide network access and interconnections for basic telephone service to electronic publishers at just and reasonable rates that are tariffed (so long as rates for such services are subject to regulation) and that are not higher on a per-unit basis than those charged for such services to any other electronic publisher or any separated affiliate engaged in electronic publishing.

(e) Private right of action.

(1) Damages. Any person claiming that any act or practice of any Bell operating company, affiliate, or separated affiliate constitutes a violation of this section may file a complaint with the Commission or bring suit as provided in section 207 of this Act, and such Bell operating company, affiliate, or separated affiliate shall be liable as provided in section 206 of this Act; except that damages may not be awarded for a violation that is discovered by a compliance review as required by subsection (b)(7) of this section and corrected within 90 days.

(2) Cease and desist orders. In addition to the provisions of paragraph (1), any person claiming that any act or practice of any Bell operating

company, affiliate, or separated affiliate constitutes a violation of this section may make application to the Commission for an order to cease and desist such violation or may make application in any district court of the United States of competent jurisdiction for an order enjoining such acts or practices or for an order compelling compliance with such requirement.

(f) Separated affiliate reporting requirement. Any separated affiliate under this section shall file with the Commission annual reports in a form substantially equivalent to the Form 10-K required by regulations of the Securities and Exchange Commission.

(g) Effective dates.

(1) Transition. Any electronic publishing service being offered to the public by a Bell operating company or affiliate on the date of enactment of the Telecommunications Act of 1996 shall have one year from such date of enactment to comply with the requirements of this section.

(2) Sunset. The provisions of this section shall not apply to conduct occurring after 4 years after the date of enactment of the Telecommunications Act of 1996.

(h) Definition of electronic publishing.

(1) In general. The term "electronic publishing" means the dissemination, provision, publication, or sale to an unaffiliated entity or person, of any one or more of the following: news (including sports); entertainment (other than interactive games); business, financial, legal, consumer, or credit materials; editorials, columns, or features; advertising; photos or images; archival or research material; legal notices or public records; scientific, educational, instructional, technical, professional, trade, or other literary materials; or other like or similar information.

(2) Exceptions. The term "electronic publishing" shall not include the following services:

(A) Information access, as that term is defined by the AT&T Consent Decree.

(B) The transmission of information as a common carrier.

(C) The transmission of information as part of a gateway to an information service that does not involve the generation or alteration of the content of information, including data transmission, address translation, protocol conversion, billing management, introductory information content, and navigational systems that enable users to access electronic publishing services, which do not affect the presentation of such electronic publishing services to users.

(D) Voice storage and retrieval services, including voice messaging and electronic mail services.

(E) Data processing or transaction processing services that do not involve the generation or alteration of the content of information.

(F) Electronic billing or advertising of a Bell operating company's regulated telecommunications services.

(G) Language translation or data format conversion.

(H) The provision of information necessary for the management, control, or operation of a telephone company telecommunications system.

(I) The provision of directory assistance that provides names, addresses, and telephone numbers and does not include advertising.

(J) Caller identification services.

(K) Repair and provisioning databases and credit card and billing validation for telephone company operations.

(L) 911-E and other emergency assistance databases.

(M) Any other network service of a type that is like or similar to these network services and that does not involve the generation or alteration of the content of information.

(N) Any upgrades to these network services that do not involve the generation or alteration of the content of information.

(O) Video programming or full motion video entertainment on demand.

(i) Additional definitions. As used in this section—

(1) The term "affiliate" means any entity that, directly or indirectly, owns or controls, is owned or controlled by, or is under common ownership or control with, a Bell operating company. Such term shall not include a separated affiliate.

(2) The term "basic telephone service" means any wireline telephone exchange service, or wireline telephone exchange service facility, provided by a Bell operating company in a telephone exchange area, except that such term does not include—

(A) a competitive wireline telephone exchange service provided in a telephone exchange area where another entity provides a wireline telephone exchange service that was provided on January 1, 1984, or

(B) a commercial mobile service.

(3) The term "basic telephone service information" means network and customer information of a Bell operating company and other

information acquired by a Bell operating company as a result of its engaging in the provision of basic telephone service.

(4) The term "control" has the meaning that it has in 17 C.F.R. 240.12b-2, the regulations promulgated by the Securities and Exchange Commission pursuant to the Securities Exchange Act of 1934 or any successor provision to such section.

(5) The term "electronic publishing joint venture" means a joint venture owned by a Bell operating company or affiliate that engages in the provision of electronic publishing which is disseminated by means of such Bell operating company's or any of its affiliates' basic telephone service.

(6) The term "entity" means any organization, and includes corporations, partnerships, sole proprietorships, associations, and joint ventures.

(7) The term "inbound telemarketing" means the marketing of property, goods, or services by telephone to a customer or potential customer who initiated the call.

(8) The term "own" with respect to an entity means to have a direct or indirect equity interest (or the equivalent thereof) of more than 10 percent of an entity, or the right to more than 10 percent of the gross revenues of an entity under a revenue sharing or royalty agreement.

(9) The term "separated affiliate" means a corporation under common ownership or control with a Bell operating company that does not own or control a Bell operating company and is not owned or controlled by a Bell operating company and that engages in the provision of electronic publishing which is disseminated by means of such Bell operating company's or any of its affiliates' basic telephone service.

(10) The term "Bell operating company" has the meaning provided in section 3, except that such term includes any entity or corporation that is owned or controlled by such a company (as so defined) but does not include an electronic publishing joint venture owned by such an entity or corporation.

§ 275. Alarm Monitoring Services

(a) Delayed entry into alarm monitoring.

(1) Prohibition. No Bell operating company or affiliate thereof shall engage in the provision of alarm monitoring services before the date which is 5 years after the date of enactment of the Telecommunications Act of 1996.

(2) Existing activities. Paragraph (1) does not prohibit or limit the provision, directly or through an affiliate, of alarm monitoring services by a Bell operating company that was engaged in providing alarm monitoring

services as of November 30, 1995, directly or through an affiliate. Such Bell operating company or affiliate may not acquire any equity interest in, or obtain financial control of, any unaffiliated alarm monitoring service entity after November 30, 1995, and until 5 years after the date of enactment of the Telecommunications Act of 1996, except that this sentence shall not prohibit an exchange of customers for the customers of an unaffiliated alarm monitoring service entity.

(b) Nondiscrimination. An incumbent local exchange carrier (as defined in section 251(h)) engaged in the provision of alarm monitoring services shall—

    (1) provide nonaffiliated entities, upon reasonable request, with the network services it provides to its own alarm monitoring operations, on nondiscriminatory terms and conditions; and

    (2) not subsidize its alarm monitoring services either directly or indirectly from telephone exchange service operations.

(c) Expedited consideration of complaints. The Commission shall establish procedures for the receipt and review of complaints concerning violations of subsection (b) or the regulations thereunder that result in material financial harm to a provider of alarm monitoring service. Such procedures shall ensure that the Commission will make a final determination with respect to any such complaint within 120 days after receipt of the complaint. If the complaint contains an appropriate showing that the alleged violation occurred, as determined by the Commission in accordance with such regulations, the Commission shall, within 60 days after receipt of the complaint, order the incumbent local exchange carrier (as defined in section 251(h)) and its affiliates to cease engaging in such violation pending such final determination.

(d) Use of data. A local exchange carrier may not record or use in any fashion the occurrence or contents of calls received by providers of alarm monitoring services for the purposes of marketing such services on behalf of such local exchange carrier, or any other entity. Any regulations necessary to enforce this subsection shall be issued initially within 6 months after the date of enactment of the Telecommunications Act of 1996.

(e) Definition of alarm monitoring service. The term "alarm monitoring service" means a service that uses a device located at a residence, place of business, or other fixed premises—

(1) to receive signals from other devices located at or about such premises regarding a possible threat at such premises to life, safety, or property, from burglary, fire, vandalism, bodily injury, or other emergency, and

(2) to transmit a signal regarding such threat by means of transmission facilities of a local exchange carrier or one of its affiliates to a remote monitoring center to alert a person at such center of the need to inform the customer or another person or police, fire, rescue, security, or public safety personnel of such threat, but does not include a service that uses a medical monitoring device attached to an individual for the automatic surveillance of an ongoing medical condition.

§ 276. Provision of Payphone Service

(a) Nondiscrimination safeguards. After the effective date of the rules prescribed pursuant to subsection (b), any Bell operating company that provides payphone service—

(1) shall not subsidize its payphone service directly or indirectly from its telephone exchange service operations or its exchange access operations; and

(2) shall not prefer or discriminate in favor of its payphone service.

(b) Regulations.

(1) Contents of regulations. In order to promote competition among payphone service providers and promote the widespread deployment of payphone services to the benefit of the general public, within 9 months after the date of enactment of the Telecommunications Act of 1996, the Commission shall take all actions necessary (including any reconsideration) to prescribe regulations that—

(A) establish a per call compensation plan to ensure that all payphone service providers are fairly compensated for each and every completed intrastate and interstate call using their payphone, except that emergency calls and telecommunications relay service calls for hearing disabled individuals shall not be subject to such compensation;

(B) discontinue the intrastate and interstate carrier access charge payphone service elements and payments in effect on such date of enactment, and all intrastate and interstate payphone subsidies from basic exchange and exchange access revenues, in favor of a compensation plan as specified in subparagraph (A);

(C) prescribe a set of nonstructural safeguards for Bell operating company payphone service to implement the provisions of paragraphs (1) and (2) of subsection (a), which safeguards shall, at a minimum, include the nonstructural safeguards equal to those adopted in the Computer Inquiry-III (CC Docket No. 90-623) proceeding;

 (D) provide for Bell operating company payphone service providers to have the same right that independent payphone providers have to negotiate with the location provider on the location provider's selecting and contracting with, and, subject to the terms of any agreement with the location provider, to select and contract with, the carriers that carry interLATA calls from their payphones, unless the Commission determines in the rulemaking pursuant to this section that it is not in the public interest; and

 (E) provide for all payphone service providers to have the right to negotiate with the location provider on the location provider's selecting and contracting with, and, subject to the terms of any agreement with the location provider, to select and contract with, the carriers that carry intraLATA calls from their payphones.

 (2) Public interest telephones. In the rulemaking conducted pursuant to paragraph (1), the Commission shall determine whether public interest payphones, which are provided in the interest of public health, safety, and welfare, in locations where there would otherwise not be a payphone, should be maintained, and if so, ensure that such public interest payphones are supported fairly and equitably.

 (3) Existing contracts. Nothing in this section shall affect any existing contracts between location providers and payphone service providers or interLATA or intraLATA carriers that are in force and effect as of the date of enactment of the Telecommunications Act of 1996.

(c) State preemption. To the extent that any State requirements are inconsistent with the Commission's regulations, the Commission's regulations on such matters shall preempt such State requirements.

(d) Definition. As used in this section, the term "payphone service" means the provision of public or semi-public pay telephones, the provision of inmate telephone service in correctional institutions, and any ancillary services.

§ 309. Application for License

(a) Considerations in granting application. Subject to the provisions of this section, the Commission shall determine, in the case of each application filed with it to which section 308 applies, whether the public interest, convenience, and necessity will be served by the granting of such application, and, if the Commission, upon examination of such application and upon consideration of such other matters as the Commission may officially notice, shall find that public interest, convenience, and necessity would be served by the granting thereof, it shall grant such application.

(b) Time of granting application. Except as provided in subsection (c) of this section, no such application—
 (1) for an instrument of authorization in the case of a station in the broadcasting or common carrier services, or
 (2) for an instrument of authorization in the case of a station in any of the following categories:
  (A) industrial radio positioning stations for which frequencies are assigned on an exclusive basis,
  (B) aeronautical en route stations,
  (C) aeronautical advisory stations,
  (D) airdrome control stations,
  (E) aeronautical fixed stations, and
  (F) such other stations or classes of stations, not in the broadcasting or common carrier services, as the Commission shall by rule prescribe, shall be granted by the Commission earlier than thirty days following issuance of public notice by the Commission of the acceptance for filing of such application or of any substantial amendment thereof.

(c) Applications not affected by subsection (b). Subsection (b) of this section shall not apply—
 (1) to any minor amendment of an application to which such subsection is applicable, or
 (2) to any application for—
  (A) a minor change in the facilities of an authorized station,
  (B) consent to an involuntary assignment or transfer under section 310(b) or to an assignment or transfer thereunder which does not involve a substantial change in ownership or control,
  (C) a license under section 319(c) or, pending application for or grant of such license, any special or temporary authorization to permit interim operation to facilitate completion of authorized construction or to provide substantially the same service as would be authorized by such license,
  (D) extension of time to complete construction of authorized facilities,
  (E) an authorization of facilities for remote pickups, studio links and similar facilities for use in the operation of a broadcast station,

(F) authorizations pursuant to section 325(c) where the programs to be transmitted are special events not of a continuing nature,

(G) a special temporary authorization for nonbroadcast operation not to exceed thirty days where no application for regular operation is contemplated to be filed or not to exceed sixty days pending the filing of an application for such regular operation, or

(H) an authorization under any of the proviso clauses of section 308(a).

(d) Petition to deny application; time; contents; reply; findings.

(1) Any party in interest may file with the Commission a petition to deny any application (whether as originally filed or as amended) to which subsection (b) of this section applies at any time prior to the day of Commission grant thereof without hearing or the day of formal designation thereof for hearing; except that with respect to any classification of applications, the Commission from time to time by rule may specify a shorter period (no less than thirty days following the issuance of public notice by the Commission of the acceptance for filing of such application or of any substantial amendment thereof), which shorter period shall be reasonably related to the time when the applications would normally be reached for processing. The petitioner shall serve a copy of such petition on the applicant. The petition shall contain specific allegations of fact sufficient to show that the petitioner is a party in interest and that a grant of the application would be prima facie inconsistent with subsection (a) (or subsection (k) in the case of renewal of any broadcast station license). Such allegations of fact shall, except for those of which official notice may be taken, be supported by affidavit of a person or persons with personal knowledge thereof. The applicant shall be given the opportunity to file a reply in which allegations of fact or denials thereof shall similarly be supported by affidavit.

(2) If the Commission finds on the basis of the application, the pleadings filed, or other matters which it may officially notice that there are no substantial and material questions of fact and that a grant of the application would be consistent with subsection (a) (or subsection (k) in the case of renewal of any broadcast station license), it shall make the grant, deny the petition, and issue a concise statement of the reasons for denying the petition, which statement shall dispose of all substantial issues raised by the petition. If a substantial and material question of fact is presented or if the Commission for any reason is unable to find that grant of the application would be consistent with subsection (a) (or subsection (k) in the case of

renewal of any broadcast station license), it shall proceed as provided in subsection (e).

(e) Hearings; intervention; evidence; burden of proof. If, in the case of any application to which subsection (a) of this section applies, a substantial and material question of fact is presented or the Commission for any reason is unable to make the finding specified in such subsection, it shall formally designate the application for hearing on the ground or reasons then obtaining and shall forthwith notify the applicant and all other known parties in interest of such action and the grounds and reasons therefor, specifying with particularity the matters and things in issue but not including issues or requirements phrased generally. When the Commission has so designated an application for hearing the parties in interest, if any, who are not notified by the Commission of such action may acquire the status of a party to the proceeding thereon by filing a petition for intervention showing the basis for their interest not more than thirty days after publication of the hearing issues or any substantial amendment thereto in the Federal Register. Any hearing subsequently held upon such application shall be a full hearing in which the applicant and all other parties in interest shall be permitted to participate. The burden of proceeding with the introduction of evidence and the burden of proof shall be upon the applicant, except that with respect to any issue presented by a petition to deny or a petition to enlarge the issues, such burdens shall be as determined by the Commission.

(f) Temporary authorization of operations under subsection (b). When an application subject to subsection (b) has been filed, the Commission, notwithstanding the requirements of such subsection, may, if the grant of such application is otherwise authorized by law and if it finds that there are extraordinary circumstances requiring temporary operations in the public interest and that delay in the institution of such temporary operations would seriously prejudice the public interest, grant a temporary authorization, accompanied by a statement of its reasons therefor, to permit such temporary operations for a period not exceeding 180 days, and upon making like findings may extend such temporary authorization for additional periods not to exceed 180 days. When any such grant of a temporary authorization is made, the Commission shall give expeditious treatment to any timely filed petition to deny such application and to any petition for rehearing of such grant filed under section 405.

(g) Classification of applications. The Commission is authorized to adopt reasonable classifications of applications and amendments in order to effectuate the purposes of this section.

(h) Form and conditions of station licenses. Such station licenses as the Commission may grant shall be in such general form as it may prescribe, but each license shall contain, in addition to other provisions, a statement of the following conditions to which such license shall be subject: (1) The station license shall not vest in the licensee any right to operate the station nor any right in the use of the frequencies designated in the license beyond the term thereof nor in any other manner than authorized therein; (2) neither the license nor the right granted thereunder shall be assigned or otherwise transferred in violation of this Act; (3) every license issued under this Act shall be subject in terms to the right of use or control conferred by section 706 of this Act.

(i) Certain initial licenses and permits; random selection procedure; significant preferences; rules.

(1) General authority. Except as provided in paragraph (5), if there is more than one application for any initial license or construction permit, then the Commission shall have the authority to grant such license or permit to a qualified applicant through the use of a system of random selection.

(2) No license or construction permit shall be granted to an applicant selected pursuant to paragraph (1) unless the Commission determines the qualifications of such applicant pursuant to subsection (a) and section 308(b). When substantial and material questions of fact exist concerning such qualifications, the Commission shall conduct a hearing in order to make such determinations. For the purpose of making such determinations, the Commission may, by rule, and notwithstanding any other provision of law—

(A) adopt procedures for the submission of all or part of the evidence in written form;

(B) delegate the function of presiding at the taking of the evidence to Commission employees other than administrative law judges; and

(C) omit the determination required by subsection (a) with respect to any application other than the one selected pursuant to paragraph (1).

(3) (A) The Commission shall establish rules and procedures to ensure that, in the administration of any system of random selection under this subsection used for granting licenses or construction permits for any media of mass communications, significant preferences will be granted to applicants or groups of applicants, the grant to which of the license or permit

would increase the diversification of ownership of the media of mass communications. To further diversify the ownership of the media of mass communications, an additional significant preference shall be granted to any applicant controlled by a member or members of a minority group.

(B) The Commission shall have authority to require each qualified applicant seeking a significant preference under subparagraph (A) to submit to the Commission such information as may be necessary to enable the Commission to make a determination regarding whether such applicant shall be granted such preference. Such information shall be submitted in such form, at such times, and in accordance with such procedures, as the Commission may require.

(C) For purposes of this paragraph:

(i) The term "media of mass communications" includes television, radio, cable television, multipoint distribution service, direct broadcast satellite service, and other services, the licensed facilities of which may be substantially devoted toward providing programming or other information services within the editorial control of the licensee.

(ii) The term "minority group" includes Blacks, Hispanics, American Indians, Alaska Natives, Asians, and Pacific Islanders.

(4) (A) The Commission shall, after notice and opportunity for hearing, prescribe rules establishing a system of random selection for use by the Commission under this subsection in any instance in which the Commission, in its discretion, determines that such use is appropriate for the granting of any license or permit in accordance with paragraph (1).

(B) The Commission shall have authority to amend such rules from time to time to the extent necessary to carry out the provisions of this subsection. Any such amendment shall be made after notice and opportunity for hearing.

(C) Not later than 180 days after the date of enactment of this subparagraph [enacted Aug. 10, 1993], the Commission shall prescribe such transfer disclosures and antitrafficking restrictions and payment schedules as are necessary to prevent the unjust enrichment of recipients of licenses or permits as a result of the methods employed to issue licenses under this subsection.

(5) Termination of authority.

(A) Except as provided in subparagraph (B), the Commission shall not issue any license or permit using a system of random selection under this subsection after July 1, 1997.

(B) Subparagraph (A) of this paragraph shall not apply with respect to licenses or permits for stations described in section 397(6) of this Act.

(j) Use of competitive bidding.

(1) General authority. If, consistent with the obligations described in paragraph (6)(E), mutually exclusive applications are accepted for any initial license or construction permit, then, except as provided in paragraph (2), the Commission shall grant the license or permit to a qualified applicant through a system of competitive bidding that meets the requirements of this subsection.

(2) Exemptions. The competitive bidding authority granted by this subsection shall not apply to licenses or construction permits issued by the Commission—

(A) for public safety radio services, including private internal radio services used by State and local governments and non-government entities and including emergency road services provided by not-for-profit organizations, that—

(i) are used to protect the safety of life, health, or property; and

(ii) are not made commercially available to the public;

(B) for initial licenses or construction permits for digital television service given to existing terrestrial broadcast licensees to replace their analog television service licenses; or

(C) for stations described in section 397(6) of this Act.

(3) Design of systems of competitive bidding. For each class of licenses or permits that the Commission grants through the use of a competitive bidding system, the Commission shall, by regulation, establish a competitive bidding methodology. The Commission shall seek to design and test multiple alternative methodologies under appropriate circumstances. The Commission shall, directly or by contract, provide for the design and conduct (for purposes of testing) of competitive bidding using a contingent combinatorial bidding system that permits prospective bidders to bid on combinations or groups of licenses in a single bid and to enter multiple alternative bids within a single bidding round. In identifying classes of licenses and permits to be issued by competitive bidding, in specifying eligibility and other characteristics of such licenses and permits, and in designing the methodologies for use under this subsection, the Commission shall include safeguards to protect the public interest in the use of the spectrum and shall seek to promote the purposes specified in section 1 of this Act and the following objectives:

(A) the development and rapid deployment of new technologies, products, and services for the benefit of the public, including those residing in rural areas, without administrative or judicial delays;

(B) promoting economic opportunity and competition and ensuring that new and innovative technologies are readily accessible to the American people by avoiding excessive concentration of licenses and by disseminating licenses among a wide variety of applicants, including small businesses, rural telephone companies, and businesses owned by members of minority groups and women;

(C) recovery for the public of a portion of the value of the public spectrum resource made available for commercial use and avoidance of unjust enrichment through the methods employed to award uses of that resource;

(D) efficient and intensive use of the electromagnetic spectrum; and

(E) ensure that, in the scheduling of any competitive bidding under this subsection, an adequate period is allowed—

(i) before issuance of bidding rules, to permit notice and comment on proposed auction procedures; and

(ii) after issuance of bidding rules, to ensure that interested parties have a sufficient time to develop business plans, assess market conditions, and evaluate the availability of equipment for the relevant services.

(4) Contents of regulations. In prescribing regulations pursuant to paragraph (3), the Commission shall—

(A) consider alternative payment schedules and methods of calculation, including lump sums or guaranteed installment payments, with or without royalty payments, or other schedules or methods that promote the objectives described in paragraph (3)(B), and combinations of such schedules and methods;

(B) include performance requirements, such as appropriate deadlines and penalties for performance failures, to ensure prompt delivery of service to rural areas, to prevent stockpiling or warehousing of spectrum by licensees or permittees, and to promote investment in and rapid deployment of new technologies and services;

(C) consistent with the public interest, convenience, and necessity, the purposes of this Act, and the characteristics of the proposed service, prescribe area designations and bandwidth assignments that promote (i) an equitable distribution of licenses and services among geographic areas, (ii) economic opportunity for a wide variety of applicants, including small businesses, rural telephone companies, and businesses owned by members of minority groups and women, and (iii) investment in and rapid deployment of new technologies and services;

(D) ensure that small businesses, rural telephone companies, and businesses owned by members of minority groups and women are given the opportunity to participate in the provision of spectrum-based services, and, for such purposes, consider the use of tax certificates, bidding preferences, and other procedures;

(E) require such transfer disclosures and antitrafficking restrictions and payment schedules as may be necessary to prevent unjust enrichment as a result of the methods employed to issue licenses and permits; and

(F) prescribe methods by which a reasonable reserve price will be required, or a minimum bid will be established, to obtain any license or permit being assigned pursuant to the competitive bidding, unless the Commission determines that such a reserve price or minimum bid is not in the public interest.

(5) Bidder and licensee qualification. No person shall be permitted to participate in a system of competitive bidding pursuant to this subsection unless such bidder submits such information and assurances as the Commission may require to demonstrate that such bidder's application is acceptable for filing. No license shall be granted to an applicant selected pursuant to this subsection unless the Commission determines that the applicant is qualified pursuant to subsection (a) and sections 308(b) and 310. Consistent with the objectives described in paragraph (3), the Commission shall, by regulation, prescribe expedited procedures consistent with the procedures authorized by subsection (i)(2) for the resolution of any substantial and material issues of fact concerning qualifications.

(6) Rules of construction. Nothing in this subsection, or in the use of competitive bidding, shall—

(A) alter spectrum allocation criteria and procedures established by the other provisions of this Act;

(B) limit or otherwise affect the requirements of subsection (h) of this section, section 301, 304, 307, 310, or 706, or any other provision of this Act (other than subsections (d)(2) and (e) of this section);

(C) diminish the authority of the Commission under the other provisions of this Act to regulate or reclaim spectrum licenses;

(D) be construed to convey any rights, including any expectation of renewal of a license, that differ from the rights that apply to other licenses within the same service that were not issued pursuant to this subsection;

(E) be construed to relieve the Commission of the obligation in the public interest to continue to use engineering solutions, negotiation, threshold qualifications, service regulations, and other means in order to avoid mutual exclusivity in application and licensing proceedings;

(F) be construed to prohibit the Commission from issuing nationwide, regional, or local licenses or permits;

(G) be construed to prevent the Commission from awarding licenses to those persons who make significant contributions to the development of a new telecommunications service or technology; or

(H) be construed to relieve any applicant for a license or permit of the obligation to pay charges imposed pursuant to section 8 of this Act.

(7) Consideration of revenues in public interest determinations.

(A) Consideration prohibited. In making a decision pursuant to section 303(c) to assign a band of frequencies to a use for which licenses or permits will be issued pursuant to this subsection, and in prescribing regulations pursuant to paragraph (4)(C) of this subsection, the Commission may not base a finding of public interest, convenience, and necessity on the expectation of Federal revenues from the use of a system of competitive bidding under this subsection.

(B) Consideration limited. In prescribing regulations pursuant to paragraph (4)(A) of this subsection, the Commission may not base a finding of public interest, convenience, and necessity solely or predominantly on the expectation of Federal revenues from the use of a system of competitive bidding under this subsection.

(C) Consideration of demand for spectrum not affected. Nothing in this paragraph shall be construed to prevent the Commission from continuing to consider consumer demand for spectrum-based services.

(8) Treatment of revenues.

(A) General rule. Except as provided in subparagraph (B), all proceeds from the use of a competitive bidding system under this subsection shall be deposited in the Treasury in accordance with chapter 33 of title 31, United States Code.

(B) Retention of revenues. Notwithstanding subparagraph (A), the salaries and expenses account of the Commission shall retain as an offsetting collection such sums as may be necessary from such proceeds for the costs of developing and implementing the program required by this subsection. Such offsetting collections shall be available for obligation subject to the terms and conditions of the receiving appropriations account, and shall be deposited in such accounts on a quarterly basis. Such offsetting collections are authorized to remain available until expended. No sums may be retained under this subparagraph during any fiscal year beginning after September 30, 1998, if the annual report of the Commission under section 4(k) for the second preceding fiscal year fails to include in the itemized statement required by paragraph (3) of such section a statement of each expenditure made for purposes of conducting competitive bidding under this subsection during such second preceding fiscal year.

(C) Deposit and use of auction escrow accounts. Any deposits the Commission may require for the qualification of any person to bid in a system of competitive bidding pursuant to this subsection shall be deposited in an interest bearing account at a financial institution designated for purposes of this subsection by the Commission (after consultation with the Secretary of the Treasury). Within 45 days following the conclusion of the competitive bidding—

(i) the deposits of successful bidders shall be paid to the Treasury;

(ii) the deposits of unsuccessful bidders shall be returned to such bidders; and

(iii) the interest accrued to the account shall be transferred to the Telecommunications Development Fund established pursuant to section 714 of this Act.

(9) Use of former government spectrum. The Commission shall, not later than 5 years after the date of enactment of this subsection [Aug. 10, 1993], issue licenses and permits pursuant to this subsection for the use of bands of frequencies that—

(A) in the aggregate span not less than 10 megahertz; and

(B) have been reassigned from Government use pursuant to part B of the National Telecommunications and Information Administration Organization Act.

(10) Authority contingent on availability of additional spectrum.

(A) Initial conditions. The Commission's authority to issue licenses or permits under this subsection shall not take effect unless—

(i) the Secretary of Commerce has submitted to the Commission the report required by section 113(d)(1) of the National Telecommunications and Information Administration Organization Act;

(ii) such report recommends for immediate reallocation bands of frequencies that, in the aggregate, span not less than 50 megahertz;

(iii) such bands of frequencies meet the criteria required by section 113(a) of such Act; and

(iv) the Commission has completed the rulemaking required by section 332(c)(1)(D) of this Act.

(B) Subsequent conditions. The Commission's authority to issue licenses or permits under this subsection on and after 2 years after the date of the enactment of this subsection [Aug. 10, 1993] shall cease to be effective if—

(i) the Secretary of Commerce has failed to submit the report required by section 113(a) of the National Telecommunications and Information Administration Organization Act;

(ii) the President has failed to withdraw and limit assignments of frequencies as required by paragraphs (1) and (2) of section 114(a) of such Act;

(iii) the Commission has failed to issue the regulations required by section 115(a) of such Act;

(iv) the Commission has failed to complete and submit to Congress, not later than 18 months after the date of enactment of this subsection, a study of current and future spectrum needs of State and local government public safety agencies through the year 2010, and a specific plan to ensure that adequate frequencies are made available to public safety licensees; or the Commission has failed under section 332(c)(3) to grant or deny within the time required by such section any petition that a State has filed within 90 days after the date of enactment of this subsection;

until such failure has been corrected.

(11) Termination. The authority of the Commission to grant a license or permit under this subsection shall expire September 30, 2007.

(12) Evaluation. Not later than September 30, 1997, the Commission shall conduct a public inquiry and submit to the Congress a report—

(A) containing a statement of the revenues obtained, and a projection of the future revenues, from the use of competitive bidding systems under this subsection;

(B) describing the methodologies established by the Commission pursuant to paragraphs (3) and (4);

(C) comparing the relative advantages and disadvantages of such methodologies in terms of attaining the objectives described in such paragraphs;

(D) evaluating whether and to what extent—

(i) competitive bidding significantly improved the efficiency and effectiveness of the process for granting radio spectrum licenses;

(ii) competitive bidding facilitated the introduction of new spectrum-based technologies and the entry of new companies into the telecommunications market;

(iii) competitive bidding methodologies have secured prompt delivery of service to rural areas and have adequately addressed the needs of rural spectrum users; and

(iv) small businesses, rural telephone companies, and businesses owned by members of minority groups and women were able to participate successfully in the competitive bidding process; and

(E) recommending any statutory changes that are needed to improve the competitive bidding process.

(13) Recovery of value of public spectrum in connection with pioneer preferences.

(A) In general. Notwithstanding paragraph (6)(G), the Commission shall not award licenses pursuant to a preferential treatment accorded by the Commission to persons who make significant contributions to the development of a new telecommunications service or technology, except in accordance with the requirements of this paragraph.

(B) Recovery of value. The Commission shall recover for the public a portion of the value of the public spectrum resource made available to such person by requiring such person, as a condition for receipt of the license, to agree to pay a sum determined by—

(i) identifying the winning bids for the licenses that the Commission determines are most reasonably comparable in terms of bandwidth, scope of service area, usage restrictions, and other technical characteristics to the license awarded to such person, and excluding licenses that the Commission determines are subject to bidding anomalies due to the award of preferential treatment;

(ii) dividing each such winning bid by the population of its service area (hereinafter referred to as the per capita bid amount);

(iii) computing the average of the per capita bid amounts for the licenses identified under clause (i);

(iv) reducing such average amount by 15 percent; and

(v) multiplying the amount determined under clause (iv) by the population of the service area of the license obtained by such person.

(C) Installments permitted. The Commission shall require such person to pay the sum required by subparagraph (B) in a lump sum or in guaranteed installment payments, with or without royalty payments, over a period of not more than 5 years.

(D) Rulemaking on pioneer preferences. Except with respect to pending applications described in clause (iv) of this subparagraph, the Commission shall prescribe regulations specifying the procedures and criteria by which the Commission will evaluate applications for preferential treatment in its licensing processes (by precluding the filing of mutually exclusive applications) for persons who make significant contributions to the development of a new service or to the development of new technologies that substantially enhance an existing service. Such regulations shall—

(i) specify the procedures and criteria by which the significance of such contributions will be determined, after an opportunity for review and verification by experts in the radio sciences drawn from among persons who are not employees of the Commission or by any applicant for such preferential treatment;

(ii) include such other procedures as may be necessary to prevent unjust enrichment by ensuring that the value of any such contribution justifies any reduction in the amounts paid for comparable licenses under this subsection;

(iii) be prescribed not later than 6 months after the date of enactment of this paragraph;

(iv) not apply to applications that have been accepted for filing on or before September 1, 1994; and

(v) cease to be effective on the date of the expiration of the Commission's authority under subparagraph (F).

(E) Implementation with respect to pending applications. In applying this paragraph to any broadband licenses in the personal communications service awarded pursuant to the preferential treatment accorded by the Federal Communications Commission in the Third Report and Order in General Docket 90-314 (FCC 93-550, released February 3, 1994)—

(i) the Commission shall not reconsider the award of preferences in such Third Report and Order, and the Commission shall not delay the grant of licenses based on such awards more than 15 days following the date of enactment of this paragraph, and the award of such preferences and licenses shall not be subject to administrative or judicial review;

(ii) the Commission shall not alter the bandwidth or service areas designated for such licenses in such Third Report and Order;

(iii) except as provided in clause (v), the Commission shall use, as the most reasonably comparable licenses for purposes of subparagraph (B)(i), the broadband licenses in the personal communications service for blocks A and B for the 20 largest markets (ranked by population) in which no applicant has obtained preferential treatment;

(iv) for purposes of subparagraph (C), the Commission shall permit guaranteed installment payments over a period of 5 years, subject to—

(I) the payment only of interest on unpaid balances during the first 2 years, commencing not later than 30 days after the award of the license (including any preferential treatment used in making such award) is final and no longer subject to administrative or judicial review, except that no such payment shall be required prior to the date of completion of the auction of the comparable licenses described in clause (iii); and

(II) payment of the unpaid balance and interest thereon after the end of such 2 years in accordance with the regulations prescribed by the Commission; and

(v) the Commission shall recover with respect to broadband licenses in the personal communications service an amount under this paragraph that is equal to not less than $400,000,000, and if such amount is less than $400,000,000, the Commission shall recover an amount equal to $400,000,000 by allocating such amount among the holders of such licenses based on the population of the license areas held by each licensee.

The Commission shall not include in any amounts required to be collected under clause (v) the interest on unpaid balances required to be collected under clause (iv).

(F) Expiration. The authority of the Commission to provide preferential treatment in licensing procedures (by precluding the filing of mutually exclusive applications) to persons who make significant contributions to the development of a new service or to the development of new technologies that substantially enhance an existing service shall expire on the date of enactment of the Balanced Budget Act of 1997.

(G) Effective date. This paragraph shall be effective on the date of its enactment and apply to any licenses issued on or after August 1, 1994, by the Federal Communications Commission pursuant to any licensing procedure that provides preferential treatment (by precluding the filing of mutually exclusive applications) to persons who make significant contributions to the development of a new service or to the development of new technologies that substantially enhance an existing service.

(14) Auction of recaptured broadcast television spectrum.

(A) Limitations on terms of terrestrial television broadcast licenses. A television broadcast license that authorizes analog television service may not be renewed to authorize such service for a period that extends beyond December 31, 2006.

(B) Extension. The Commission shall extend the date described in subparagraph (A) for any station that requests such extension in any television market if the Commission finds that—

(i) one or more of the stations in such market that are licensed to or affiliated with one of the four largest national television networks are not broadcasting a digital television service signal, and the Commission finds that each such station has exercised due diligence and satisfies the conditions for an extension of the Commission's applicable construction deadlines for digital television service in that market;

(ii) digital-to-analog converter technology is not generally available in such market; or

(iii) in any market in which an extension is not available under clause (i) or (ii), 15 percent or more of the television households in such market—

(I) do not subscribe to a multichannel video programming distributor (as defined in section 602) that carries one of the digital television service programming channels of each of the television stations broadcasting such a channel in such market; and

(II) do not have either—

(a) at least one television receiver capable of receiving the digital television service signals of the television stations licensed in such market; or

(b) at least one television receiver of analog television service signals equipped with digital-to-analog converter technology capable of receiving the digital television service signals of the television stations licensed in such market.

(C) Spectrum reversion and resale.

(i) The Commission shall—

(I) ensure that, as licenses for analog television service expire pursuant to subparagraph (A) or (B), each licensee shall cease using electromagnetic spectrum assigned to such service according to the Commission's direction; and

(II) reclaim and organize the electromagnetic spectrum in a manner consistent with the objectives described in paragraph (3) of this subsection.

(ii) Licensees for new services occupying spectrum reclaimed pursuant to clause (i) shall be assigned in accordance with this subsection. The Commission shall complete the assignment of such licenses, and report to the Congress the total revenues from such competitive bidding, by September 30, 2002.

(D) Certain limitations on qualified bidders prohibited. In prescribing any regulations relating to the qualification of bidders for spectrum reclaimed pursuant to subparagraph (C)(i), the Commission, for any license that may be used for any digital television service where the grade A contour of the station is projected to encompass the entirety of a city with a population in excess of 400,000 (as determined using the 1990 decennial census), shall not—

(i) preclude any party from being a qualified bidder for such spectrum on the basis of—

(I) the Commission's duopoly rule (47 C.F.R. 73.3555(b)); or

(II) the Commission's newspaper cross-ownership rule (47 C.F.R. 73.3555(d)); or

(ii) apply either such rule to preclude such a party that is a winning bidder in a competitive bidding for such spectrum from using such spectrum for digital television service.

(k) Broadcast station renewal procedures.

(1) Standards for renewal. If the licensee of a broadcast station submits an application to the Commission for renewal of such license, the Commission shall grant the application if it finds, with respect to that station, during the preceding term of its license—

(A) the station has served the public interest, convenience, and necessity;

(B) there have been no serious violations by the licensee of this Act or the rules and regulations of the Commission; and

(C) there have been no other violations by the licensee of this Act or the rules and regulations of the Commission which, taken together, would constitute a pattern of abuse.

(2) Consequence of failure to meet standard. If any licensee of a broadcast station fails to meet the requirements of this subsection, the Commission may deny the application for renewal in accordance with paragraph (3), or grant such application on terms and conditions as are appropriate, including renewal for a term less than the maximum otherwise permitted.

(3) Standards for denial. If the Commission determines, after notice and opportunity for a hearing as provided in subsection (e), that a licensee has failed to meet the requirements specified in paragraph (1) and that no mitigating factors justify the imposition of lesser sanctions, the Commission shall—

(A) issue an order denying the renewal application filed by such licensee under section 308; and

(B) only thereafter accept and consider such applications for a construction permit as may be filed under section 308 specifying the channel or broadcasting facilities of the former licensee.

(4) Competitor consideration prohibited. In making the determinations specified in paragraph (1) or (2), the Commission shall not consider whether the public interest, convenience, and necessity might be served by the grant of a license to a person other than the renewal applicant.

(l) Applicability of competitive bidding to pending comparative licensing cases. With respect to competing applications for initial licenses or construction permits for commercial radio or television stations that were filed with the Commission before July 1, 1997, the Commission shall—

(1) have the authority to conduct a competitive bidding proceeding pursuant to subsection (j) to assign such license or permit;

(2) treat the persons filing such applications as the only persons eligible to be qualified bidders for purposes of such proceeding; and

(3) waive any provisions of its regulations necessary to permit such persons to enter an agreement to procure the removal of a conflict between their applications during the 180-day period beginning on the date of enactment of the Balanced Budget Act of 1997.

§ 310. License Ownership Restrictions

(a) Grant to or holding by foreign government or representative. The station license required under this Act shall not be granted to or held by any foreign government or the representative thereof.

(b) Grant to or holding by alien or representative, foreign corporation, etc. No broadcast or common carrier or aeronautical en route or aeronautical fixed radio station license shall be granted to or held by—

(1) any alien or the representative of any alien;

(2) any corporation organized under the laws of any foreign government;

(3) any corporation of which more than one-fifth of the capital stock is owned of record or voted by aliens or their representatives or by a foreign government or representative thereof or by any corporation organized under the laws of a foreign country;

(4) any corporation directly or indirectly controlled by any other corporation of which more than one-fourth of the capital stock is owned of record or voted by aliens, their representatives, or by a foreign government or representative thereof, or by any corporation organized under the laws of a foreign country, if the Commission finds that the public interest will be served by the refusal or revocation of such license.

(c) Authorization for aliens licensed by foreign governments; multilateral or bilateral agreement to which United States and foreign country are parties as prerequisite. In addition to amateur station licenses which the Commission may issue to aliens pursuant to this Act, the Commission may issue authorizations, under such conditions and terms as it may prescribe, to permit an alien licensed by his government as an amateur radio operator to operate his amateur radio station licensed by his government in the United States, its possessions, and the Commonwealth of Puerto Rico provided there is in effect a multilateral or bilateral agreement, to which the United States and the alien's government are parties, for such operation on a reciprocal basis by United States amateur radio operators. Other provisions of this Act and of the Administrative Procedure Act shall not be applicable to any request or application for or modification, suspension, or cancellation of any such authorization.

(d) Assignment and transfer of construction permit or station license. No construction permit or station license, or any rights thereunder, shall be transferred, assigned, or disposed of in any manner, voluntarily or involuntarily, directly or indirectly, or by transfer of control of any corporation holding such permit or license, to any person except upon application to the Commission and upon finding by the Commission that the public interest, convenience, and necessity will be served thereby. Any such application shall be disposed of as if the proposed transferee or assignee were making application under section 308 for the permit or license in question; but in acting thereon the Commission may not consider whether the public interest, convenience, and necessity might be served by the transfer, assignment, or disposal of the permit or license to a person other than the proposed transferee or assignee.

(e) Administration of regional concentration rules for broadcast stations.
(1) In the case of any broadcast station, and any ownership interest therein, which is excluded from the regional concentration rules by reason of the savings provision for existing facilities provided by the First Report and Order adopted March 9, 1977 (docket No. 20548; *42 Fed. Reg. 16145),* the exclusion shall not terminate solely by reason of changes made in the technical facilities of the station to improve its service.
(2) For purposes of this subsection, the term "regional concentration rules" means the provisions of sections 73.35, 73.240, and 73.636 of title 47, Code of Federal Regulations (as in effect June 1, 1983), which prohibit any party from directly or indirectly owning, operating, or controlling three broadcast stations in one or several services where any two of such stations are within 100 miles of the third (measured city-to-city), and where there is a primary service contour overlap of any of the stations.

About the Author

Charles H. Kennedy is a partner in the law firm of Morrison & Foerster, LLP, and a member of the adjunct faculty of the Columbus School of Law, Catholic University of America. He is the author or coauthor of four books on communications law and cyberlaw, and advises a wide range of clients on the problems posed by domestic and international regulation of telecommunications and Internet-based services. Mr. Kennedy is a graduate of The University of Chicago Law School, where he was an editor of *The University of Chicago Law Review*. His email address is ckennedy@mofo.com.

Index

2-PIC rule, 49
911 service, 67, 190

Access
    defined, 47
    equal, 46–49, 289
    special, 50, 52
    switched, 50, 52
Access software provider, 297
Affiliates, 253
    defined, 336
    joint marketing, 325
    separate, 322–23, 335, 337
Age of competition (1996–present), *xxvi–xxviii*
Age of hybrid regulation (1968–1996), *xxi–xxiii*
Age of monopoly (1913–1968), *xviii–xxi*
Aggregators
    defined, 289
    requirements for, 291
Alarm monitoring, 70–71
    definition of, 338–39
    delayed entry into, 337–38
    services, 337–39

Alternate operator services (AOSs)
    See Operator service providers (OSPs)
Amateur station, 253
Answering services, 28
Antitrust laws, nonbasic services pricing under, 29–31
Arbitration
    interconnection, 44–45
    procedures for, 303–5
AT&T, 104, 105
    judicial separation of, 106
    nondominant classification, 106, 108–9
    price-cap regulation, 106–8
    purpose of divestiture, 62–63
AT&T Consent Decree, 47, 72
    access definition, 47
    defined, 253
Audio programming services, 321
Average cost, 223
    curves, 226
    as low as suppliers can make it, 227
    price equal to, 227
Average fixed cost (AFC), 224
Average variable cost (AVC), 224, 225

Barriers, removal, 308–9
Basic service elements (BSEs), 73
Basic telephone service, 336
Bell operating companies (BOCs), 61–78
   access to 911, directory assistance and operator services, 67
   access to databases and signaling, 67
   access to network elements, 65–66
   access to poles, ducts, conduits and rights-of-way, 66
   access to telephone numbers, 67
   alarm monitoring and payphones, 70–71
   COCOTs disparity, 119
   comparably efficient interconnection, 74–75
   Computer III rules, 71–78
   defined, 61, 253
   electronic publishing, 332–37
   electronic publishing restriction, 70
   equipment procurement and sales, 330–31
   interconnection with data networks and enhanced service providers, 71–78
   interconnection with local competitors, 65
   interLATA restriction, 62–71
   interLATA services, 315–22
   local loop transmission, 66
   manufacturing, 325–31
   manufacturing restriction, 68–70
   network engineering, 70
   number portability and dialing parity, 68
   ONA model, 73–74
   reciprocal compensation arrangements, 68
   resale, 68
   restrictions in Telecommunications Act of 1996, 62–78
   separate affiliates, 71
   unbundled local switching, 67
   unbundled local transport, 66–67
Below-cost pricing, 243
Ben Ezra, Weinstein, and Company, Inc. v. America Online, Inc., 143–44

Bensuan Restaurant Corp. v. King, 160
Blocking, 295–97
Blumenthal v. Drudge and America Online, Inc., 143
Broadcasting, 253
Broadcast stations, 253
   administration of regional concentration rules for, 358
   renewal procedures, 356–57
Cable service, 254
Cable system, 254
Call splashing, 289
Carrier common line (CCL) charge, 50
Carrier market power, 244–46
   Oregon statute, 245, 245–46
   Texas statute, 246
Carrier pricing, 237–42
   alternatives, 238
   marginal cost and, 237–39
Carriers
   common, 254
   contracts of, 278
   dominant, 202, 209
   information confidentiality, 282
   interconnection duties, 297
   liability for damages, 276
C block auctions, 128
Chain broadcasting, 254
Channel mileage, 52
Channel terminations, 52
Charges, 273–76
   just and reasonable, 275–76
   schedules of, 274–75
   service and, 273
Child On-line Protection Act (COPA), 152
   defined, 153
   enforcement, 153
   judicial rejections, 153, 154
CLEC certification requirements, 86–89
   area specification, 87–88
   construction plans, 89
   election of facilities-based or resale authority, 86–87
   environmental impact statements, 89
   financial ability to serve, 88–89
   satisfaction, 85

services specification, 87
tariff filings, 89
See also Competing local exchange carriers (CLECs)
CMRS providers
compensation of, 133
defined, 53
emergency E911 services, 131
ILEC services provided to, 53–54
interconnection authority, 54
nonpaging, 134
requirements, 130–31
services, 53
tariffs and, 130
as telecommunications carriers, 133
Commercial mobile radio service (CMRS), 129–31
customers, 53
interconnection obligations, 132
nonpaging, 134
services, 130
See also CMRS providers
Common carriers, 254
Communications Act, 251–358
Section 151. Purposes of chapter; Federal Communications Commission created, 251–52
Section 152. Application of chapter, 252
Section 153. Definitions, 253–60
Section 154. Federal Communications Commission, 260–70
Section 155. Commission, 270–73
Section 201. Service and charges, 273
Section 202. Discriminations and preferences, 273–74
Section 203. Schedules of charges, 274–75
Section 205. Commission authorized to prescribe just and reasonable charges: penalties for violations, 275–76
Section 206. Carriers' liability for damages, 276
Section 207. Recovery of damages, 276
Section 208. Complaints to Commission: investigations; duration of investigation; appeal of order concluding investigation, 277
Section 209. Orders of payment of money, 277
Section 211. Contracts of carriers: filing with Commission, 278
Section 214. Extension of lines or discontinuities of service; certificate of public convenience and necessity, 199–202, 278–82
Section 222. Privacy of customer information, 282–85
Section 224. Pole attachments, 285–89
Section 226. Telephone operator services, 289–95
Section 230. Protection for private blocking and screening of offensive material, 295–97
Section 251. Interconnection, 297–303
Section 252. Procedures for negotiation, arbitration, and approval of agreements, 303–8
Section 253. Removal of barriers to entry, 308–9
Section 254. Universal service, 309–15
Section 271. Bell operating company entry into interLATA services, 315–22
Section 272. Separate affiliate: safeguards, 322–25
Section 273. Manufacturing by Bell operating companies, 325–31
Section 274. Electronic publishing by Bell operating companies, 332–37
Section 275. Alarm monitoring services, 337–39
Section 276. Provision of payphone service, 339–40
Section 309. Application for license, 340–57
Section 310. License ownership restrictions, 357–58
Section 310(b), 203–4, 357
Communications Decency Act (CDA), 152
challenges, 153
defined, 152–53

Communications Decency Act (continued)
  federal obscenity statute, 154
  judicial rejections, 153, 154
Comparably efficient interconnection, 74–75
Competing local exchange carriers (CLECs), 85–98
  access to ILEC facilities, services and rights-of-way, 93–98
  access to unbundled network elements, 95–97
  contributions to universal service, 89–91
  cost allocation and, 241–42
  defined, 1, 85
  dialing parity and, 93
  facilities-based, 86
  facility collocation, 37
  financial ability, 88–89
  ILECs vs., 1–2
  interconnection and reciprocal compensation, 94–95
  interconnection and related obligations, 92–93
  interconnection provided to, 36–45
  number portability and, 93
  poles, conduits and rights-of-way, 97–98
  regulatory freedom, 85
  regulatory obligations, 89–93
  resale and, 92–93
  as resellers, 87
  shared line rate negotiation, 97
  state certification, 86–89
  tariffs and informational filings, 91–92
  See also CLEC certification requirements
Competition
  economic case for, 210–12
  efficiency and, 222–23
  maximizing profits and, 218
  monopoly and, 210–31
  pricing efficiency and, 213–18
  pure, 212, 227, 235
  superiority, 210
  supply/demand and, 213–18
  telecommunications law and, 210–31

Competitive bidding, 346–56
  applicability of, 357
  auction of recaptured broadcast television spectrum, 354–56
  bidder and licensee qualifications, 348
  contents of regulations, 347–48
  design of systems of, 346–47
  evaluation, 351–52
  former government spectrum use, 350
  recovery of value of public spectrum, 352–54
  revenues in public interest determinations, 349
  rules of construction, 348–49
  treatment of revenues, 349–50
Competitive markets, 212–18
  characteristics, 212
  defined, 212
  in long-run equilibrium, 220–22, 227
  monopolistic markets vs., 230
  output level and, 219
  pricing efficiency and, 213–18
CompuServe, Inc. v. Patterson, 159–60
Computer Fraud and Abuse Act (CFAA), 157
  defined, 165
  penalties, 166
  unauthorized access and, 165–67
Computer III rules, 71–78
  accounting safeguards, 76
  comparably efficient interconnection, 74–75
  concerns, 75
  ONA model, 73–74
  other safeguards, 75–76
  post-1996 reform of, 78
  state regulation of enhanced services, 77–78
  See also Bell operating companies (BOCs)
Computers
  access to, 163–70
  CFAA and, 165–67
  ECPA and, 164–65
  privacy, 167–68
  tort remedies, 168–70

See also Internet; Internet service
    providers (ISPs)
Conduits
    access to, 66, 97
    attachment to, 288–89
    owner modifications, 288
Connecting carriers, 254
Construction permits, 254
Consumer welfare, 210–12
    defined, 210
    description, 211
    monopolies and, 230
Contracts of carriers, 278
Contributory infringement, 146
COPPA, 174
Copyright infringement
    basics, 145–46
    contributory, 146
    direct, 146
    DMCA and, 148–49
    liability, 145–49
    vicarious, 146
Corporations, 254
Cost of capital, 10
"Cost-plus" regulation, 26
Costs
    allocating, among regulated services, 239–41
    allocating, when carriers face competition, 241–42
    allocating to local exchange service, 8–9
    average, 223, 226
    average fixed (AFC), 224
    average variable (AVC), 224, 225
    changes in, 222
    exogenous, 16
    fixed, 223, 225
    local loop, 51
    long-run incremental (LRIC), 242
    marginal, 218–23
    minimizing, in long run, 225–27
    minimizing, in short run, 223–25
    monopoly, 229
    nontraffic-sensitive (NTS), 50, 51
    output and, 218–23
    schedule, 220
    total, 219, 223, 225
    traffic-sensitive (TS), 50
    types of, 223
    variable, 223, 224, 225
Cross-subsidization, 30
    price-cap regulation and, 30
    problem, 27
Cubby, Inc. et al. v. CompuServe, Inc., et al., 144
Customer-owned coin-operated telephone providers (COCOTs), 116
    BOC disparity, 119
    defined, 116
    economic/technical arrangements, 118
    phones, 118
    Telecommunications Act of 1996 and, 118–19
Customer premises equipment, 254
Customer proprietary network information (CPNI), 75–76
    aggregated, 76
    defined, 75
Customers
    CMRS, 53
    information for emergency services, 284
    information privacy, 282–85
"Cyberpiracy," 151

Damages
    award of, 277
    carrier liability, 276
    recovery of, 276
Databases, access to, 67
Defamation
    liability, 141–45
    successful action of, 141
Demand
    changes in, 222
    increase, effect of, 218
    market curve, 214
    market price and, 213–18
    market schedule, 214
    See also Supply
Depreciation, 10–12
    concept, 11
    regulation of, 12
Dial Equipment Minutes (DEM), 186, 187

Dialing parity, 37
  BOC compliance, 68
  CLECs and, 93
  defined, 254
  interLATA service, 319–20
  mandate, 49
Digital Millennium Copyright Act
  (DMCA), 148–49
  defined, 148
  notice and takedown requirements, 148
  value, 148–49
Digital subscriber line (DSL)
    technology, 96
Direct infringement, 146
Directory assistance, access to, 67
Discriminatory rates, 14–15
  Economist's view, 14–15
  regulator's view, 14
Dominant carriers, 202, 209
Dual tone multifrequency (DTMF), 190
Ducts
  access to, 66
  owner modifications, 288
Due Process test, 159

Economic profit, 213, 221
The Economics of Regulation, 246
Economies of scale, 226
Effective competitive opportunities
    (ECO) test, 202
Efficiency
  competitive firms and, 222–23
  pricing, 211, 213–18, 237–42
  productive, 211, 236
  regulation of rates/earnings and,
    234–37
Electronic commerce
  on-line contract enforcement, 171–74
  personal information protection,
    174–75
  problems with, 170–75
  See also Internet
Electronic Communications Privacy Act
    (ECPA), 162–63
  interception and, 162–63
  stored communication and, 164
  unauthorized access and, 164–65

Electronic publishing, 332–37
  BOC requirement, 334
  definition of, 335–36
  effective dates, 335
  joint marketing, 333–34
  joint venture requirements, 332–33
  private right of action, 334–35
  restriction, 70, 332
  separate affiliate reporting
    requirement, 335
Electronic Signatures in Global and National Commerce Act (ESign Act), 172
End-user common line (EUCL) charge, 50
Enhanced 911 (E911) service, 131, 190
Enhanced service providers (ESPs), 73
  access parity, 74
  "common interconnection charge," 75
  competition ability, 74
  ISPs classified as, 161
  status advantages, 161
  "two-mile rule," 75
Equal access, 46–49, 289
  defined, 46–47
  implementation, 47–49
Equilibrium price, 215, 217, 218
European Union Data Directive, 174, 175
Exchange access, 254
Exclusive franchise, *xix*
Execunet, 105
Explicit subsidies, 186–88
  before 1996 Act, 186–87
  criticism, 188
  DEM, 186, 187
  Lifeline program, 187–88, 194
  Link Up, 187–88, 194
  for low-income subscribers, 187–88
  LTS program, 186, 187
  USF, 186–87
  See also Universal service

False light theory, 170
Federal Communications Commission
    (FCC), 5
  annual reports to Congress, 269
  auction authority, 128
  authority, 261–62

authorization to prescribe
  just/reasonable charges, 275
C block auction rules, 128
chairman, 270
commissioners, 260–61
communication in safety of
  life/property, 270
compensation of appointees, 269
complaints to, 277
conduct of proceedings: hearings,
  268–69
creation, 251–52
delegation of functions, 270–72
dominant carriers and, 202, 209
duties and powers, 268
employees and assistants, 263–66
expenditures, 266–68
filing with, 278
international settlements policy, 205
investigations, 277
ISPs and, 161–62
licensing authority, 125
licensing in international arena, 200
Managing Director, 272–73
meetings, 272
organization of staff, 270
price-cap approach, 6, 27–28
principle office, 262–63
publication of reports, 269
quorum, 268
record of reports, 269
spectrum allocation, 125
terms of office, 262
USP, 205
Federal Trade Commission (FTC), 174
Fiber-optic technology, 234
Fleet Call, 129
Foreign communication, 254–55
Foreign ownership restrictions, 203–4
Fully distributed cost methods, 240

Geographic averaging, 13
Great Lakes Agreement, 255

Harbor/port, 255

Implicit subsidies, 186
Inbound telemarketing, 337

Incentive regulation. *See* Price-cap
  regulation
Incidental interLATA service, 64, 320–21
Incumbent local exchange carriers (ILECs)
  CLECs vs., 1–2
  cost of capital, 10
  defined, 1, 302–3
  facilities/services access, 93–98
  fully competitive services, 28
  interconnection obligations, 298–99
  local exchange service rates, 6–18
  nonbasic service rates, 23–31
  predatory pricing and, 31
  "revenue requirement," 6
  services provided to CLECs, 36–45
  services provided to CMRS providers,
    53–54
  services provided to ISPs, 54–55
  services provided to IXCs, 45–53
Indexing, 15–17
  defined, 15–16
  formula, 15–16, 17
Information content provider, 297
Information service, 255
Inset Systems, Inc. v. Instruction Set, Inc.,
  160
Interactive computer service, 297
Interception, communication, 162–63, 168
Interconnection, 35–55
  agreements, negotiating, 37–44
  arbitration, 44–45
  Communications Act and, 297–303
  comparably efficient, 74–75
  equal access and, 46–49
  exemptions, suspensions, and
    modifications, 301–2
  future of, 55
  implementation, 299–300
  inconsistencies, 55
  judicial review, 44–45
  with local competitors, 65
  mediation, 44–45
  mobile telephone licensees and wireline
    carriers, 131–34
  numbering administration, 300–301
  obligations, 36
  provided to CLECs, 36–45

Interconnection (continued)
  provided to CMRS providers, 53–54
  provided to ISPs, 54–55
  provided to IXCs, 45–53
  reciprocal compensation and, 41–43
  requirements enforcement, 302
  resale of services and, 44
  Telecommunications Act of 1996 and, 44–45
  unbundled network elements and, 43
  See also Incumbent local exchange carriers (ILECs)
Interexchange carriers (IXCs), 5, 103–10
  charge for services provided to, 49–53
  defined, 103
  ILEC serviced provided to, 45–53
  origins and growth, 104–9
  regulation, 109–10
  universal service support, 192
InterLATA restriction, 62–71
  alarm monitoring and payphones, 70–71
  electronic publishing restriction, 70
  escaping, 64–68
  explained, 64
  manufacturing restriction, 68–70
  separate affiliates, 71
  See also Bell operating companies (BOCs)
InterLATA service
  BOC entry into, 315–22
  calls, 49
  defined, 64, 255
  "incidental," 64, 320–21
  in-region, 65, 315–19
  prohibition, 63
  termination, 64
  toll dialing parity, 319–20
International services, 199–206
  dominant carrier safeguards, 202
  foreign-ownership restrictions, 203–4
  Section 214 process, 199–202
  settlements policy, 204–5
International settlements policy, 204–5
International simple resale (ISR), 206
International Telecommunications Union (ITU), 125

Internet
  access to computers and stored information, 163–70
  communications, interception of, 162–63, 168
  defined, 297
  electronic commerce problems, 170–75
  law, 139–40
  privacy and data security on, 162–70
  regulation of, 161–62
  services, 139
Internet Corporation for Assigned Names and Numbers (ICANN), 151
Internet service providers (ISPs), 139–75
  activities jurisdiction, 158–60
  copyright infringement liability, 145–49
  defamation liability, 141–45
  as enhanced service providers (ESPs), 161
  FCC and, 161–62
  ILEC services provided to, 54–55
  liability for harmful content, 140–58
  obscenity/indecency liability, 152–57
  "passive," 142
  reciprocal compensation and, 95
  as share of CLEC's customer base, 94
  spamming liability, 157–58
  trademark infringement liability, 149–51
Interstate communication, 255
IntraLATA calls, 49
Intrastate rate base, 10
Intrusion tort, 169
Investigations, 277

Jurisdiction
  cases, 159–60
  Due Process test, 159
  over ISP activities, 158–60
  types of, 158

Katz v. United States, 167–68

Land station, 255
Liability (Internet), 140–58
  copyright infringement, 145–49
  defamation, 141–45

obscenity and indecency, 152–57
spamming, 157–58
trademark infringement, 149–51
License application, 340–57
broadcast station renewal procedures, 356–57
classification of, 344
competitive bidding and, 346–56
granting considerations, 340–41
hearings; intervention; evidence; burden of proof, 343
petition to deny, 342–43
time of granting, 341–42
Licensee, 255
Licenses
assignment and transfer of, 358
initial, 344–45
ownership restrictions, 357–58
station, form and conditions of, 344
Lifeline program, 187–88, 194, 314
Link Up, 187–88, 194
Local access and transport areas (LATAs), 62–64
defined, 63, 256–57
"shadow," 63
size, 63
See also InterLATA restriction; interLATA service
Local exchange carriers (LECs), 1, 92–93
dialing parity and, 93
interconnection duties, 298
number portability and, 93
poles, ducts, conduits, rights-of-way access, 93
reciprocal compensation arrangements, 93
resale and, 92–93
See also Incumbent local exchange carriers (ILECs)
Local exchange service rates, 6–18
cost allocation and, 8–9
with price-cap regulation, 15–18
with rate-of-return regulation, 6–15
Local loop
costs, 51
transmission, 66
Long-run equilibrium, 213

competitive markets in, 220–22
defined, 213
Long-run incremental cost (LRIC), 242
Long-Term Support (LTS) program, 186, 187
Low-Earth orbit (LEO) satellites, 134

Manufacturing, 325–31
administration and enforcement authority, 331
authorization, 325
collaboration, 325–26
defined, 69–70
equipment, 69
information requirements, 326
limitations for standard-setting organizations, 326–30
restriction, 68–69
rules and regulations, 331
Marginal cost, 218–23
carrier pricing and, 237–39
defined, 244
price equal to, 227
Market price
equilibrium, 215, 217, 218
supply/demand and, 213–18
MCI, 104, 105
MCI v. AT&T, 243
Message storage services, 28
Minimum contacts analysis, 159
Misappropriation tort, 169–70
Mobile satellite service (MSS), 123, 134–35
complexities, 135
satellites for, 134
spectrum, 134–35
Mobile telephone companies, 123–35
economic terms of interconnection, 133–34
interconnection with wireline carriers, 131–34
licensing, 125–29
physical terms of interconnection, 131–33
regulation of, 129–31
spectrum assignment, 126–29
Mobile telephone switching office (MTSO), 124, 132

Mobile telephone technology, 123–25
  MSS, 123
  MTSO, 124
  PCS, 123
  SMR, 123
Modification of Final Judgment
    (MFJ), 61, 64, 65
  BOCs and, 61
  interLATA restrictions, 64
  restriction on BOC manufacturing,
    68–69
Monopolies
  competition and, 210–31
  competitive markets vs., 230
  consumer welfare and, 230
  cost and revenue schedule, 229
  market, 228–31
  natural, *xix*, 230–31
  profit-maximizing output, 229
  pure, 212
  regulated, 235
  telecommunications law and, 211

Natural monopolies, *xix*, 230–31
  defined, 231
  identifying, 232–34
  legal implications, 233–34
  market illustration, 230
  puzzle, 232–34
  telecommunication issues, 233
  See also Monopolies
Negotiation
  interconnection agreement, 37–44
  OSP contract, 119
  procedures for, 303
  shared line rate, 97
Network elements. *See* Unbundled network
    elements (UNEs)
Nonbasic service rates, 23–31
  with antitrust laws, 29–31
  with price caps, 27–29
  with rate-of-return regulation, 24–27
  regulation question, 24
Nonbasic services, types of, 23, 28
Nondiscrimination safeguards, 323
Nonstructural safeguards, *xxiv*
Nontraffic-sensitive (NTS) costs, 50, 51

North American Numbering Plan
    (NANP), 90
Numbering administration, 300–301
Number portability, 37
  BOCs and, 68
  CLECs and, 93

Obscenity/indecency
  indecent material on Internet, 152–54
  liability, 152–57
  obscenity/child pornography on
    Internet, 154–57
Off-premises switching (OPS), 24–27
  costs, 27
  defined, 24
  lowering price of, 241
  premium, 25, 26
On-line contracts, 171–74
  claims, 171–72
  clickwrap, 173
  enforcing, 173–74
  ESign Act and, 172
  laws, 172
  making, 171–72
  signatures, 171–72
  uniform acts and, 172
  See also Electronic commerce
Open Network Architecture (ONA)
    model, 73–74
Operating expenses, calculating, 9–10
Operator service providers (OSPs), 117,
    120–21
  contract negotiation, 119
  defined, 120, 290
  presubscribed, 119, 289–90
  requirements, 290–91
Operator services, 289–95
  access to, 67
  defined, 289
  definitions, 289–90
  determination of rate compliance,
    293–95
  general rulemaking, 292
  rulemaking on access and
    compensation, 292–93
  statutory construction, 295

Optical line terminating multiplexer
    (OLTM), 41
Optimal scale, 227

Payphones, 71, 115–20
    call types, 115
    providers, 116–18
    safeguards, 339–40
    service provision, 339–40
PenCLEC
    interconnection agreement, 37–44
    needs, 38–40
    negotiation, 40–41
    reciprocal compensation, 41–43
    unbundled network element access, 43
Peninsula Telephone example, 2–3,
        239–40
    illustration, 3
    interconnection agreement negotiation,
        37–44
    local exchange service rates under
        price-cap regulation, 15–18
    local exchange service rates under
        rate-of-return regulation, 6–15
    nonbasic services pricing under antitrust
        laws, 29–31
    nonbasic services pricing under price
        caps, 27–29
    nonbasic services pricing under rate-of-
        return regulation, 24–27
    See also Incumbent local exchange
        carriers (ILECs)
Personal communications service (PCS),
        53, 123, 129
Personal information, protecting, 174–75
Playboy v. Frena, 147, 150
Poles
    access to, 66, 97
    attachments, 285–89
    owner modifications, 288
Predatory pricing, 29–30, 242–44
    ILECs and, 31
    MCI v. AT&T, 243–44
Presubscribed interexchange carrier charge
        (PICC), 52
Price-cap regulation, 236–37
    of AT&T's rates, 106–8

cross-subsidies and, 30
defined, 6
incentives, 15
indexing, 15–17
local exchange service rates under,
    15–18
nonbasic service pricing with, 27–29
in practice, 15
productive efficiency and, 236
sharing, 15–17
Pricing efficiency, 211
    by carriers, 237–42
    competition and, 213–18
    defined, 211, 213
    incentive regimes and, 237
    promotion of, 235
    pure competition and, 235
Pricing standards, 305–6
Privacy
    common-law rights, 168–70
    informational, 167
    torts, 168–70
    under U.S. Constitution, 167–68
Private branch exchanges (PBXs), 24
Private mobile radio service (PMRS),
        53, 129
Productive efficiency, 211
    defined, 211
    price-cap regulation and, 236
Profit
    defined, 218
    economic, 213, 221
    maximizing, 218
Protection of Children Against Sexual
        Exploitation Act of 1977, 156
Public disclosure tort, 170
Public utilities commission (PUC), 5–6
    "cost-plus" regulation, 26
    price-cap regulations, 28
    rate setting, 5, 6
    rules, 6
Pure competition, 212
    character of, 227
    pricing efficiency and, 235
    See also Competition

Quarantine, *xxiv*

Radio communication, 257
Radio officer, 257
Radio station, 257
Radiotelegraph, 257–58
Rate-of-return calculation, 7
Rate-of-return regulation, 235
    capital assets and, 236
    cost allocation, 8–9
    defined, 6
    depreciation and, 10–12
    discriminatory rates, 14–15
    local exchange service rates under, 6–15
    nonbasic service pricing with, 24–27
    rate structure, 12–14
    revenue requirement calculation, 9
Rates
    discriminatory, 14–15
    intrastate, 5, 53
    local exchange service, 6–18
    nonbasic service, 23–31
Rate structure, 12–14
    for basic exchange service, 14
    designing, 13
    geographic averaging, 13
    value of service pricing, 13
Reciprocal compensation, 37
    BOC arrangements, 68
    ISP traffic and, 95
Recovery of damages, 276
Regional concentration rules, 358
Regulation
    access charge, 51
    "cost-plus," 26
    of depreciation, 12
    differential, *xxv–xxvi*
    interexchange carriers (IXCs), 109–10
    Internet, 161–62
    mobile telephone licensees, 129–31
    of new entrants, *xxv*
    price-cap, 15–18, 236–37
    rate-of-return, 6–15, 235
    traditional, 235–36
Regulatory control, 5
Religious Technology Center v. Netcom Online Communication Services, Inc., 147
Resale

BOCs and, 68
CLECs and, 92–93
    interconnection and, 44
Revenue schedule, 220, 229
Rights-of-way
    access to, 66, 97
    attachment to, 288–89
    owner modifications, 288
Rural telephone company, 258

Safeguards
    accounting, 76
    for BOC payphone service, 339–40
    dominant carrier, 202
    nondiscrimination, 323
Safety convention, 258
Schools and Libraries program, 194–95
    controversy, 195
    defined, 194
    See also Universal service
Screening, 295–97
Section 214, 278–82
    applications of foreign carriers, 201–2
    applications of U.S. carriers, 200–201
    approval or disapproval; injunction, 279
    defined, 199
    dominant carrier safeguards, 202
    exceptions, 278–79
    notification of Secretary of Defense, 279
    order of Commission; hearing; penalty, 279–80
    process, 199–202
    provision of universal service, 280–82
    streamlined treatment, 201
    See also Communications Act
Section 310(b), 203–4, 357
    application discretion, 203
    application of, 204
    defined, 203
    See also Communications Act
Sega Enterprises Ltd. v. MAPHIA, 146–47
Sharing, 15–17
    defined, 15
    functioning, 17
    requirements, 29
Ship/vessel, 258–59
Short-run equilibrium, 221

Signaling, access to, 67
Smith v. Maryland, 168
Spamming liability, 157–58
Special access
 charges, 52
 defined, 50
 services, 52
Specialized mobile radio service
  (SMR), 123, 129
Spectrum assignment, 126–29
State commissions, 259
 approval by, 306–7
 review, 307
*Stratton-Oakmont, Inc., et al., v. Prodigy
  Services Company, et al.*, 144–45
Structural separation, *xxiv*
Subscriber line charge (SLC), 50
Subsidies
 explicit, 186–88, 194
 implicit, 186
Supply
 market curve, 217
 market price and, 213–18
 market schedule, 216
 weekly curve, 216
 weekly schedule, 215
 See also Demand
Switched access
 charges, 52
 defined, 50

Telecommunications
 carrier, 259
 defined, 259
 equipment, 259
 service, 259
Telecommunications Act of 1996,
  xxvii–xxviii
 BOC restrictions in, 62–78
 COCOTs and, 118–19
 ILEC/CLEC interconnection, 44–45
 universal service requirements, 188–89
Telecommunications law
 competition and monopoly, 210–31
 defined, *xvii–xviii*
 economic background, 209–46
 issues, 231–46

Telecommunications Relay Service
  (TRS), 90, 91
Telephone exchange service, 259–60
Telephone numbers, access to, 67
Telephone Operator Consumer Services
 Improvement Act (TOCSIA), 120
Television service, 260
Torts
 false light, 170
 intrusion, 169
 misappropriation, 170
 privacy, 168–70
 public disclosure, 170
Total cost, 219, 223, 225
Total element long-run incremental cost
  (TELRIC), 42
Trademark infringement
 considerations, 149–50
 domain name disputes, 151
 liability, 149–51
Trademark protection, 149
Traffic-sensitive (TS) costs, 50
Transmission of energy by radio, 260
"Two-mile rule," 75
Two-tier rate schemes, 239

Unbundled local switching, 67
Unbundled local transport, 66–67
Unbundled network elements
  (UNEs), 43, 86
 BOCs and, 65–66
 CLEC access to, 95–97
Uniform Computer Information
  Transactions Act (UCITA), 172
Uniform Electronic Transactions Act
  (UETA), 172
Uniform Settlements Policy (USP), 205
United States v. Maxwell, 168
United States v. Miller, 168
Universal service, 185–95
 CLECs contributions to, 89–91
 Communications Act and, 309–15
 consumer protection, 314
 contributor exemption, 193
 defined, *xx–xxi*
 defining, 190
 goal, 185

Universal service (continued)
   Lifeline program and, 194, 314
   Link-Up program and, 194
   mandatory contributors, 192
   policy difficulty, 185
   post-1996 rules, 189–95
   principles, 310–11
   provision of, 280–82
   relinquishment of, 281
   requirements of 1996 Act, 188–89
   review procedures, 309–10
   Schools and Libraries program, 194–95
   support, 312
   support availability, 189
   support contributions, 192–93
   support eligibility, 190–92
   support payment calculation, 193–94
   telecommunications services for certain providers, 312–14

   See also Explicit subsidies
Universal Service Fund (USF), 186–87

Value of service pricing, 13
Variable costs, 223, 224, 225
Vicarious infringement, 146
Video programming services, 321

Watts v. Network Solutions, Inc., 150–51
Wire communication, 260
World Trade Organization (WTO), 200
   foreign carrier applications and, 201–2
   Telecommunications Agreement, 200, 201

Yellow Pages, 28

Zeran v. America Online, Inc., 143

Recent Titles in the Artech House Telecommunications Library

Vinton G. Cerf, Senior Series Editor

Access Networks: Technology and V5 Interfacing, Alex Gillespie

Achieving Global Information Networking, Eve L. Varma, Thierry Stephant, et al.

Advanced High-Frequency Radio Communications, Eric E. Johnson, Robert I. Desourdis, Jr., et al.

ATM Switches, Edwin R. Coover

ATM Switching Systems, Thomas M. Chen and Stephen S. Liu

Broadband Access Technology, Interfaces, and Management, Alex Gillespie

Broadband Networking: ATM, SDH, and SONET, Mike Sexton and Andy Reid

Broadband Telecommunications Technology, Second Edition, Byeong Lee, Minho Kang, and Jonghee Lee

The Business Case for Web-Based Training, Tammy Whalen and David Wright

Communication and Computing for Distributed Multimedia Systems, Guojun Lu

Communications Technology Guide for Business, Richard Downey, Seán Boland, and Phillip Walsh

Community Networks: Lessons from Blacksburg, Virginia, Second Edition, Andrew M. Cohill and Andrea Kavanaugh, editors

Component-Based Network System Engineering, Mark Norris, Rob Davis, and Alan Pengelly

Computer Telephony Integration, Second Edition, Rob Walters

Desktop Encyclopedia of the Internet, Nathan J. Muller

Digital Modulation Techniques, Fuqin Xiong

E-Commerce Systems Architecture and Applications, Wasim E. Rajput

Error-Control Block Codes for Communications Engineers,
  L. H. Charles Lee

FAX: Facsimile Technology and Systems, Third Edition,
  Kenneth R. McConnell, Dennis Bodson, and Stephen Urban

Fundamentals of Network Security, John E. Canavan

Guide to ATM Systems and Technology, Mohammad A. Rahman

A Guide to the TCP/IP Protocol Suite, Floyd Wilder

*Information Superhighways Revisited: The Economics of
  Multimedia,* Bruce Egan

Internet E-mail: Protocols, Standards, and Implementation,
  Lawrence Hughes

Introduction to Telecommunications Network Engineering,
  Tarmo Anttalainen

Introduction to Telephones and Telephone Systems, Third Edition,
  A. Michael Noll

An Introduction to U.S. Telecommunications Law, Second Edition
  Charles H. Kennedy

IP Convergence: The Next Revolution in Telecommunications,
  Nathan J. Muller

The Law and Regulation of Telecommunications Carriers,
  Henk Brands and Evan T. Leo

*Managing Internet-Driven Change in International
  Telecommunications,* Rob Frieden

*Marketing Telecommunications Services: New Approaches for a
  Changing Environment,* Karen G. Strouse

*Multimedia Communications Networks: Technologies and
  Services,* Mallikarjun Tatipamula and Bhumip Khashnabish,
  editors

Performance Evaluation of Communication Networks,
  Gary N. Higginbottom

Performance of TCP/IP over ATM Networks, Mahbub Hassan and
  Mohammed Atiquzzaman

Practical Guide for Implementing Secure Intranets and Extranets,
  Kaustubh M. Phaltankar

Practical Multiservice LANs: ATM and RF Broadband,
  Ernest O. Tunmann

Principles of Modern Communications Technology,
  A. Michael Noll

Protocol Management in Computer Networking, Philippe Byrnes

Pulse Code Modulation Systems Design, William N. Waggener

Service Level Management for Enterprise Networks, Lundy Lewis

SIP: Understanding the Session Initiation Protocol,
  Alan B. Johnston

Smart Card Security and Applications, Second Edition,
  Mike Hendry

SNMP-Based ATM Network Management, Heng Pan

Strategic Management in Telecommunications, James K. Shaw

Strategies for Success in the New Telecommunications
  Marketplace, Karen G. Strouse

Successful Business Strategies Using Telecommunications Services,
  Martin F. Bartholomew

Telecommunications Department Management, Robert A. Gable

Telecommunications Deregulation and the Information Economy,
  Second Edition, James K. Shaw

Telephone Switching Systems, Richard A. Thompson

Understanding Modern Telecommunications and the Information
  Superhighway, John G. Nellist and Elliott M. Gilbert

Understanding Networking Technology: Concepts, Terms, and
  Trends, Second Edition, Mark Norris

Videoconferencing and Videotelephony: Technology and
  Standards, Second Edition, Richard Schaphorst

Visual Telephony, Edward A. Daly and Kathleen J. Hansell

Wide-Area Data Network Performance Engineering,
  Robert G. Cole and Ravi Ramaswamy

Winning Telco Customers Using Marketing Databases,
  Rob Mattison

World-Class Telecommunications Service Development,
  Ellen P. Ward

For further information on these and other Artech House titles, including previously considered out-of-print books now available through our In-Print-Forever® (IPF®) program, contact:

Artech House
685 Canton Street
Norwood, MA 02062
Phone: 781-769-9750
Fax: 781-769-6334
e-mail: artech@artechhouse.com

Artech House
46 Gillingham Street
London SW1V 1AH UK
Phone: +44 (0)20 7596-8750
Fax: +44 (0)20 7630-0166
e-mail: artech-uk@artechhouse.com

Find us on the World Wide Web at:
www.artechhouse.com